Remote Sensing Data Analysis in R

Remote Sensing Data Analysis in R

Alka Rani
Scientist, Division of Soil Physics
ICAR – Indian Institute of Soil Science
Bhopal, Madhya Pradesh

Nirmal Kumar
Scientist, Division of Remote Sensing Applications
ICAR – National Bureau of Soil Survey and Land Use Planning
Nagpur, Maharashtra

S.K. Singh
Director
ICAR – National Bureau of Soil Survey and Land Use Planning
Nagpur, Maharashtra

N.K. Sinha
Scientist, Division of Soil Physics
ICAR – Indian Institute of Soil Science
Bhopal, Madhya Pradesh

R.K. Jena
Scientist, ICAR – National Bureau of Soil Survey and Land Use Planning
Regional Centre, Jorhat

Himesh Patra
Scientist (Geoinformatics)
Crop Data Technology Private Limited
Nagpur, Maharashtra

CRC Press is an imprint of the
Taylor & Francis Group, an informa business

First published 2021
by CRC Press
4 Park Square, Milton Park, Abingdon, Oxon OX14 4RN

and by CRC Press
6000 Broken Sound Parkway NW, Suite 300, Boca Raton, FL 33487-2742

CRC Press is an imprint of Taylor & Francis Group, an Informa business

British Library Cataloguing-in-Publication Data
A catalogue record for this book is available from the British Library

Library of Congress Cataloging-in-Publication Data
A catalog record has been requested

ISBN: 978-0-367-72562-4 (hbk)

Preface

Remote sensing and GIS are undeniably important methodological tools for studies related to inventory, surveying, and monitoring of natural resources. However, the highly priced proprietary softwares and unavailability of public satellite data are barriers to their use by broader community, particularly in developing countries. To maximize the scientific and societal benefits of the spatial data, open data and softwares are being promoted. Several satellite data from different sources with a wide array of spatial, spectral, temporal, and radiometric resolutions are now being made available free to all categories of users. To analyze the satellite data of widely differing characteristics and other spatial and non-spatial data from different sources, rapid advancements have been observed in free/open-source software (FOSS) GIS.

R has become the world-wide language for statistics, predictive analytics, and data visualization. Not only statistics, it has a wide range of tools for spatial analysis as well as digital image analysis and can handle any process needed for image analysis with more flexibilities. It offers the widest range available of methodologies available for digital image processing, from the most basic to the most complex. As an open source project, it's freely available for a range of platforms, including Windows, Mac OS X, and Linux. It's under constant development, with new procedures added daily. Additionally, R is supported by a large and diverse community of data scientists and programmers who gladly offer their help and advice to users. Further, some of its most powerful features come from the thousands of extensions (packages) provided by contributing authors.

It can be hard for new users to get a handle on what R is and what it can do. Although the codes for different digital image analysis procedures are available on the web, the new users will find it difficult to work with all the analysis as codes for all type of image analysis is not available at one place.

Remote Sensing Data Analysis in R is a guide book containing codes for most of the operations which are being performed for analysing any satellite data for deriving meaningful information. The goal of this book is to provide hands on experience in performing all the activities from the loading of raster and vector data, mapping or visualisation of data, pre-processing, calculation of indices, classification and advanced machine learning algorithms on remote sensing data in R. The reader will be able to acquire skills to carry out most of the operations of raster data analysis – more flexibly - in open-source freely available software *i.e.* R which are generally available in the paid digital image processing softwares.

Who should read this book?

This book should appeal to anyone who deals with satellite data and digital image analysis. The book assumes a light background of remote sensing and no background in statistical programming and R language to the users.

Roadmap

This book is designed to give you a guided tour of the R platform for satellite data and digital image analysis, with a focus on those methods most immediately applicable for loading, manipulating, visualizing, classifying and understanding data. There are 18 chapters which mainly deals with "Introduction to R", 'Pre – processing", "Enhancements, filtering and transformations", "Classification", "Digital terrain analysis" and "Thematic mapping".

Chapter 1 begins with the procedure of downloading and installation of R and R-Studio. It further discusses about the libraries or packages required for remote sensing image analysis. How to set the working directory and the data provided in the DVD for performing the given examples is also discussed. The freely available source for vector is also discussed.

Chapter 2 tells about the sources of the remote sensing data from various satellites and sensors at different spatial, temporal, spectral and radiometric resolutions from where they can be freely downloaded.

Chapter 3 demonstrates loading and plotting the single and multiband raster data along with saving the raster output in the disk using R.

Chapter 4 covers the radiometric conversion techniques *i.e.* converting the digital number data in raster band to reflectance or radiance, particularly for Landsat image.

Chapter 5 discusses about creating the vector file from tables, loading of the available vector data, and its plotting in R.

Chapter 6 presents the method of reading and assigning projection to the spatial data, and its reprojection to the desired projection.

Chapter 7 tells about the procedure for spectral or spatial sub-setting of raster data in R.

Chapter 8 depicts the commonly used vector data analysis like buffer, union, interesect etc.

Chapter 9 describes the procedure of preparing mosaic of the several raster bands.

Chapter 10 demonstrates the resampling of raster data to higher or lower resolution pixels. The radiometric resampling is also discussed in the chapter.

Chapter 11 briefs about the R codes for computing band statistics, cell statistics, plotting of histogram from the band data, comparison between the spectra of two bands, extracting and plotting the raster values from the sample points.

Chapter 12 covers various image enhancement techniques like linear, histogram, min-max, etc.

Chapter 13 demonstrates the application of filters like high-pass, low-pass, etc. on the raster image.

Chapter 14 tells about the several methods of image transformations like computation of spectral indices, tasselled cap transformation, principal component analysis and pan sharpening.

Chapter 15 presents the several procedures for unsupervised classification of raster image like K-means, Clara and unsupervised random forest.

Chapter 16 discusses about both conventional method and machine learning algorithms for supervised classification of raster image. Conventional approach includes maximum likelihood method whereas machine learning methods involve logistic regressions, support vector machine, decision trees, and random forest approach.

Chapter 17 discusses the derivation of primary terrain variables like slope, aspect, hill shades, etc. by digital terrain analysis of the DEM image. It also includes the derivation of secondary terrain variables like compound topographic index and stream power index by digital terrain analysis of the DEM image.

Chapter 18 demonstrates the methods to prepare thematic maps from the raster and vector data in R.

Practice data for the users

We are providing a DVD containing all the datasets used, chapter wise, in the book for effortless practice of the methods discussed in the book to the users.

Authors

Contents

Chapter 1

Download and Installation of R

1.1. R and R studio

R has become the world-wide language for statistics, predictive analytics, and data visualization. R is not limited to statistics; it has a wide range of tools for spatial analysis and digital image analysis and can handle any process needed for image analysis flexibly. It offers a wide range of methodologies for digital image processing, from the most basic to the most complex. As an open source project, it's freely available for a range of platforms, including Windows, Mac OS X, and Linux. An integrated development environment (IDE) for R, R Studio, allows interactive execution of R functions with user interface. RStudio is available in two formats - RStudio Desktop and RStudio Server - both available in free and fee-based (commercial) editions.

1.1.1. Download and install R

The R software setup files can be downloaded from any of the widely distributed Comprehensive R Archive Network (CRAN). To download R, please choose your preferred CRAN mirror, a location close to you. Follow the following steps:

1. Go to url: https://www.r-project.org/

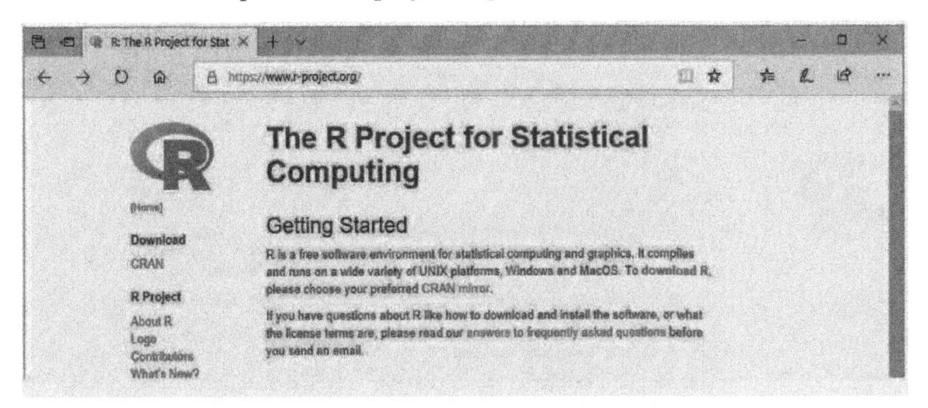

2. Click on CRAN or download R. It will take you to https://cran.r-project.org/mirrors.html

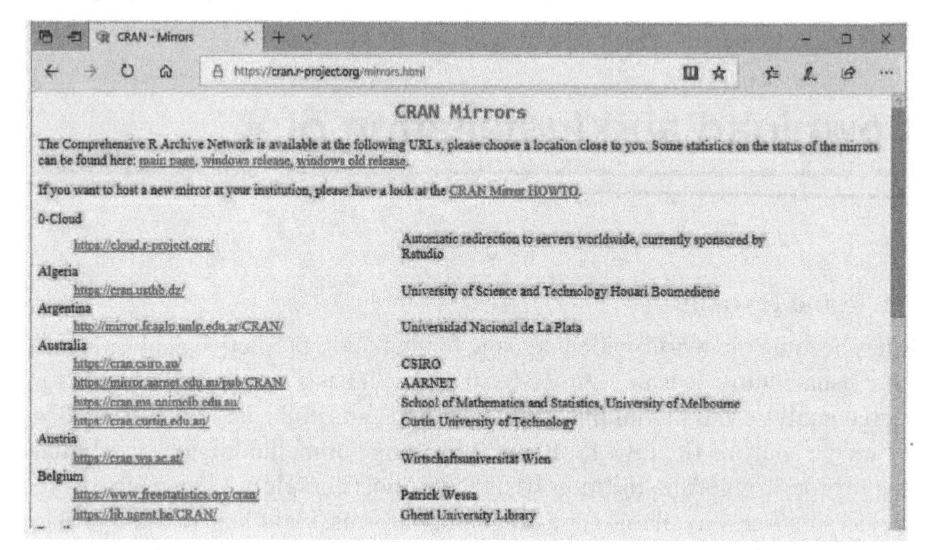

3. All the CRAN URLs are listed alphabetically here. Choose any CRAN near to you.

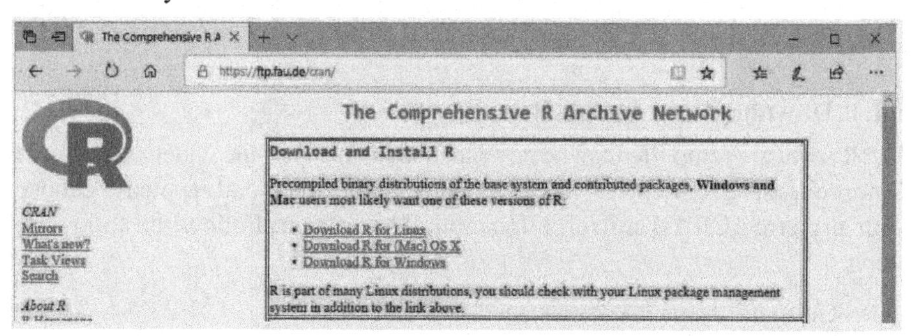

4. Choose the version depending on your operating system (Windows here).

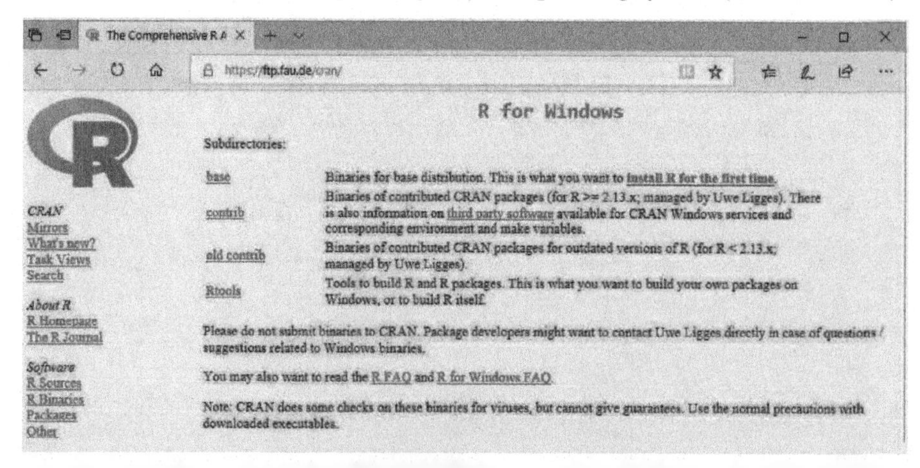

5. To download the Base R, click on install R for the first time.

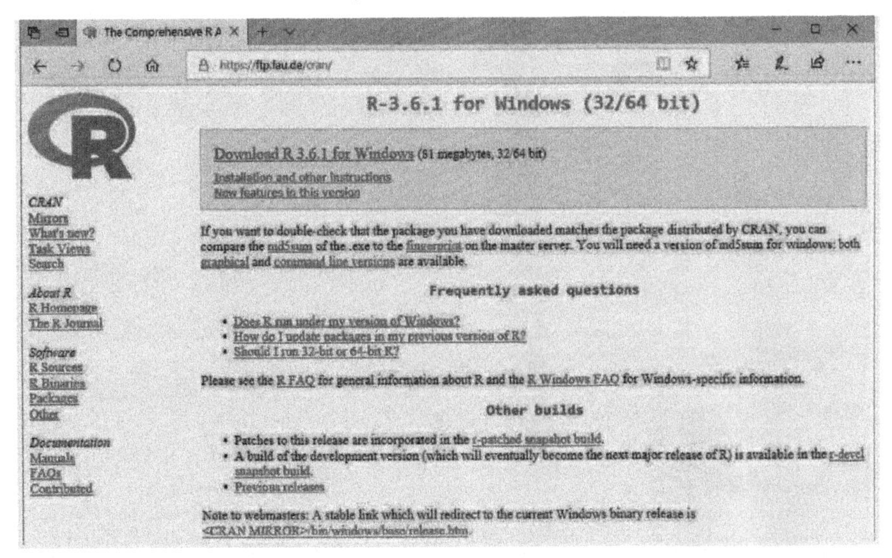

6. Click on download R version for windows and save the installer.
7. Run the installer and follow the instructions to install it.

1.1.2. Download and install R Studio

Download and Install the R Studio once the base R is installed. Go to https://www.rstudio.com/products/rstudio/download/ and click on download R Studio Desktop with Open Source license.

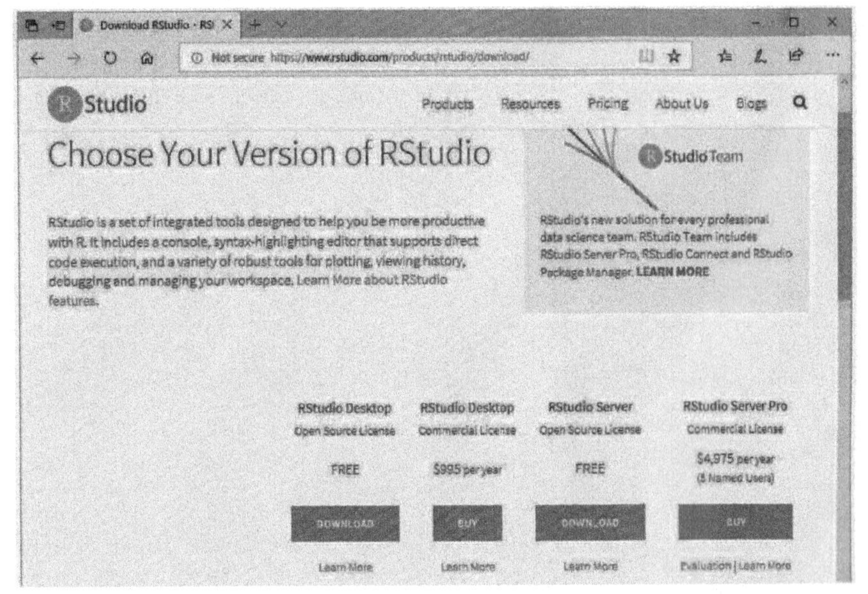

Select the version based on the operating system (Windows 7+ for 64 bit). Save the Installer.

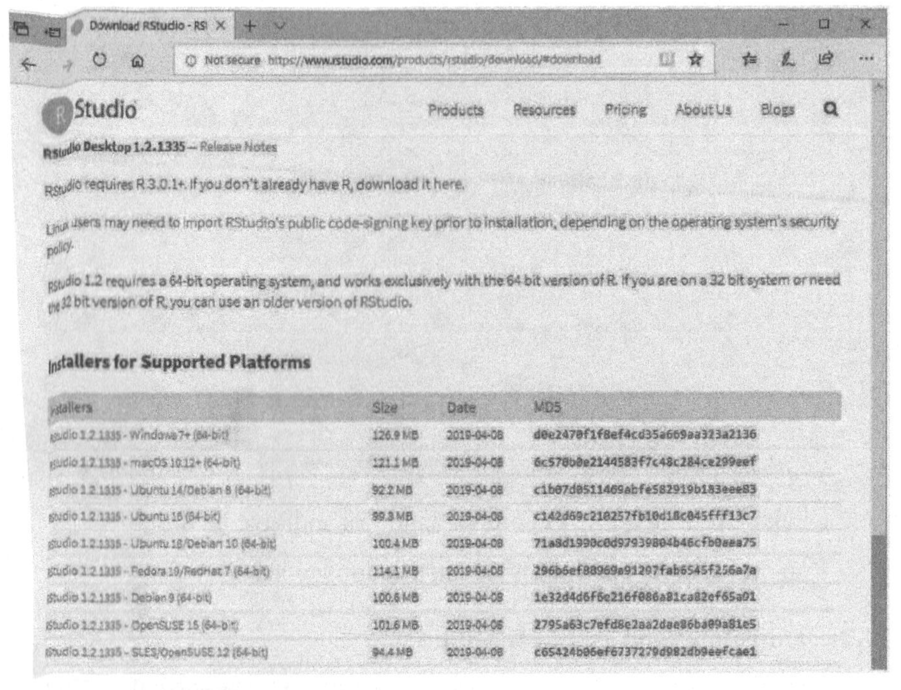

Double click the installer and follow the instructions to install the R Studio.

1.1.3. Starting up R Studio

Once the base R and R Studio have been installed, click on RStudio icon to open it.

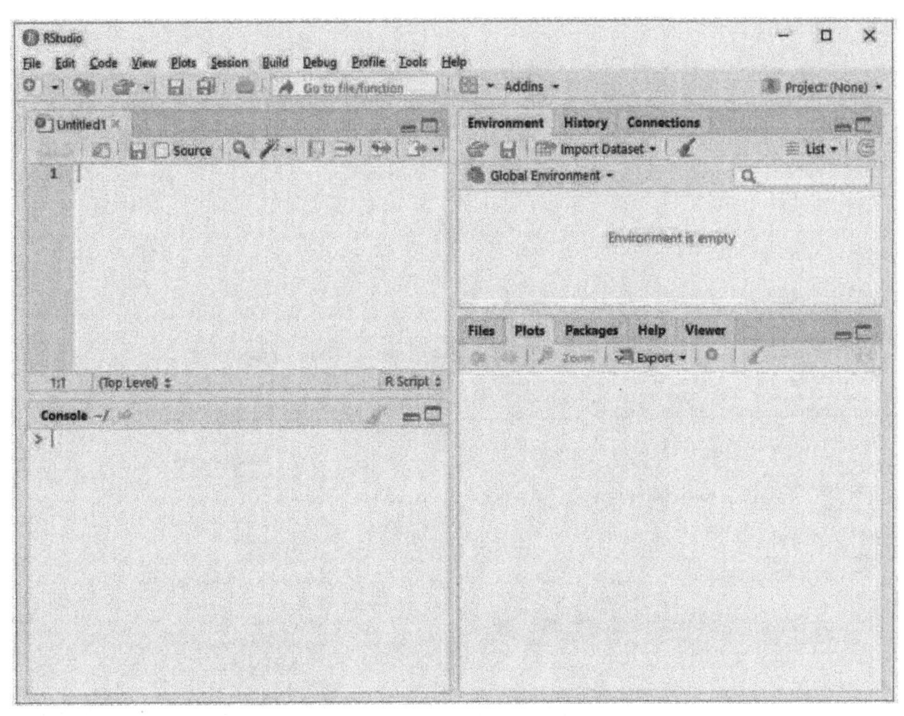

1.1.4. Installing packages

All R functions and datasets are stored in packages. To use the functions, the packages are needed to be loaded. The base R is provided with a few of built in packages, called standard and recommended packages. Many more packages are available through the CRAN family of Internet sites (via https://CRAN.R-project.org) and elsewhere. This strategy helps in low memory use and faster processing. The R environment is under constant development, with new procedures added daily. To install any library, click on the ***packages*** tab and click on ***Install***. In the install Packages windows type the name of the package and click on ***Install***.

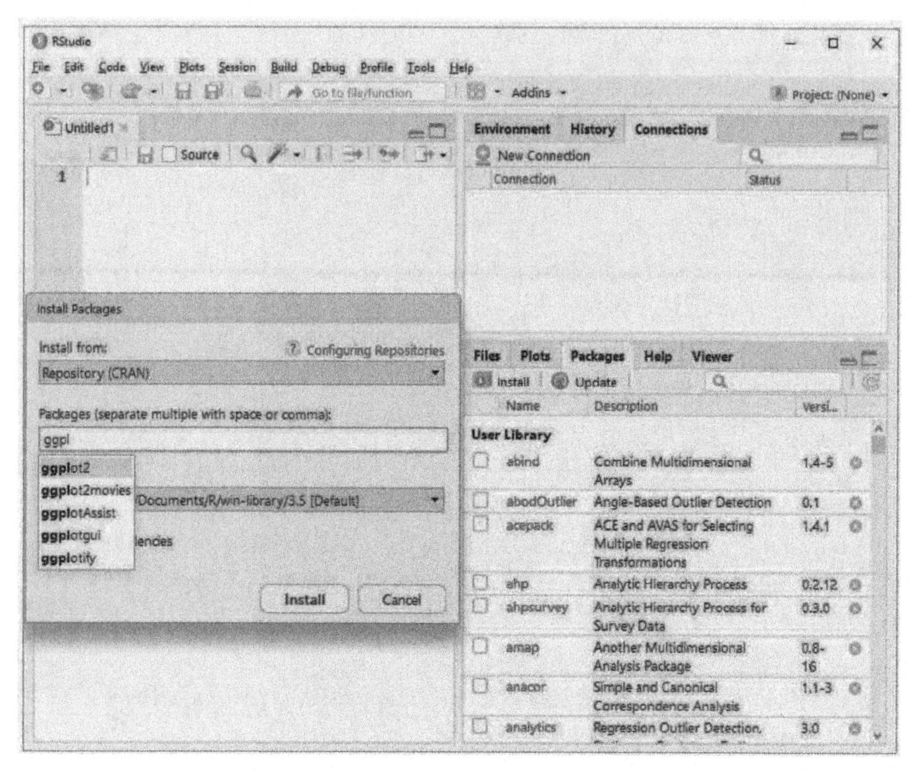

If you have any package downloaded and saved on your computer, select *Package Archive File (.zip; .tar.gz)* in the **install from** dropdown. Click on ***Browse*** and show the file directory.

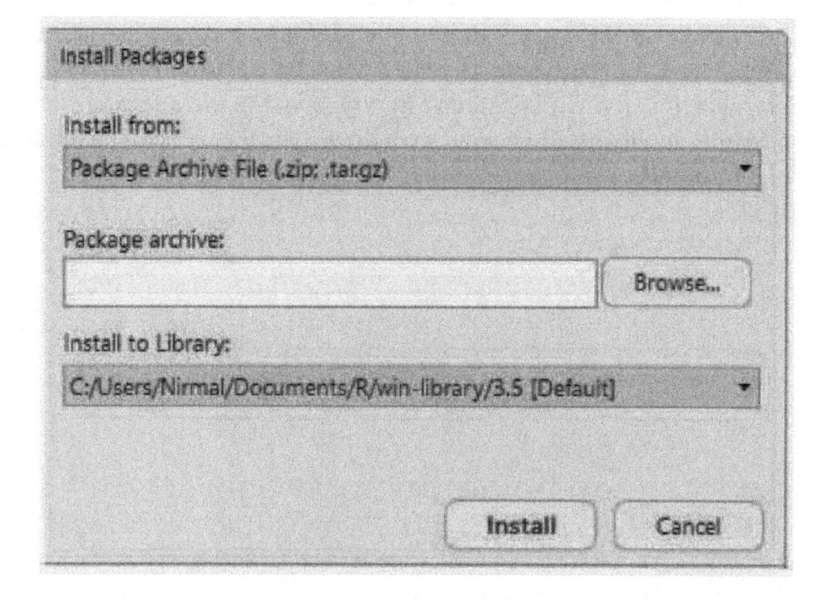

1.1.5. Setting up working directory

R is always pointed at a directory on your computer. You can find out which directory by running the getwd (get working directory) function; this function has no arguments. To change your working directory, use setwd and specify the path to the desired folder. Type the following codes.

```
getwd()       # will give default directory when run first
time
## [1] "C:/Users/Nirmal/Documents"
setwd("C:/Users/Nirmal/Desktop/New folder")   #Sets the
directory to a folder named 'New Folder' on desktop
getwd()       # will show the newly assigned directory
## [1] "C:/Users/Nirmal/Desktop/New folder"
```

Chapter 2

Data Availability and Downloading

2.1. Open source satellite data

The satellite images have been invaluable information sources for Natural Resource Management. They provide synoptic view of the Earth periodically helping in inventory, surveying, and monitoring in agriculture, hydrology, geology, mineralogy, land cover, land use and environment. Free global satellite image data of spatial resolutions of 1m to 1 km and spectral resolution of single band to hyperspectral are available from an increasing number of sources (Table 2.1).

2.2. Open Access Vector Data

Natural Earth is a public domain map dataset available at 1:10m, 1:50m, and 1:110 million scales. Natural Earth was built through a collaboration of many volunteers and is supported by NACIS (North American Cartographic Information Society), and is free for use in any type of project in any manner, including modifying the content and design, electronic dissemination, and offset printing. The data can be downloaded from http://www.naturalearthdata.com/downloads/.

Another open source shapefiles for administrative boundaries are available with http://gadm.org/data.html. World wise or country wise administrative boundaries with sub-divisions are freely available for academic and other non-commercial use.

Yet another open source vector database is OpenStreetMap®, licensed under the Open Data Commons Open Database License (ODbL) by the OpenStreetMap Foundation (OSMF). OpenStreetMap (OSM) is a collaborative project to create a free editable map of the world. Users are free to copy, distribute, transmit and adapt the data, as long as credit is given to OpenStreetMap and its contributors. If the users alter or build upon this data, the result may be distributed only under the same license.

Table 2.1: A list of freely available satellite data

Sensor	Satellite	Resolution			Availability
		Spatial	Spectral	Temporal	
Multispectral data					
Landsat MSS	Landsat 1-3 (Bands 4-7) Landsat 4-5 (Bands 1-4)	60* m	4 bands: VNIR	18 days 16 days	July 1972 to October 1992 & June 2012 to January 2013
Landsat TM	Landsat 4-5	30 m (band 1-5, 7) 120** m (band 6)	7 bands: VNIR, SWIR, TIR	16 days	July 1982 to May 2012
Landsat ETM+	Landsat 7	30 m (band 1-5, 7) 60** m (band 6) 15 m (band 8)	8 bands: VNIR, SWIR, TIR, PAN	16 days	July 1999 to May 2003 (SLC on) May 30, 2003 till date (SLC off)
Landsat OLI/TIRS	Landsat 8	30 m (band 1-7, 9) 100 m (band 10, 11) 15 m (band 8)	11 bands: VNIR, SWIR, TIR, PAN	16 days	March, 2013 till date
ALI	EO 1	30 m(band 1-7); 10m (band 8)	10 bands: VNIR, SWIR, PAN	16 days	May 2001 to March 2017
MODIS	Terra and Aqua	250 m (band 1, 2) 500 m (band 3-7) 1 km (band 8-36)	36 bands: VNIR, SWIR, TIR	1-2 days	February 2000 till date
AVHRR	NOAA	1.1 km	6 bands: VNIR, SWIR, TIR		1981 till date
LISS-III	Resourcesat 1	24 m	4 bands: VNIR, SWIR	24 days	Selected dates
AWiFS	Resourcesat 1	56 m	4 bands: VNIR, SWIR	5 days	
MIS	Sentinel 2A	10 m (band 2-4, 8)	13 bands: VNIR, SWIR	3 to 5 days	From October, 2015

Contd.

Hyperspectral Data					
		20 m (band 5-7, 8a, 11, 12)60 m (band 1, 9, 10)			
Hyperion	EO 1	30 m	220 bands: VNIR, SWIR		May 2001 to March 2017
HySI	IMS1	500 m	64 bands: VNIR	24 days	Selected dates
Digital elevation models					
SRTM	SRTM	30 m, 90 m, 250 m	C-band SAR	-	-
ASTER	ASTER	30 m	Optical	-	-
Cartosat	Cartosat	30 m	Optical	-	-
ALOS PALSAR	ALOS	30 m	L-band SAR	-	-
ALOS PALSAR	ALOS	12.5 m	L-band SAR		
High resolution data					
OrbView 3	GeoEye	1 m	PAN	3 day	Selected number of data

2.1. Downloading satellite data

Most of the data are available to download with https://earthexplorer.usgs.gov/. We will be discussing how to download satellite data from earth explorer. Follow the steps:

Go to https://earthexplorer.usgs.gov/ and click on ***Register.***

1. After completion of register, login.

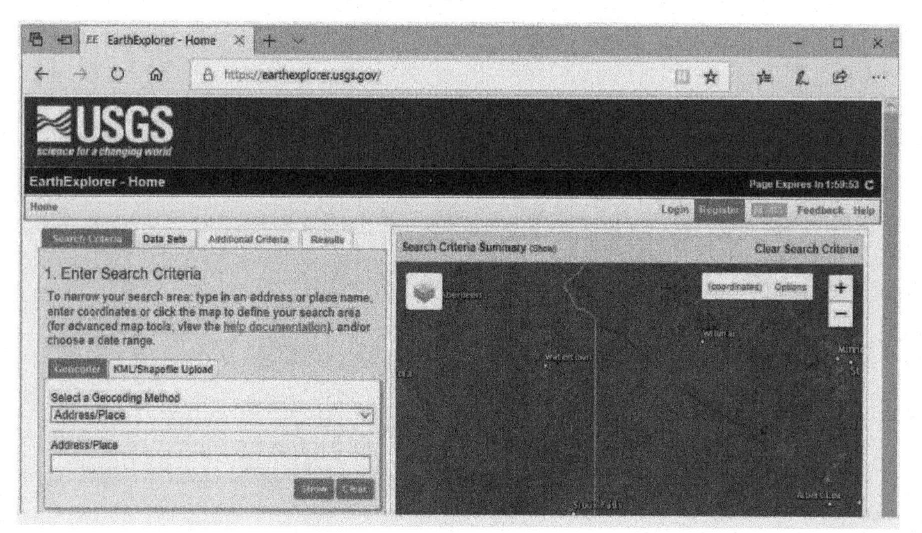

2. The next step is to define the search criteria such as the region of interest (ROI), preferred dates, and number of scenes preferred. These preferences can be set in the **'Search Criteria'** tab. The ROI can be defined by

a. Typing Address/ Place name

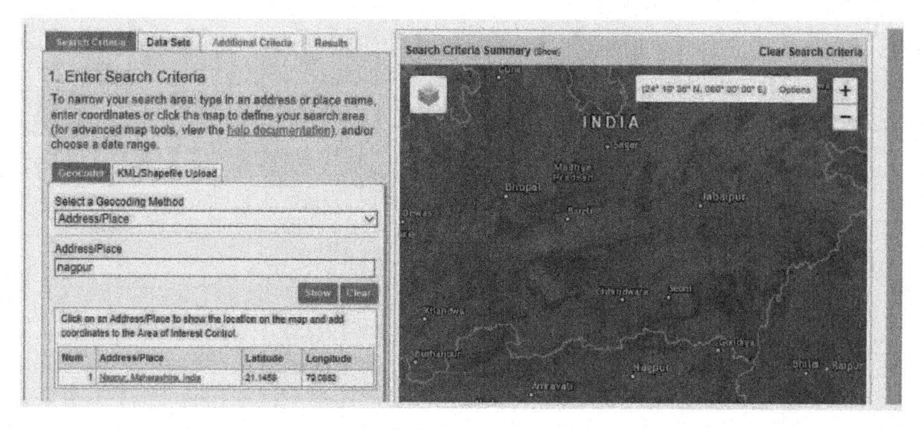

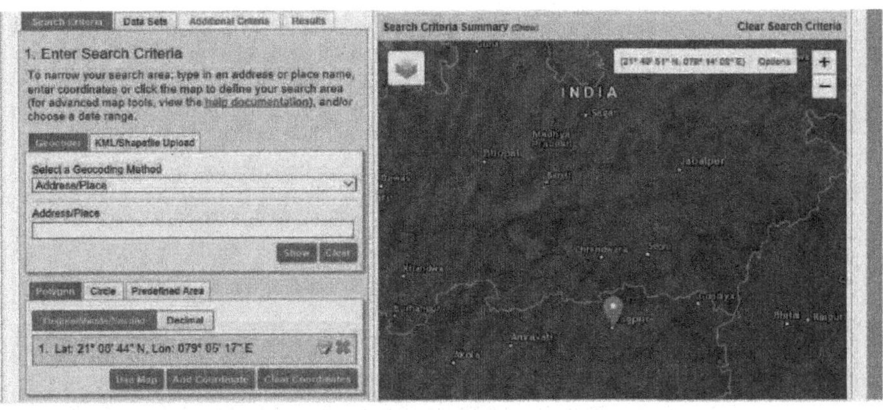

b. By interactively drawing ROI by clicking on the map,

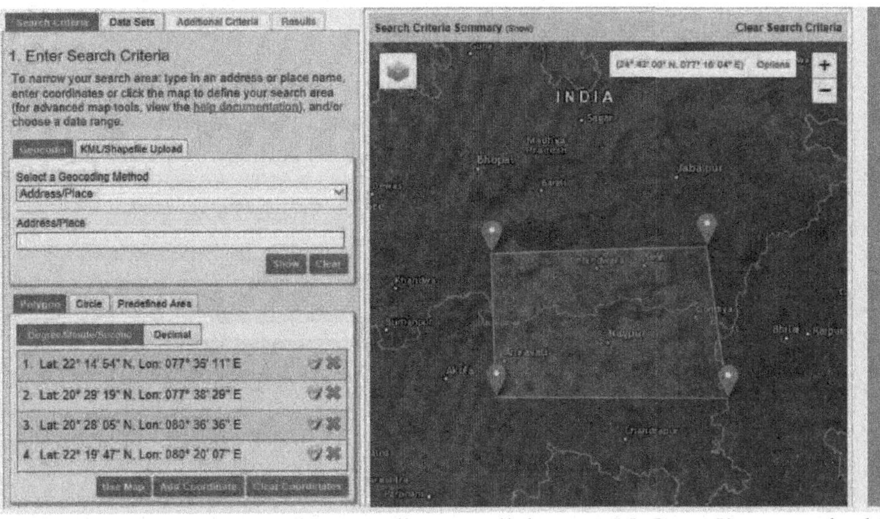

For getting the precise coordinates, click on **Add Coordinate** and add the coordinates.

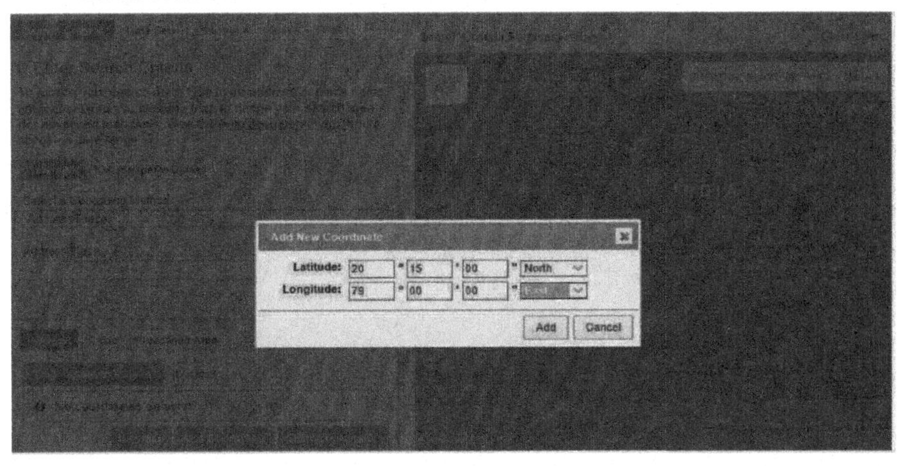

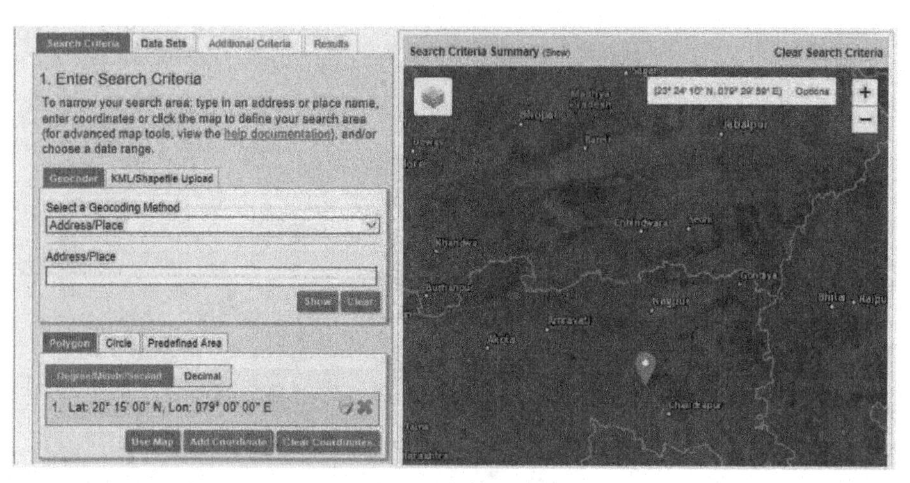

c. By providing a .kml or .shp file of the ROI.

Click on **KML/Shapefile Upload**, select **Shapefile** and click on **Select File**. Locate the shapefile and click **Open**. The shapefile must be in archived format and having not more than 30 nodes.

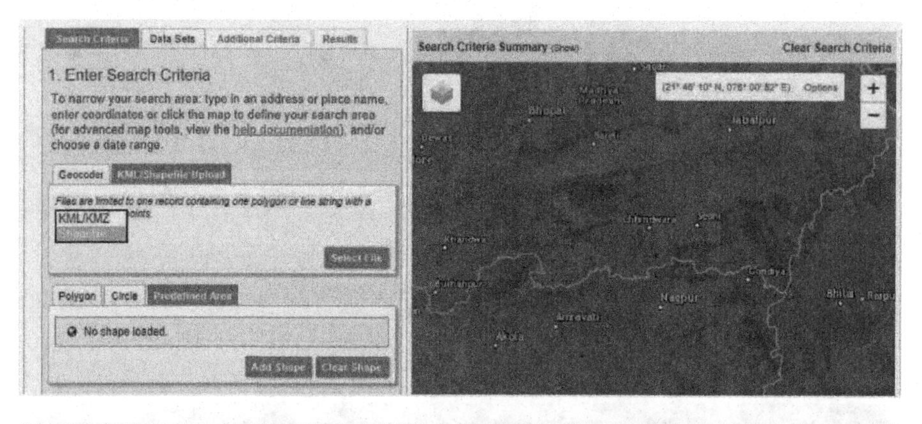

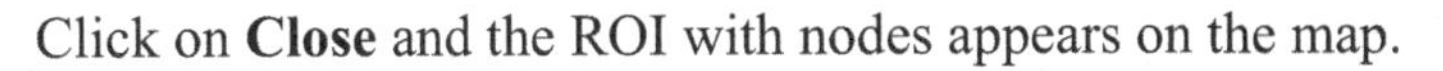

Click on **Close** and the ROI with nodes appears on the map.

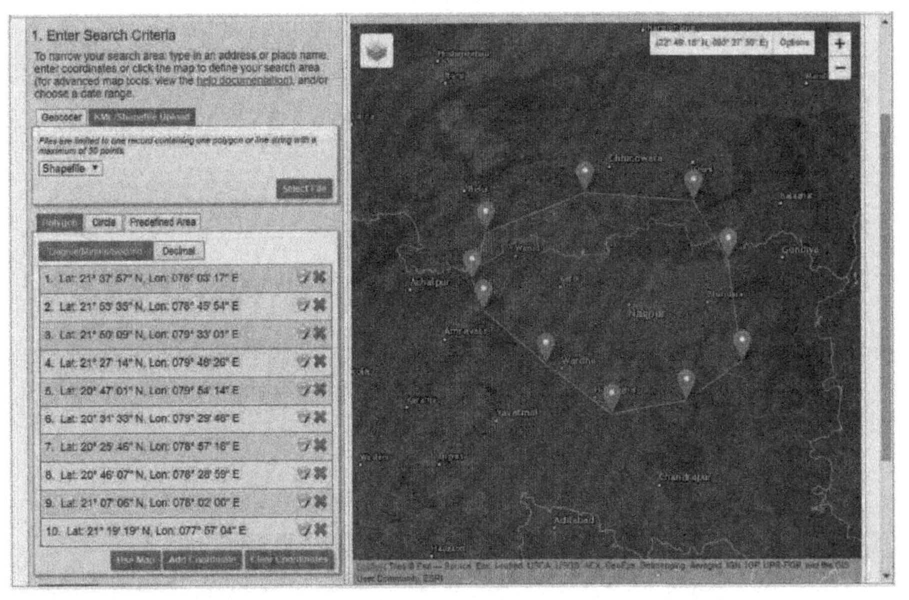

The date's criterion can be given by clicking on the "**Date Range**" tab.

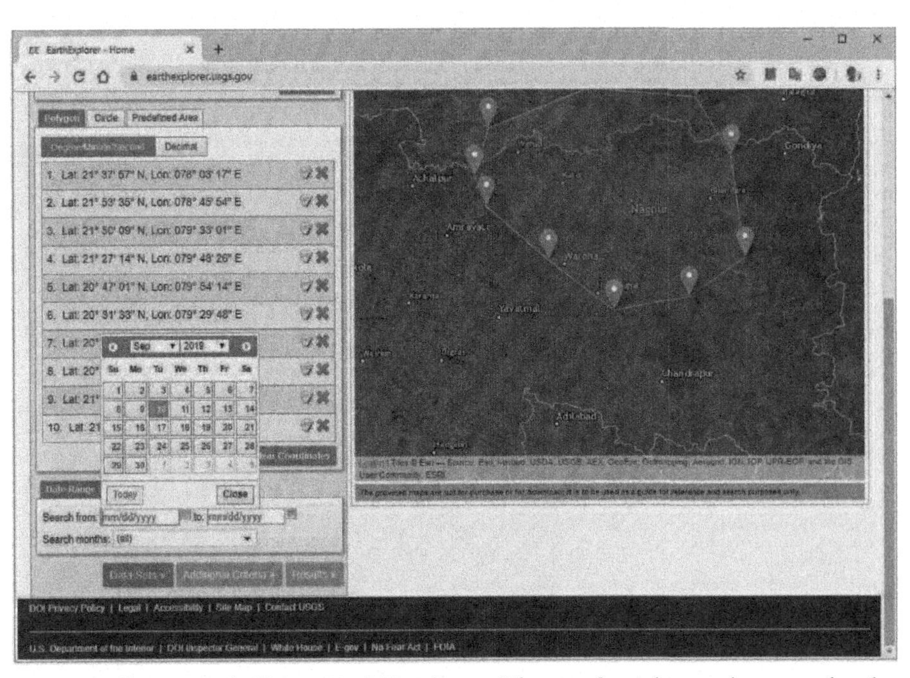

Define search **From** and **To** dates. The preferred months can also be defined.

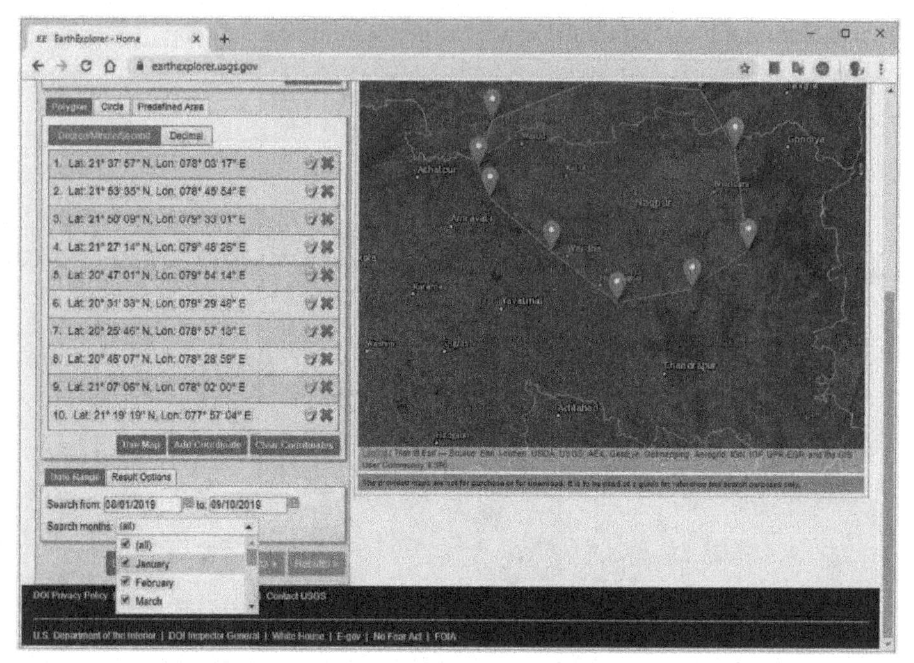

4. After defining the search criteria, click on **"Data Sets"** tab for selection of data. A list of datasets available for free are listed here, including Landsat, Sentinel, SRTM DEM, MODIS, AVHRR, etc. we will be downloading the latest Landsat-8 OLI/TIRS data.

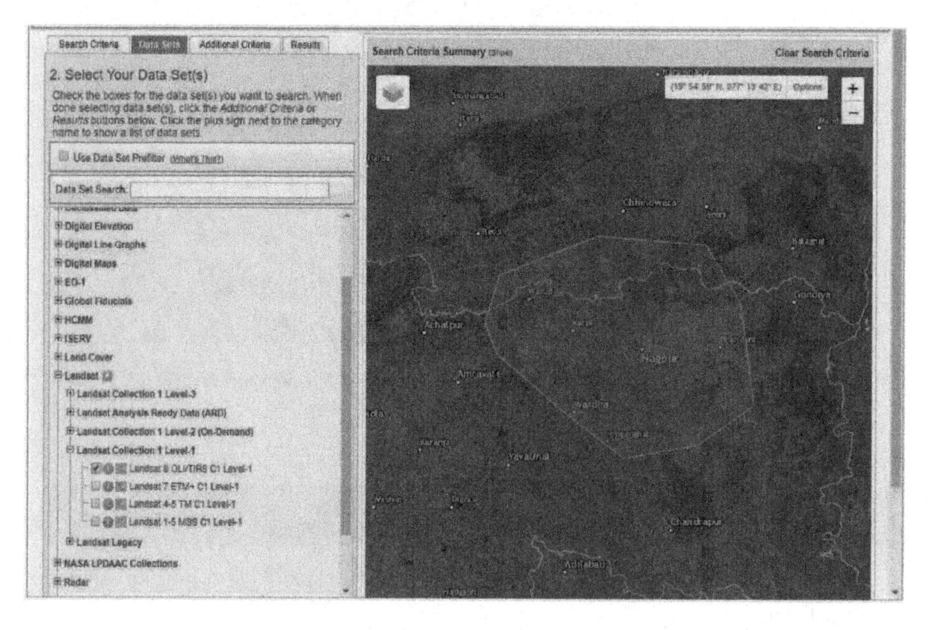

5. To visualize the available scenes with defined search criteria, click on **Results** tab.

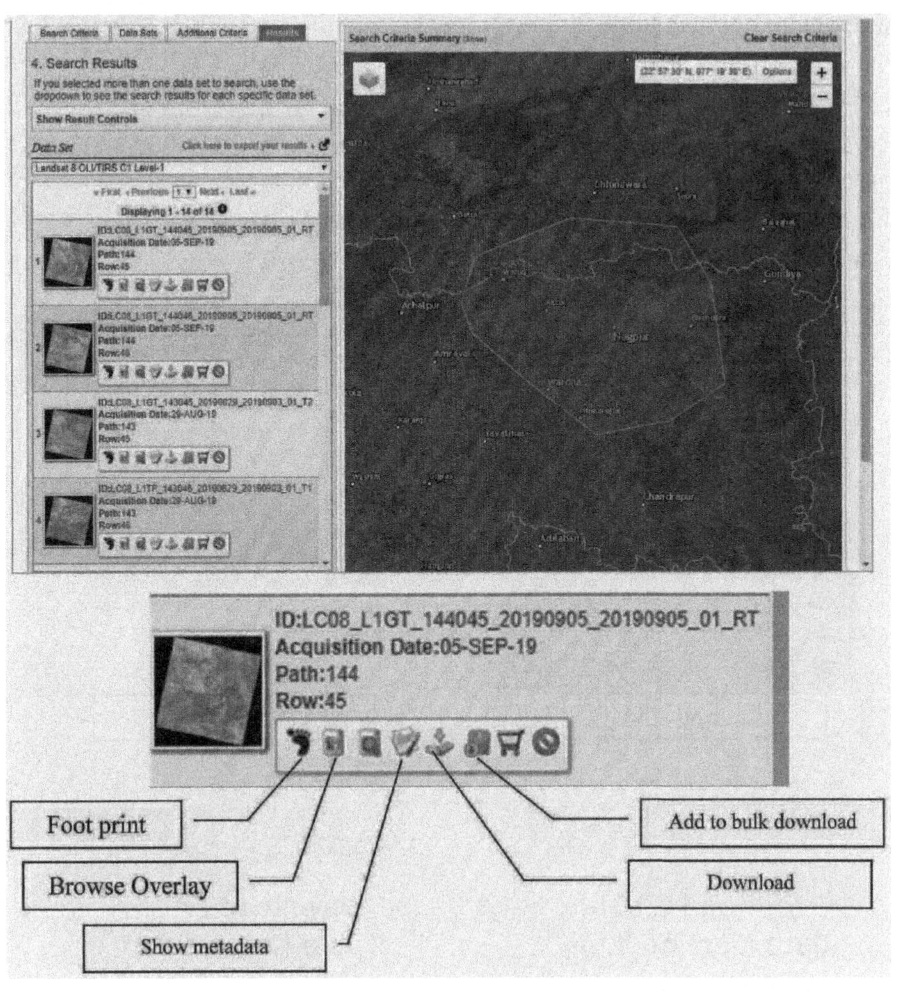

6. Click on Download icon of the selected scene. And select the Level-1 GeoTiff Product to download the .tif files of all the bands.

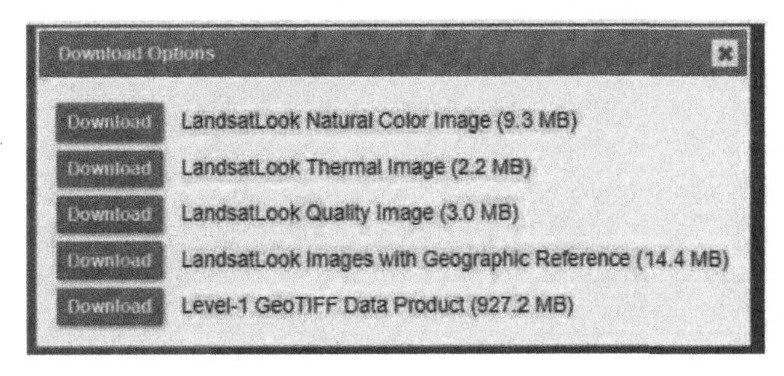

2.4. Downloading administrative boundaries data

In this section we will download the administrative boundaries shapefiles from http://gadm.org/data.html.

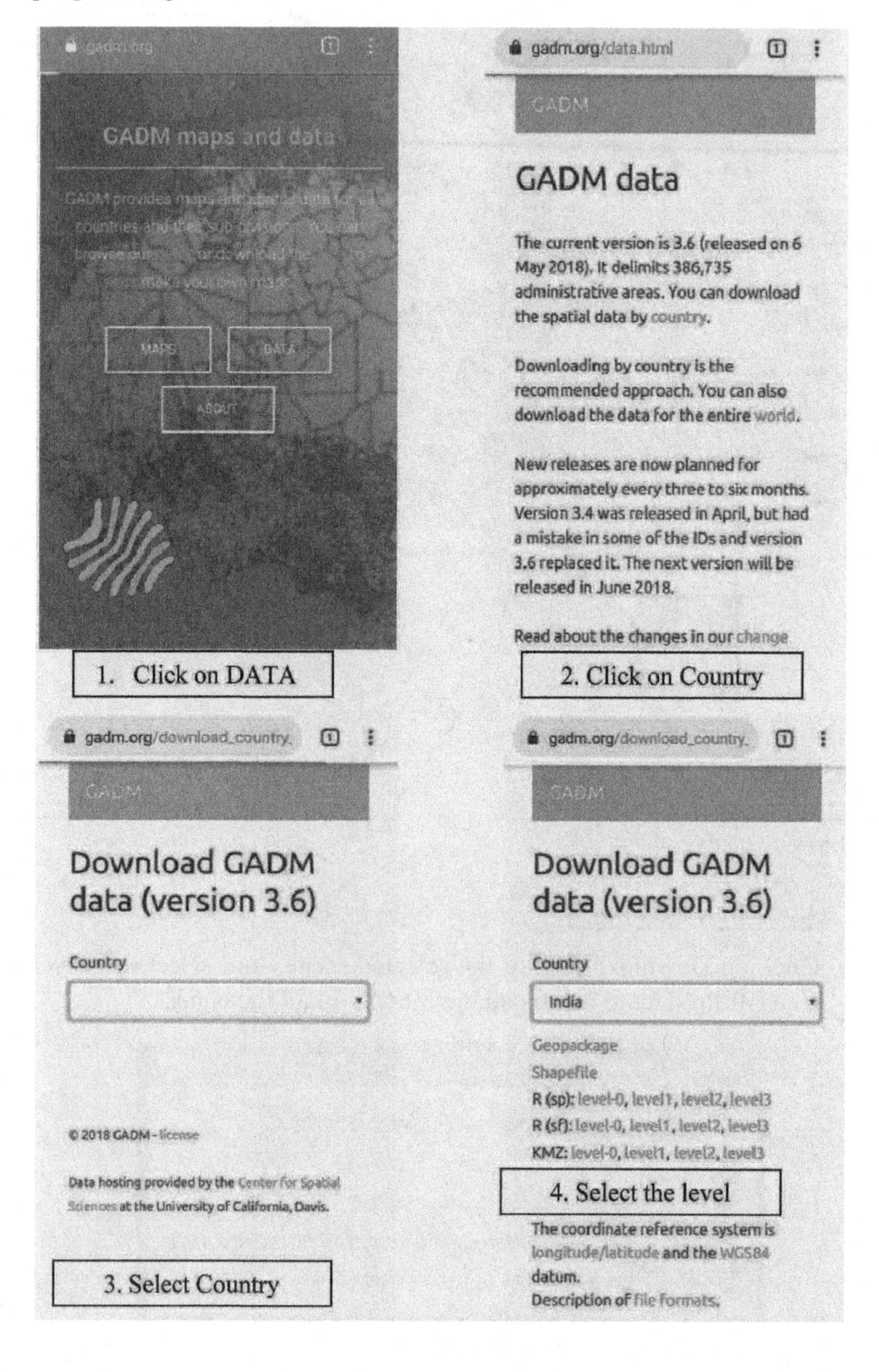

Chapter 3

Raster Data in R

The satellite data is available in raster format. Raster data consists of two – dimensional array of pixels or grids which stores the values representing features. All the pixels are of same size and shape with their particular coordinate location. A raster data format is shown below.

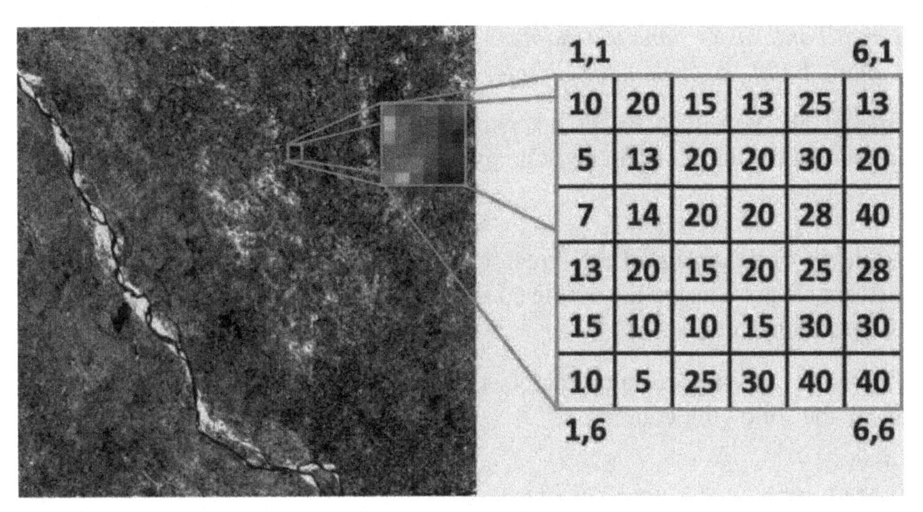

Fig. 3.1: Raster Data

In this chapter, we are going to use Landsat 8 Operational Land Imager (OLI) and Thermal Infrared Sensor (TIRS) data downloaded from USGS Earth Explorer in GeoTIFF format. Landsat 8 dataset has 11 bands with following specifications (Table 3.1):

Table 3.1: Specification of Landsat 8 images

Bands	Wavelength (μm)	Resolution (m)
Band 1 – Coastal aerosol	0.43 – 0.45	30
Band 2 – Blue	0.45 – 0.51	30
Band 3 – Green	0.53 – 0.59	30
Band 4 – Red	0.64 – 0.67	30

Contd.

Band 5 – Near Infrared	0.85 – 0.88	30
Band 6 – SWIR 1	1.57 – 1.65	30
Band 7 – SWIR 2	2.11 – 2.29	30
Band 8 – Panchromatic	0.50 – 0.68	15
Band 9 – Cirrus	1.36 – 1.38	30
Band 10 – Thermal Infrared (TIRS) 1	10.60 – 11.19	100
Band 11 – Thermal Infrared (TIRS) 2	11.50 – 12.51	100

This chapter discusses how to load a raster, both single band and multiple bands, in R. Plotting of single raster, natural colour composite, and false colour composites are also discussed. Further, how to know the properties of the raster and writing the output raster to the hard disk is discussed. Following datasets will be used in the current chapter:

Landsat data: LC08_L1TP_144042_20150419_20170409_01_T1 with all bands separately saved in: Chapter3\data\Landsat_data

A multiband raster named: Chapter3\data\Landsat_multi.tif
A single band raster named: Chapter3\data\Landsat_NIR.tif

First of all, you need to set the working directory to the path having the downloaded data. This can be done by the following code:

```
>setwd('E:\\......\\Chapter3\\data')
```

While writing the path of the directory, always use double backslashes (\\) or forward slash (/) instead of single (\) back slash because single backslash (\) is used to escape character.

Check the current working directory or whether the directory is appropriately set by the following code:

```
>getwd()
## [1] "E:/................./Chapter3/data"
```

For loading the raster data `raster` package is required.

```
#Import package
>library(raster)
```

3.1. Loading single band raster

The single band of the raster data can be loaded using the following code. The raster command if used with a multiband data, will load single band - only the first band. To display the attributes of the loaded raster layer type the name of the raster created

```
#Load single band
>img =raster('Landsat_NIR.TIF')
```

```
# display the attributes of the loaded raster
>img
## class      : RasterLayer
## dimensions : 7801, 7661, 59763461  (nrow, ncol, ncell)
## resolution : 30, 30  (x, y)
## extent     : 308985, 538815, 2757585, 2991615  (xmin, xmax, ymin,
ymax)
## crs          : +proj=utm +zone=44 +datum=WGS84 +units=m +no_defs
+ellps=WGS84 +towgs84=0,0,0
## source     : E:/................/data/Landsat_NIR.TIF
## names      : Landsat_NIR
## values     : 0, 65535  (min, max)
```

The class, number of rows, columns and cells, spatial resolution, extent of the image, Coordinate Reference System (CRS) projection, source, as well as minimum and maximum values of the raster image is displayed.

3.2. Plotting Single Band

Single band can be plotted with the `plot` command as given below:

```
#plotting single band
>plot(img)
```

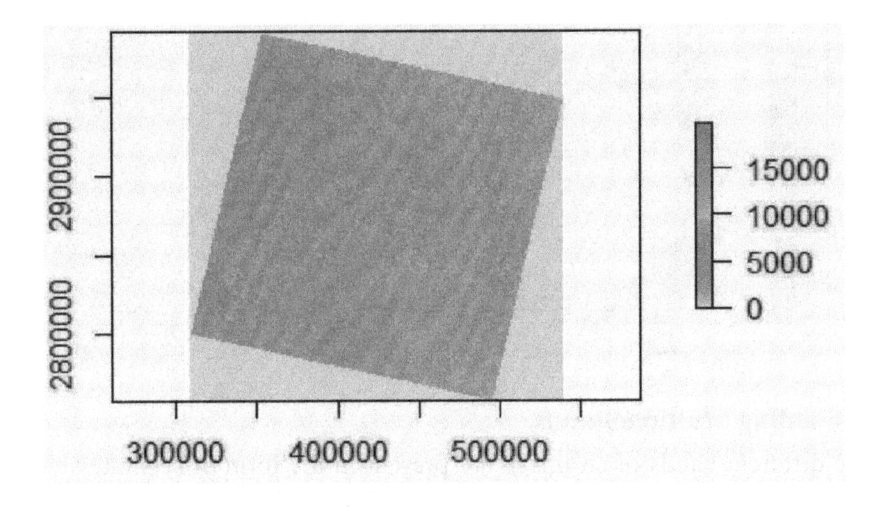

By using simple `plot` command, the default colour is applied to the output. The single band can also be plotted using specific colour. Generally, grey colour is used to display single band having different levels based on radiance value.

```
#Plotting single band with different levels of grey shades
>plot(img, col =gray(0:100/100))
```

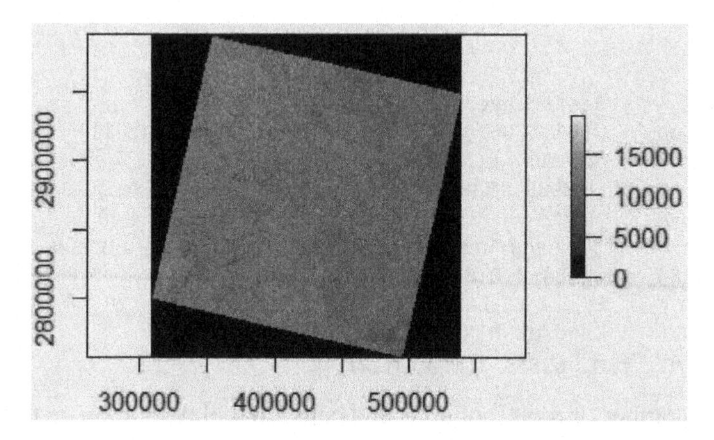

The distribution of pixel values in the raster layer can be plotted in the form of a histogram by using the following codes.

```
# the distribution of values in the single band raster
>hist(img, main="Distribution of pixel values", col="purple",
maxpixels=ncell(img))
```

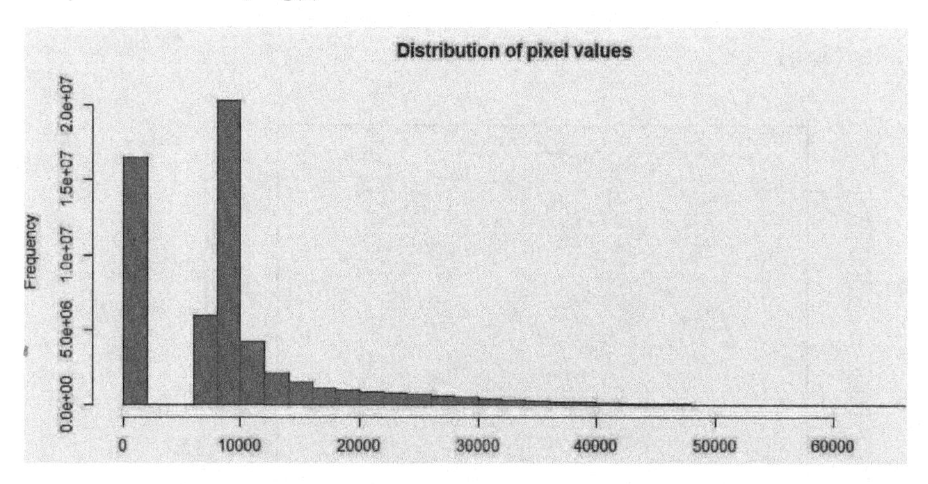

3.3. Loading Multiple Bands

If the different bands of an image are present in the form of separate single raster layers, then `stack` command available in the `raster` package is used to load these bands in the form of a single raster image with multiple bands. The `stack` command only links the files present in the source path of the computer as single raster image, but doesn't actually form or load a multiple band raster file. For loading a raster image with multiple bands in it, `brick` command is used. The `brick` command can also be applied on stacked image in order to form a single raster image containing all bands.

If the multiple bands of the same scene are available as separate rasters:

```
#Create the list of source files containing all bands as separate raster layers with same path and naming convention, but band number is different
>landsat =paste0('E:\\........\\LC08_L1TP_144042_20150419_20170409_01_T1_B', 1:7,".TIF")
>landsat# to display the created list containing path and file names
## [1] "E:\\..........\\LC08_L1TP_144042_20150419_20170409_01_T1_B1.TIF"
## [2] "E:\\..........\\LC08_L1TP_144042_20150419_20170409_01_T1_B2.TIF"
## [3] "E:\\..........\\LC08_L1TP_144042_20150419_20170409_01_T1_B3.TIF"
## [4] "E:\\..........\\LC08_L1TP_144042_20150419_20170409_01_T1_B4.TIF"
## [5] "E:\\..........\\LC08_L1TP_144042_20150419_20170409_01_T1_B5.TIF"
## [6] "E:\\..........\\LC08_L1TP_144042_20150419_20170409_01_T1_B6.TIF"
## [7] "E:\\..........\\LC08_L1TP_144042_20150419_20170409_01_T1_B7.TIF"
#creating raster stack
>landsat_stack =stack(landsat)
# to display the attributes of created raster stack
>landsat_stack
## class      : RasterStack
## dimensions : 7801, 7661, 59763461, 7  (nrow, ncol, ncell, nlayers)
## resolution : 30, 30  (x, y)
## extent     : 308985, 538815, 2757585, 2991615  (xmin, xmax, ymin, ymax)
## crs          : +proj=utm +zone=44 +datum=WGS84 +units=m +no_defs +ellps=WGS84 +towgs84=0,0,0
## names      : LC08_L1TP//9_01_T1_B1, LC08_L1TP//9_01_T1_B2, LC08_L1TP//9_01_T1_B3, LC08_L1TP//9_01_T1_B4, LC08_L1TP//9_01_T1_B5, LC08_L1TP//9_01_T1_B6, LC08_L1TP//9_01_T1_B7
## min values : 0,  0, 0,  0,  0,  0,  0
## max values : 65535, 65535, 65535, 65535, 65535, 65535, 65535
#creating raster brick from the stacked layer
>landsat_brick =brick(landsat_stack)
# to display the attributes of raster brick
>landsat_brick
## class      : RasterBrick
## dimensions : 7801, 7661, 59763461, 7  (nrow, ncol, ncell, nlayers)
## resolution : 30, 30  (x, y)
## extent     : 308985, 538815, 2757585, 2991615  (xmin, xmax, ymin, ymax)
## crs          : +proj=utm +zone=44 +datum=WGS84 +units=m +no_defs +ellps=WGS84 +towgs84=0,0,0
## source     : C:/................/raster/r_tmp_2019-06-10_105510_60600_24680.grd
## names      : LC08_L1TP//9_01_T1_B1, LC08_L1TP//9_01_T1_B2, LC08_L1TP//9_01_T1_B3, LC08_L1TP//9_01_T1_B4, LC08_L1TP//9_01_T1_B5, LC08_L1TP//9_01_T1_B6, LC08_L1TP//9_01_T1_B7
```

```
## min values :  0,    0,    0,    0,     0,    0,    0
## max values :  20530, 21049, 22192, 23430, 27718, 57702, 65535
```

You can see that by using raster brick, a raster image containing 7 raster layers is created in the memory as you may see the source path in the properties of raster brick. This temporary multiband raster layer in the memory is not created in case of raster stack.

Note:`raster` command cannot be used for loading multiband raster image unlike single band image because it considers only first layer and ignores remaining layer. It is clear from the following example where number of layers is not present in the attributes of the loaded multiband raster image.

```
#Loading multiband raster using raster command
>r =raster(landsat_stack)
>r #to get attributes of the Loaded Layer
## class      : RasterLayer
## dimensions : 7801, 7661, 59763461  (nrow, ncol, ncell)# no nLayers
## resolution : 30, 30  (x, y)
## extent     : 308985, 538815, 2757585, 2991615  (xmin, xmax, ymin,
ymax)
## crs         : +proj=utm +zone=44 +datum=WGS84 +units=m +no_defs
+ellps=WGS84 +towgs84=0,0,0
```

If the multiple bands of the same scene are already stacked in a single image, type the following command to load the raster:

```
#reading multiband raster file (one raster file containing all
bands)
>brick('Landsat_multi.tif')
## class      : RasterBrick
## dimensions : 7801, 7661, 59763461, 7  (nrow, ncol, ncell, nlayers)
## resolution : 30, 30  (x, y)
## extent     : 308985, 538815, 2757585, 2991615  (xmin, xmax, ymin,
ymax)
## crs         : +proj=utm +zone=44 +datum=WGS84 +units=m +no_defs
+ellps=WGS84 +towgs84=0,0,0
## source     : E:/............................/Landsat_multi.tif
## names       : Landsat_multi.1, Landsat_multi.2, Landsat_multi.3,
Landsat_multi.4, Landsat_multi.5, Landsat_multi.6, Landsat_multi.7
## min values : 0,      0,      0,      0,      0,      0,    0
## max values : 20530, 21049, 22192, 23430, 27718, 57702, 65535
```

3.4. Attributes of the loaded raster layer

The specific attributes of the loaded multiband or single band raster image like number of cells, dimensions, extent, number of rows, columns and bands, spatial resolution, projection, minimum and maximum pixel values, and size of the image can be known by using the following codes.

```
#to get number of cells
>ncell(landsat_brick)
## [1] 59763461
#to get number of rows
>nrow(landsat_brick)
## [1] 7801
#to get number of columns
>ncol(landsat_brick)
## [1] 7661
#to get number of bands in image
>nlayers(landsat_brick)
## [1] 7
#to get dimensions of raster
>dim(landsat_brick)
## [1] 7801 7661    7
#to get spatial resolution
>xres(landsat_brick)
## [1] 30
>yres(landsat_brick)
## [1] 30
>res(landsat_brick)
## [1] 30 30
#to get Coordinate Reference System (CRS)
>crs(landsat_brick)
>landsat_brick@crs#alternative code
## CRS arguments:
## +proj=utm +zone=44 +datum=WGS84 +units=m +no_defs +ellps=WGS84
## +towgs84=0,0,0
#view extent of raster
>landsat_brick@extent
> extent(landsat_brick) #alternative code
## class       : Extent
## xmin        : 308985
## xmax        : 538815
## ymin        : 2757585
## ymax        : 2991615
#to get minimum value of each band
>cellStats(landsat_brick, stat ='min')
## LC08_L1TP_144042_20150419_20170409_01_T1_B1
##                                           0
## LC08_L1TP_144042_20150419_20170409_01_T1_B2
##                                           0
## LC08_L1TP_144042_20150419_20170409_01_T1_B3
##                                           0
## LC08_L1TP_144042_20150419_20170409_01_T1_B4
```

```
##                                           0
## LC08_L1TP_144042_20150419_20170409_01_T1_B5
##                                           0
## LC08_L1TP_144042_20150419_20170409_01_T1_B6
##                                           0
## LC08_L1TP_144042_20150419_20170409_01_T1_B7
##                                           0
#to get maximum value of each band
>cellStats(landsat_brick, stat ='max')
## LC08_L1TP_144042_20150419_20170409_01_T1_B1
##                                       20530
## LC08_L1TP_144042_20150419_20170409_01_T1_B2
##                                       21049
## LC08_L1TP_144042_20150419_20170409_01_T1_B3
##                                       22192
## LC08_L1TP_144042_20150419_20170409_01_T1_B4
##                                       23430
## LC08_L1TP_144042_20150419_20170409_01_T1_B5
##                                       27718
## LC08_L1TP_144042_20150419_20170409_01_T1_B6
##                                       57702
## LC08_L1TP_144042_20150419_20170409_01_T1_B7
##                                       65535
#view size of image file on directory
object.size(landsat_brick)
## 14384 bytes
object.size(landsat_stack)
## 95704 bytes
```

2.5. Naming of bands in raster brick

In the above examples, you can see that the whole file name is displayed for the band name. We can specify the band names corresponding to its wavelength to increase the understandability. In the following example, we are setting the names of bands of the raster brick created from the Landsat 8 OLI image.

```
>names(landsat_brick) #check the original names of bands
## [1] "LC08_L1TP_144042_20150419_20170409_01_T1_B1"
## [2] "LC08_L1TP_144042_20150419_20170409_01_T1_B2"
## [3] "LC08_L1TP_144042_20150419_20170409_01_T1_B3"
## [4] "LC08_L1TP_144042_20150419_20170409_01_T1_B4"
## [5] "LC08_L1TP_144042_20150419_20170409_01_T1_B5"
## [6] "LC08_L1TP_144042_20150419_20170409_01_T1_B6"
## [7] "LC08_L1TP_144042_20150419_20170409_01_T1_B7"
```

```
#set the band names
>names(landsat_brick) =c('Coastal aerosol', 'Blue', 'Green',
'Red', 'NIR', 'SWIR-1', 'SWIR-2')
>names(landsat_brick) #display the newly set band names
## [1] "Coastal.aerosol" "Blue"            "Green"           "Red"
## [5] "NIR"             "SWIR.1"          "SWIR.2"
```

3.6. Plotting multiple band raster

The multiple band raster can be plotted in many ways as explained in the further section.

3.6.1. Plotting single band from a raster brick or stack

A single band from a multiband raster brick or stack can be plotted by mentioning the band number in `[[band number]]` notation as shown below.

```
#Plotting single band from a brick
> plot(landsat_brick[[1]], #plots first band
col =gray(0:100/100), axes =TRUE, main ="First band of Landsat")
```

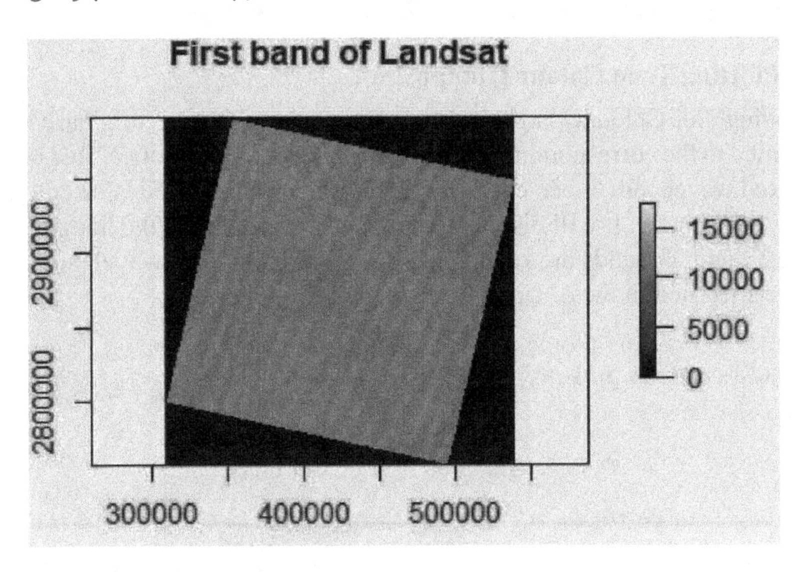

3.6.2. Plotting all bands separately

All bands can be plotted separately by using the following code.

```
#Plotting all bands separately
>plot(landsat_brick, col =gray(0:100/100), axes =TRUE, main
=names(landsat_brick))
```

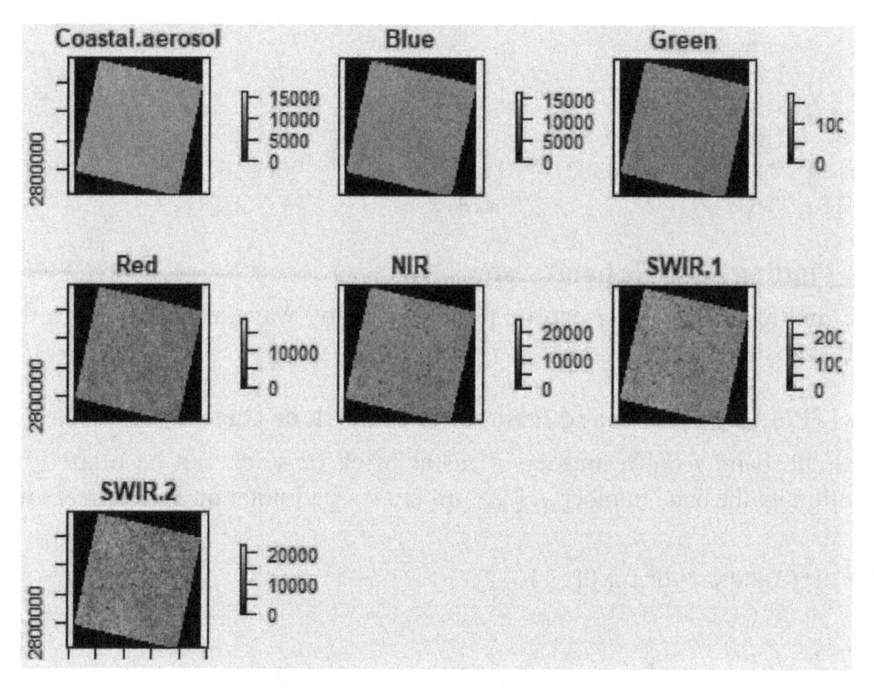

3.6.3. Plotting True Colour Composite

For plotting True Colour Composite of a multiband raster image, original colours are allotted to the corresponding wavelength bands in RGB plot *i.e.* Red colour to the Red waveband, Green colour to the Green waveband and Blue colour to the Blue waveband. It will display the image as it appears to the human eyes. As 2nd, 3rd and 4th bands are related to Blue, Green and Red wavelengths, so, they are specified in the order of RGB in the following code.

```
#Plot True Colour Composite
>plotRGB(landsat_brick, r =4, g = 3, b = 2,axes = T, stretch
='lin', main ="Landsat True colour composite")
```

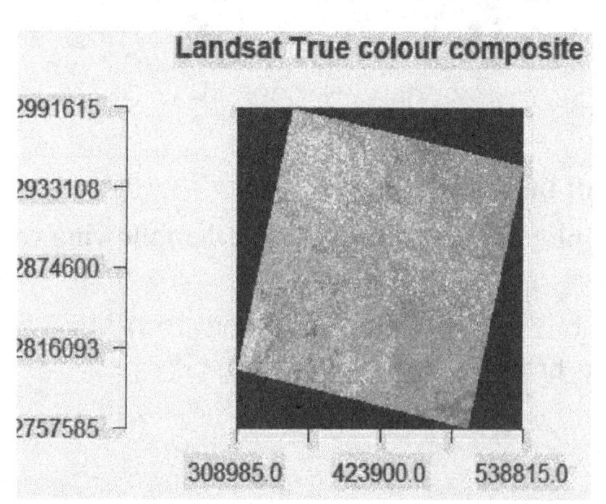

3.6.4. Plotting False Colour Composite

For plotting standard False Colour Composite of the multiband raster image, red colour is assigned to Near Infrared (NIR) band, Green colour to Red band and Blue colour to Green band. This False Colour Composite (FCC) can readily detect the vegetation as it appears in red colour because vegetation has higher reflectance in the NIR band. Water appears in dark-bluish colour whereas bare soils, roads and buildings appear in various shades of cyan, blue or grey colour depending on their composition. The 3rd, 4th and 5th bands in Landsat 8 image corresponds to Green, Red and NIR wavelengths, respectively therefore, these bands were selected while plotting the standard FCC.

```
#Plots False Colour Composite
>plotRGB(landsat_brick,r =5, g = 4, b = 3,axes = T, stretch
='lin', main ="Landsat False colour composite")
```

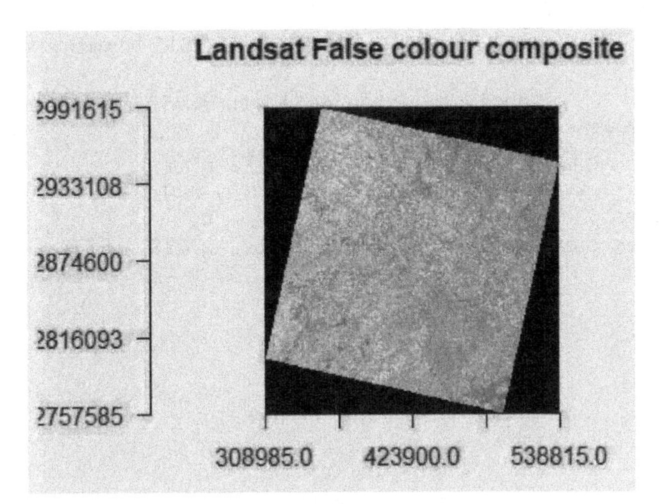

3.7. Writing output raster to the disk

The stacked file can be saved to disk and a raster file can be created by using `writeRaster` function. This is similar to 'composite band' or 'layer stacking'. The raster can be saved either in GeoTiff format or raster-grd format. While saving in GeoTiff format, band or layer names are lost whereas layer or band names are preserved in 'raster-grd' format, but 'raster-grd' format is not amenable to many programs or software like GeoTiff format.

```
#writing raster file in geotiff format - bands names lost
>writeRaster(landsat_stack, filename ="Landsat_multi.tif",
overwrite =TRUE)
>brick('Landsat_multi.tif')
## class      : RasterBrick
## dimensions : 7801, 7661, 59763461, 7  (nrow, ncol, ncell, nlayers)
```

```
## resolution : 30, 30  (x, y)
## extent     : 308985, 538815, 2757585, 2991615  (xmin, xmax, ymin, ymax)
## crs         : +proj=utm +zone=44 +datum=WGS84 +units=m +no_defs +ellps=WGS84 +towgs84=0,0,0
## source     : E:/............................../Landsat_multi.tif
## names       : Landsat_multi.1, Landsat_multi.2, Landsat_multi.3, Landsat_multi.4, Landsat_multi.5, Landsat_multi.6, Landsat_multi.7
## min values : 0,      0,      0,      0,      0,      0,     0
## max values : 20530, 21049, 22192,  23430,  27718,  57702,  65535
#writing raster file in raster-grd format - bands names not lost
>writeRaster(landsat_stack, filename ="Landsat_multi.grd", overwrite =TRUE)
>brick('Landsat_multi.grd')
## class      : RasterBrick
## dimensions : 7801, 7661, 59763461, 7  (nrow, ncol, ncell, nlayers)
## resolution : 30, 30  (x, y)
## extent     : 308985, 538815, 2757585, 2991615  (xmin, xmax, ymin, ymax)
## crs         : +proj=utm +zone=44 +datum=WGS84 +units=m +no_defs +ellps=WGS84 +towgs84=0,0,0
## source     : E:/............................../Landsat_multi.grd
## names      : Coastal.aerosol, Blue, Green, Red, NIR, SWIR.1, SWIR.2
## min values : 0,     0,     0,     0,     0,      0,      0
## max values : 20530, 21049, 22192, 23430, 27718,  57702,  65535
```

Chapter 4

Radiometric Calibration

Sensors onboard the satellite records the intensity of emitted or reflected electromagnetic radiations in the form of digital number (DN) for each pixel. These digital numbers can be transformed to more meaningful intensity units like radiance or reflectance. This process of conversion of digital numbers to radiance or reflectance is known as radiometric calibration. Sensor specific information is required to conduct radiometric calibration.

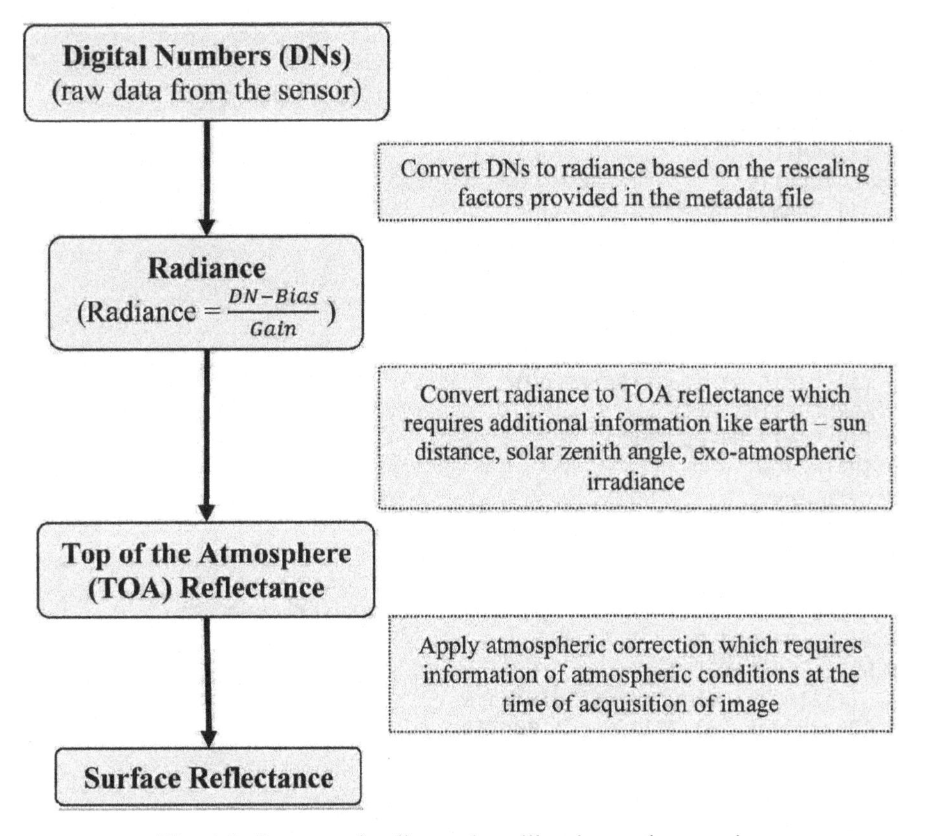

Fig. 4.1: Process of radiometric calibration and correction

In this chapter, we are using the Landsat 8 OLI and TIRS data for performing the task of radiometric calibration. The data required for radiometric calibration present in the metadata of the image file. `RStoolbox` package has functions to do radiometric calibration of the Landsat 8 data specifically. The details about `RStoolbox` package can be read from https://www.rdocumentation.org/packages/RStoolbox/versions/0.2.4/topics/RStoolbox link.

4.1. Reading metadata of the Landsat image

First of all, we need to load the `RStoolbox` and `raster` packages for performing the radiometric calibration process. After loading the packages, we need to read the metadata of the downloaded Landsat 8 OLI and TIRS image (the data used is same as in Chapter 3).

```
#Load packages
>library(RStoolbox)   #For radiometric calibration of Landsat data
>library(raster)             #For analysis of raster data
#Read metadata file from the path
>y =readMeta('E:\\.....\\LC08_L1TP_144042_20150419_20170409_01_T1_MTL.txt')
>y                       #Display the metadata of the acquired image
## $METADATA_FILE
## [1] "E:/................../LC08_L1TP_144042_20150419_20170409_01_T1_MTL.txt"
## $METADATA_FORMAT
## [1] "MTL"
## $SATELLITE
## [1] "LANDSAT8"
## $SENSOR
## [1] "OLI_TIRS"
## $SCENE_ID
## [1] "LC81440422015109LGN01"
## $ACQUISITION_DATE
## [1] "2015-04-19 05:06:39 GMT"
## $PROCESSING_DATE
## [1] "2017-04-09 GMT"
## $PATH_ROW
## path  row
##  144   42
## $PROJECTION
## CRS arguments:
##  +proj=utm +zone=44 +units=m +datum=WGS84 +ellps=WGS84
## +towgs84=0,0,0
## $SOLAR_PARAMETERS
##    azimuth  elevation   distance
## 120.363749  63.721163   1.004268
## $DATA
##                                                FILES  BANDS QUANTITY
## B1_dn   LC08_L1TP_144042_20150419_20170409_01_T1_B1.TIF  B1_dn       dn
## B2_dn   LC08_L1TP_144042_20150419_20170409_01_T1_B2.TIF  B2_dn       dn
```

```
## B3_dn   LC08_L1TP_144042_20150419_20170409_01_T1_B3.TIF  B3_dn      dn
## B4_dn   LC08_L1TP_144042_20150419_20170409_01_T1_B4.TIF  B4_dn      dn
## B5_dn   LC08_L1TP_144042_20150419_20170409_01_T1_B5.TIF  B5_dn      dn
## B6_dn   LC08_L1TP_144042_20150419_20170409_01_T1_B6.TIF  B6_dn      dn
## B7_dn   LC08_L1TP_144042_20150419_20170409_01_T1_B7.TIF  B7_dn      dn
## B9_dn   LC08_L1TP_144042_20150419_20170409_01_T1_B9.TIF  B9_dn      dn
## B10_dn LC08_L1TP_144042_20150419_20170409_01_T1_B10.TIF B10_dn     dn
## B11_dn LC08_L1TP_144042_20150419_20170409_01_T1_B11.TIF B11_dn     dn
## B8_dn   LC08_L1TP_144042_20150419_20170409_01_T1_B8.TIF  B8_dn      dn
## QA_dn   LC08_L1TP_144042_20150419_20170409_01_T1_BQA.TIF  QA_dndn
##         CATEGORY NA_VALUE SATURATE_VALUE SCALE_FACTOR DATA_TYPE
## B1_dn      image       NA             NA            1        NA
## B2_dn      image       NA             NA            1        NA
## B3_dn      image       NA             NA            1        NA
## B4_dn      image       NA             NA            1        NA
## B5_dn      image       NA             NA            1        NA
## B6_dn      image       NA             NA            1        NA
## B7_dn      image       NA             NA            1        NA
## B9_dn      image       NA             NA            1        NA
## B10_dn     image       NA             NA            1        NA
## B11_dn     image       NA             NA            1        NA
## B8_dn        pan       NA             NA            1        NA
## QA_dnqa        NA             NA            1        NA
##         SPATIAL_RESOLUTION RADIOMETRIC_RESOLUTION
## B1_dn                   30                     16
## B2_dn                   30                     16
## B3_dn                   30                     16
## B4_dn                   30                     16
## B5_dn                   30                     16
## B6_dn                   30                     16
## B7_dn                   30                     16
## B9_dn                   30                     16
## B10_dn                  30                     16
## B11_dn                  30                     16
## B8_dn                   15                     16
## QA_dn                   30                     16
## $CALRAD
##            offset       gain
## B1_dn  -62.24627 0.01244900
## B2_dn  -63.74094 0.01274800
## B3_dn  -58.73673 0.01174700
## B4_dn  -49.53012 0.00990600
## B5_dn  -30.30996 0.00606200
## B6_dn   -7.53781 0.00150760
## B7_dn   -2.54065 0.00050813
## B8_dn  -56.05446 0.01121100
## B9_dn  -11.84582 0.00236920
## B10_dn   0.10000 0.00033420
## B11_dn   0.10000 0.00033420
## $CALREF
```

```
##        offset  gain
## B1_dn    -0.1 2e-05
## B2_dn    -0.1 2e-05
## B3_dn    -0.1 2e-05
## B4_dn    -0.1 2e-05
## B5_dn    -0.1 2e-05
## B6_dn    -0.1 2e-05
## B7_dn    -0.1 2e-05
## B8_dn    -0.1 2e-05
## B9_dn    -0.1 2e-05
## $CALBT
##                K1        K2
## B10_dn  774.8853  480.8883
## B11_dn 1321.0789 1201.1442
##
## attr(,"class")
## [1] "ImageMetaData" "RStoolbox"
```

So, `readMeta` code reads the metadata file present as MTL.txt format in the Landsat image file. This code displays metadata information like satellite name, sensor name, scene ID, image acquisition date, image processing date, path and row number, projection, etc. along with the information required for radiometric calibration like solar azimuth angle, solar elevation, earth – sun distance, bands available, spatial and radiometric resolution of each band, offset and gain values of each band for conversion to radiance, reflectance and brightness temperature (in case of thermal bands).

4.2. Importing separate files into single stack from metadata

In `RStoolbox` package, `stackMeta` function reads Landsat MTL or XML metadata files and loads single Landsat Tiffs into a raster stack. This function creates stack of same resolution images by default, but if `allResolutions` parameter is set `TRUE`, then stack of all bands irrespective of their resolutions can be created. Importing Landsat images through `stackMeta` function is necessary to perform the radiometric calibration. The example of using this function is given below.

```
#Import Separate Landsat Files Into Single Stack
>lsat
=stackMeta('E:\\..........\\LC08_L1TP_144042_20150419_20170409_01_T1_MTL.txt',
quantity ="dn", category ="image", allResolutions =FALSE)

#Define no data value in the image (DN of no data value in
Landsat is 0)
>NAvalue(lsat) =0
```

```
#To display the attributes of landsat stack
>lsat
## class      : RasterStack
## dimensions : 7801, 7661, 59763461, 10  (nrow, ncol, ncell, nlayers)
## resolution : 30, 30  (x, y)
## extent     : 308985, 538815, 2757585, 2991615  (xmin, xmax, ymin,
ymax)
## crs         : +proj=utm +zone=44 +datum=WGS84 +units=m +no_defs
+ellps=WGS84 +towgs84=0,0,0
## names      : B1_dn, B2_dn, B3_dn, B4_dn, B5_dn, B6_dn, B7_dn, B9_dn,
B10_dn, B11_dn
## min values : 0,     0,     0,     0,     0,     0,     0,     0,     0,
0
## max values : 65535, 65535, 65535, 65535, 65535, 65535, 65535, 65535,
65535,  65535
```

4.3. Conversion from Digital Number to Radiance

The digital numbers of all the raster layers in the stack of the Landsat image can be converted to radiance values with unit W/m²/srad/μm by using "rad" method in the **radCor** function of the RStoolbox package. The bands for which we want to compute radiance can be defined in bandSet argument. Information about the atmosphere or cloudy condition during the acquisition of the image can be provided in the atmosphereargument. The example of usage of this function is explained further.

```
#Conversion of DN values to the radiance values
>landsat_rad =radCor(lsat, metaData = y, method =”rad”,
bandSet =c(“B1_dn”,”B2_dn”,”B3_dn”,”B4_dn”,
“B5_dn”,”B6_dn”,”B7_dn”),
atmosphere =”veryClear”)

#Check the attributes of landsat_rad
>landsat_rad
## class      : RasterStack
## dimensions : 7801, 7661, 59763461, 7  (nrow, ncol, ncell, nlayers)
## resolution : 30, 30  (x, y)
## extent     : 308985, 538815, 2757585, 2991615  (xmin, xmax, ymin,
ymax)
## crs         : +proj=utm +zone=44 +datum=WGS84 +units=m +no_defs
+ellps=WGS84 +towgs84=0,0,0
## names      :    B1_tra,     B2_tra,     B3_tra,     B4_tra,     B5_tra,
B6_tra,     B7_tra
## min values :72.0037460, 61.6866320, 42.8865670, 28.7173740, 15.1611020,
2.3294320,  0.5426828
## max values : 193.33170,  204.59171,  201.95269,  182.56746,  137.71656,
79.45373,   30.75965
```

```
#Plot RGB FCC of landsat_rad
>plotRGB(landsat_rad[[5:3]], axes = T, stretch ='lin')
```

4.4. Conversion from Digital Number to Top of Atmosphere (TOA) reflectance

The digital numbers of all the raster layers in the stack of the Landsat image can be converted to apparent reflectance by using "apref" method in the **radCor** function as explained earlier.

```
#Conversion of DN values to top of the atmosphere reflectance
>landsat_rf =radCor(lsat, metaData = y, method ="apref",
bandSet =c("B1_dn","B2_dn","B3_dn","B4_dn",
"B5_dn","B6_dn","B7_dn"),
atmosphere ="veryClear")
#To display the attributes of landsat_rf
>landsat_rf
## class      : RasterStack
## dimensions : 7801, 7661, 59763461, 7  (nrow, ncol, ncell, nlayers)
## resolution : 30, 30  (x, y)
## extent     : 308985, 538815, 2757585, 2991615  (xmin, xmax, ymin,
ymax)
## crs        : +proj=utm +zone=44 +datum=WGS84 +units=m +no_defs
+ellps=WGS84 +towgs84=0,0,0
## names      :   B1_tre,    B2_tre,    B3_tre,    B4_tre,    B5_tre,
B6_tre,    B7_tre
## min values :0.12901355, 0.10793509, 0.08143646, 0.06466291, 0.05578542,
0.03446161, 0.02382200
## max values : 0.3464005,  0.3579769,  0.3834718,  0.4110857,  0.5067306,
1.0000000,  1.0000000
```

4.5. Conversion from Digital Number to surface reflectance

Similarly, the digital numbers of all the raster layers in the stack of the Landsat image can be converted to surface reflectance by using "dos" method in the radCor function. This "dos" method computes at the surface reflectance by using Dark Object subtraction technique for atmospheric scattering correction given by Chavez (1989).

```
#conversion of DN values to at the surface reflectance
>landsat_srf =radCor(lsat, metaData = y, method =”dos”,
bandSet =c(“B1_dn”,”B2_dn”,”B3_dn”,”B4_dn”,
“B5_dn”,”B6_dn”,”B7_dn”),
atmosphere =”veryClear”)
#Check the attributes of landsat_srf
>landsat_srf
## class      : RasterStack
## dimensions : 7801, 7661, 59763461, 7  (nrow, ncol, ncell, nlayers)
## resolution : 30, 30  (x, y)
## extent     : 308985, 538815, 2757585, 2991615  (xmin, xmax, ymin,
ymax)
## crs         : +proj=utm +zone=44 +datum=WGS84 +units=m +no_defs
+ellps=WGS84 +towgs84=0,0,0
## names      :   B1_sre,    B2_sre,    B3_sre,    B4_sre,    B5_sre,
B6_sre,     B7_sre
## min values :0.11987894, 0.10874178, 0.08323355, 0.06549114, 0.05629004,
0.03325126, 0.02243862
## max values : 0.3460845, 0.3606562, 0.3919465, 0.4163525, 0.5113131,
1.0000000,  1.0000000
```

In this section, we have discussed about the R codes for performing radiometric calibration of Landsat data only. The radiometric conversions for other sensors can be done by applying their corresponding formulae mentioned in their respective literatures. Any formula can be applied on raster images using R – codes, which is further illustrated in the section 14.2 *'Performing algebra on raster images'* of the Chapter 14.

Chapter 5

Vector Data in R

Vector data consists of points, lines and polygons with x, y coordinates known as vertices representing the geographic location *i.e.* latitude and longitude. Vector data depicts the shape of the spatial feature. Points consist of single x, y locations. Examples of points are tower location, sampling location, etc. Lines represent 1 – dimensional features like roads, streams, etc., and are composed of many or at least two connected points or vertices. Polygons denote 2 – dimensional features like boundaries with three or more vertices that are connected and closed. The vector data are represented in the following figure (fig. 5.1).

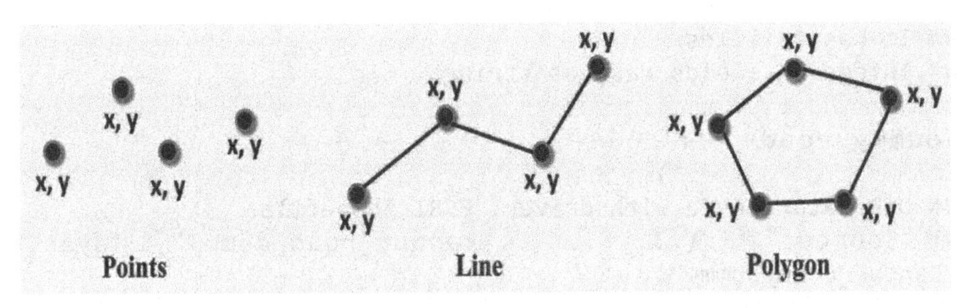

Fig. 5.1: Representation of vector data

For working with vector data in R, many R packages are required like `sp`, `raster` and `rgdal`. Package `ggplot2` is also required for creating attractive plots of vector data.

```
#Import packages
>library(sp)          #Classes and methods for spatial data
>library(raster)             #Analysis of raster data
>library(rgdal)              #Bindings for GDAL
>library(ggplot2)     #For data visualization
```

The data used in the chapter are:

Chapter5\data\India_state\Admin2.shp

Chapter5\data\kanpur\Kanpur.shp

Chapter5\data\kanpur_points\kanpur_points.shp

Chapter5\data\kanpur_road_dummy\Kanpur_road_dummy.shp

Chapter5\data\state_shpfile\Maharashtra.shp

Chapter5\data\data_for_creating_vector.csv

5.1. Importing ESRI shapefile

ESRI shapefile in the form of points, lines or polygons can be imported in R by using `readOGR` function of `rgdal` package. In `readOGR` function, you have to mention the path of the directory containing ESRI shapefile in the first argument, and then the name of the shapefile without any extension in the second argument. The R codes for importing vector data in the form of points, lines and polygon are given below.

```
#Importing point vector data
>kanpur_pts =readOGR('E:\\ …………\\kanpur_points',
'kanpur_points')
## OGR data source with driver: ESRI Shapefile
## Source: "E:\........\kanpur_points", layer: "kanpur_points"
## with 10 features
## It has 2 fields
## Integer64 fields read as strings:  id
#Importing line vector data
>dummy_road =readOGR('E:\\....\\kanpur_road_dummy',
'Kanpur_road_dummy')
## OGR data source with driver: ESRI Shapefile
## Source: "E:\...........\kanpur_road_dummy", layer:
"Kanpur_road_dummy"
## with 4 features
## It has 2 fields
## Integer64 fields read as strings:  id
#Importing polygon vector data
>ind_state =readOGR('E:\\……………………………\\India_state',
'Admin2')#Polygon data downloaded from https://map.igismap.com
## OGR data source with driver: ESRI Shapefile
## Source: "E:\.............\India_state", layer: "Admin2"
## with 36 features
## It has 1 fields
```

5.2. Shapefile metadata and attributes

Metadata of a shapefile like object type, Coordinate Reference System, extent, and attributes can be determined by using following codes in `rgdal` and `raster` packages of R.

```
#Display metadata and attribute of imported shapefile
>class(ind_state) #Object type
## [1] "SpatialPolygonsDataFrame"
## attr(,"package")
## [1] "sp"
>crs(ind_state) #Coordinate Reference System
## CRS arguments:
##   +proj=longlat +datum=WGS84 +no_defs +ellps=WGS84
+towgs84=0,0,0
>extent(ind_state) #Extent of shapefile
## class       : Extent
## xmin        : 68.18625
## xmax        : 97.41529
## ymin        : 6.755953
## ymax        : 37.07827
>ind_state#View all metadata at the same time
## class        : SpatialPolygonsDataFrame
## features     : 36
## extent       : 68.18625, 97.41529, 6.755953, 37.07827  (xmin,
xmax, ymin, ymax)
## crs        : +proj=longlat +datum=WGS84 +no_defs +ellps=WGS84
+towgs84=0,0,0
## variables    : 1
## names        :                   ST_NM
## min values   : Andaman & Nicobar Island
## max values   :             West Bengal
>ind_state$ST_NM#View attributes
##  [1] Andaman & Nicobar Island Arunanchal Pradesh
##  [3] Assam                    Bihar
##  [5] Chandigarh               Chhattisgarh
##  [7] Dadara& Nagar Havelli   Daman & Diu
##  [9] Goa                      Gujarat
## [11] Haryana                  Himachal Pradesh
## [13] Jammu & Kashmir          Jharkhand
## [15] Karnataka                Kerala
## [17] Lakshadweep              Madhya Pradesh
## [19] Maharashtra              Manipur
## [21] Meghalaya                Mizoram
## [23] Nagaland                 NCT of Delhi
## [25] Puducherry               Punjab
## [27] Rajasthan                Sikkim
## [29] Tamil Nadu               Telangana
## [31] Tripura                  Uttar Pradesh
## [33] Uttarakhand              West Bengal
## [35] Odisha                   Andhra Pradesh
```

```
## 36 Levels: Andaman & Nicobar Island Andhra Pradesh ... West
Bengal
>length(ind_state) #View feature count
## [1] 36
```

5.3. Plotting shapefiles

5.3.1. Plot single shapefile

ESRI shapefile can be plotted by using many codes from various packages. Some codes for plotting shapefile are as follows:

1. `plot` function from `raster` package
2. `spplot` function from `sp` package
3. `geom_polygon` function from `ggplot2` package

The examples of above-mentioned R codes are given below:

```
#Plotting shapefile with 'raster' package
>plot(ind_state, col="wheat", border="black", lwd=2,#Line thickness
main="Map of Indian States")
```

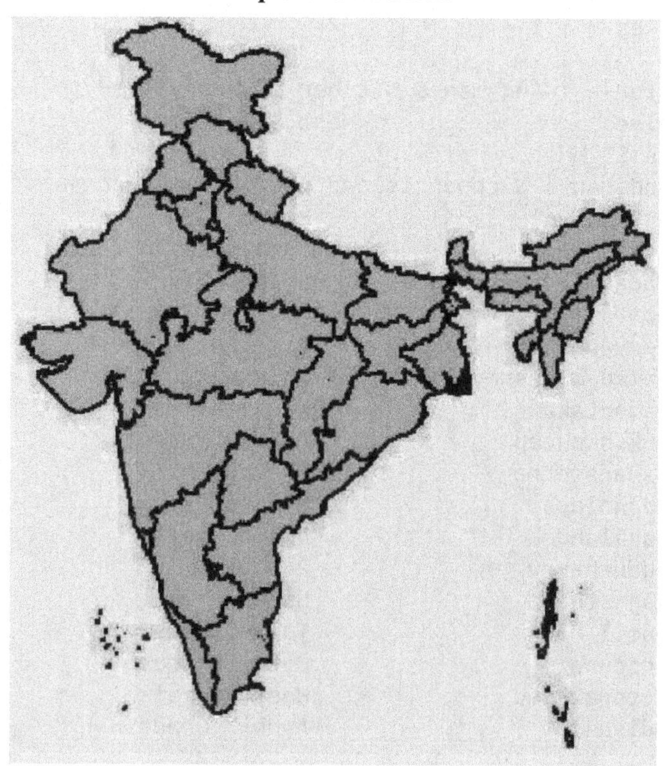

```
#Plotting shapefile with 'sp' package
>spplot(ind_state, zcol ='ST_NM') #'ST_NM' is shapefile attribute
containing state names which are displayed by various colours
```

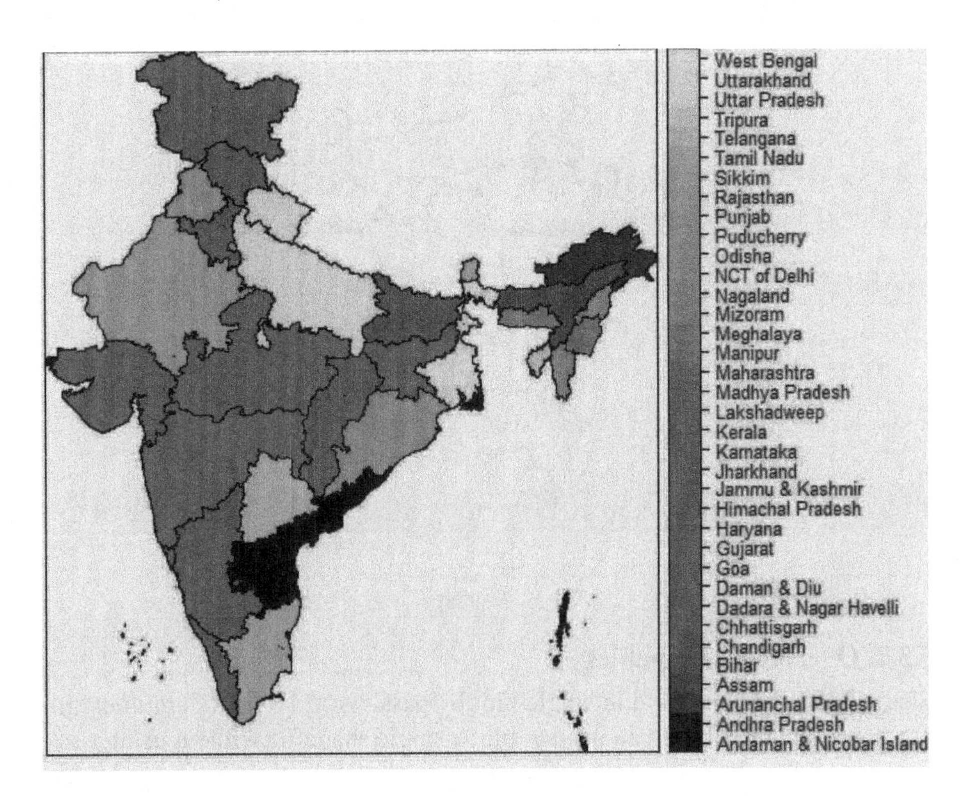


```
#Plotting shapefile with 'ggplot2' package
>ggplot() +geom_polygon(data =ind_state,
aes(x = long, y =lat, group = group),
colour ="black", fill ="white")
```

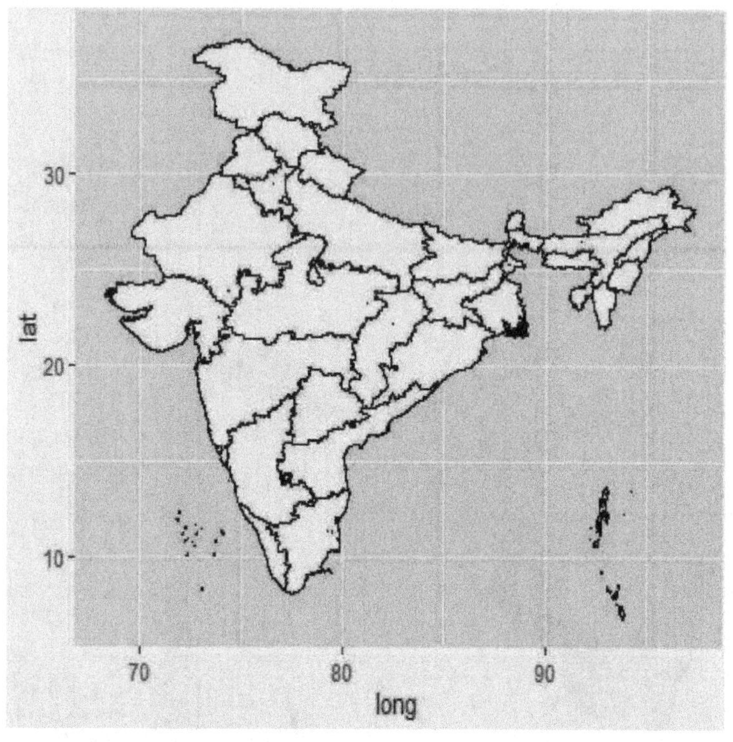

5.3.2. Overlaying shapefiles

Shapefiles can be overlaid in single plot by passing `add =TRUE` argument in the `plot` function. First you need to plot a single shapefile without using `add` function. After that, other shapefiles can be plotted in the previous plot by using `add =TRUE` argument. In the following example, we are plotting points, lines and polygons in the same plot. The polygon shapefile used in this example is of Kanpur district which is being imported prior to plotting. The points and lines shapefiles used in this example are already imported as shown in the section 5.1.

```
#Import Kanpur district polygon shapefile
>kanpur_dist =readOGR('E:\\..........\\kanpur', 'Kanpur')
## OGR data source with driver: ESRI Shapefile
## Source: "E:\.............\kanpur", layer: "Kanpur"
## with 1 features
## It has 5 fields
#Plotting multiple shapefiles
   #Plot Kanpur district polygon
>plot(kanpur_dist, col ='wheat', lwd =2, main ="Kanpur District")
>plot(dummy_road, col ="red", add =TRUE) #Plot lines
>plot(kanpur_pts, col ="blue", add =TRUE, pch =19) #Plot points
#'pch' parameter decides shape of the points
```

Kanpur District

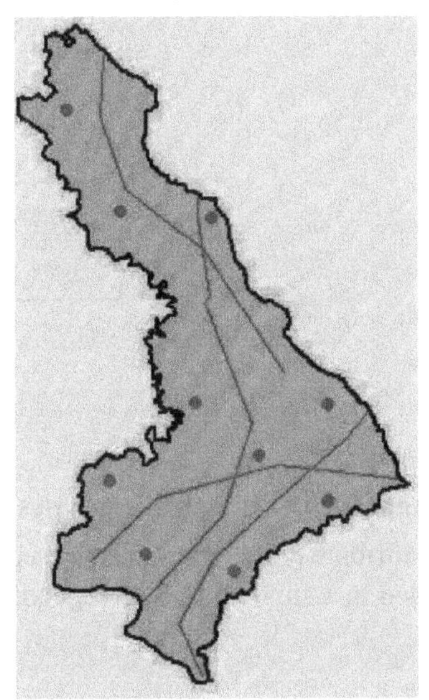

5.3.3. Plot by shapefile attributes

The shapefile has attributes of varying types or values. The plotting of shapefile with respect to its attribute type is done by providing different colour, line width, symbols, etc. to each attribute type. In the following example, a point shapefile is being plotted which has attribute named as 'Feature' with three levels *i.e.* Hospital, School and Shop as shown below.

```
>kanpur_pts$Feature#View features
## [1] Hospital Hospital School   Shop    Shop    School   Shop
## [8] Shop     School   Hospital
## Levels: Hospital School Shop
```

These three levels of the attribute 'Feature' can be plotted by giving separate colour and symbol to each level as shown below.

```
#Plot by giving separate colour and symbols to each level
>pt_col =c('brown', 'blue', 'red')#Assigning colours for each
level
>plot(kanpur_pts, col =pt_col, pch =c(15, 17, 19) #assigning
symbols,
main ="Kanpur Dummy Features")
>legend("right", legend=levels(kanpur_pts$Feature), col =pt_col,
cex =0.7, pch =c(15, 17, 19)) #Adding legends
```

Kanpur Dummy Features

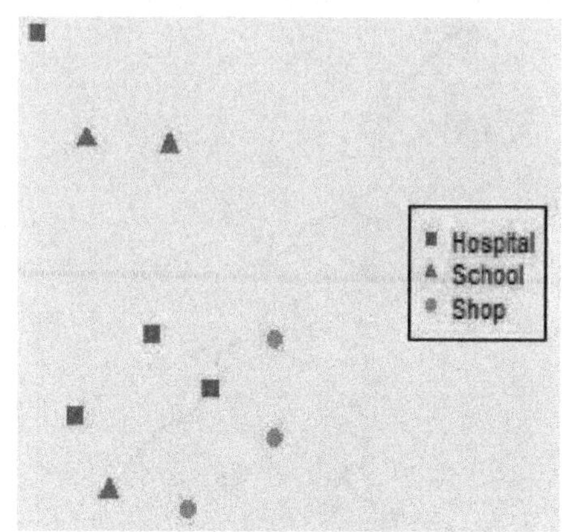

5.3.4. Plot multiple shapefiles with attribute types

Multiple shapefiles with attribute types represented by separate colours, symbols or thickness can be plotted in a single plot with legends as shown below.

```
#Plotting three shapefiles together with legends
>plot(kanpur_dist, lwd =5, main ="Kanpur Dummy Map")# Plot polygon
#Making list of colours for assigning to lines based on their
type
>road_col =c('darkgreen', 'grey', 'darkblue')
#Adding the plot of lines to the previous plot
>plot(dummy_road, col =road_col,#assigning colours based on line
type
lwd =c(4, 3, 2), #assigning different thickness based on line
type
 add =TRUE) #adding to the existing plot
#Making list of colours for assigning to points based on their
type
>pt_col =c('brown', 'blue', 'red')
#Adding the plot of points to the previous plot
>plot(kanpur_pts, col =pt_col, #assigning colours based on feature
type
pch =c(15, 17, 19), #assigning symbols based on feature type
 add =TRUE)#adding to the existing plot
#assign plot to an object for easy modification
>plot_Kanpur =recordPlot()
#create a list of all labels
>label_kanpur =c(c('Highway', 'Expressway', 'District Road'),
levels(kanpur_pts$Feature))
>label_kanpur#View list
## [1] "Highway"      "Expressway"    "District Road" "Hospital"
```

```
## [5] "School"        "Shop"
>plot_col =c(road_col, pt_col) #List of plot colours
>plot_col#View List
## [1] "darkgreen" "grey"      "darkblue"  "brown"      "blue"
"red"
>lineLegend =c(1, 1, 1, NA, NA, NA) # create Line object
>lineLegend # display created object
## [1]  1  1  1 NA NANA
>plotSym =c(NA,NA,NA, 15, 17, 19) # create List of symbols for
points
>plotSym # display created object
## [1] NA NANA 15 17 19
# render plot
>plot_Kanpur
# add Legends to the plot
>legend("right", # Alignment of the Legends
legend =label_kanpur, col =plot_col,
bty ="n", # No box around Legends
lty =lineLegend, pch =plotSym,
cex =0.8, # Character expansion factor
lwd =c(4, 3, 2, NA, NA, NA), # Line width to Legends
 inset =0.01, # Inset distance from the margin
x.intersp =1, y.intersp =1) # character interspacing factor
```

Kanpur Dummy Map

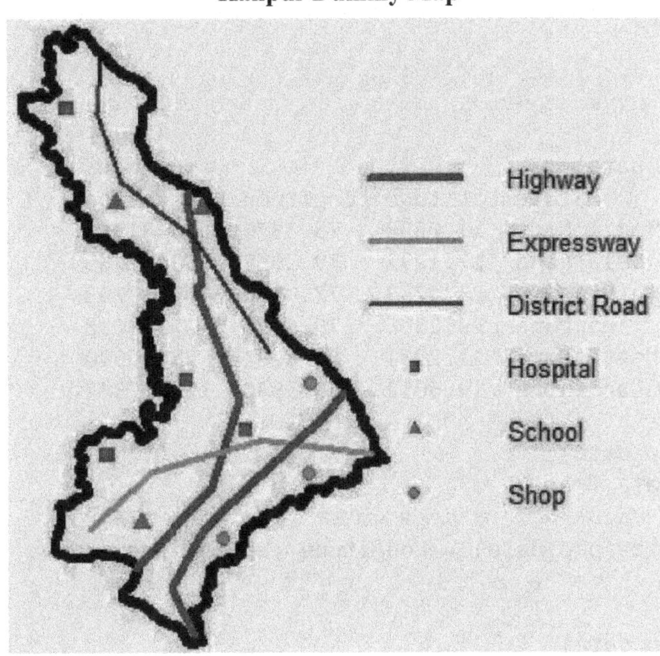

5.4. Convert table data with spatial information to point file

Table data present in '.csv' format with coordinate information in the columns can be converted to point shapefile through `sp` package by following steps:

1) Read the '.csv' file.
2) Set spatial coordinates by specifying column names containing geographic location information *i.e.* latitude and longitude in the `coordinates` function of the `sp` package in order to create spatial object.
3) Get the projection string of the desired shapefile whose projection you want to assign to the spatial object created in the previous step by using `proj4string` function.
4) Assign the projection of the desired shapefile to the created spatial object through `CRS` function.

In this way, data present in table can be converted to point file. In this section, the population data of the Indian states with their respective latitude and longitude is converted to point file. The name of the file containing data is *'data_for_creating_vector.csv'* which has four columns *viz.* State, Latitude, Longitude and Population. This file is converted to spatial data or point vector data by assigning the projection of `ind_state` spatial object imported in the section 5.1. This point file is then plotted using `spplot` function to show the distribution of points in accordance with the population data. The R code are given below.

```
#Read the csv file containing data
>pop_data =read.csv('E:/....../data_for_creating_vector.csv', header
=TRUE)
>head(pop_data) #Read first few rows of imported csv file
##              State Latitude Longitude Population
## 1       Tamil Nadu  11.0598   78.3875   76481545
## 2        Telangana  17.1232   79.2088   38472769
## 3 Madhya Pradesh   23.4733   77.9480   82342793
## 4          Haryana  29.2385   76.4319   27388008
## 5     Chhattisgarh  21.2951   81.8282   28566990
## 6      Maharashtra  19.6012   75.5530  120837347
#Check the class of imported csv file
>class(pop_data)
## [1] "data.frame"
# Convert data frame object to the spatial object
>coordinates(pop_data) =~Longitude +Latitude
# Again Check the class to confirm the creation of spatial
object
>class(pop_data)
```

```
## [1] "SpatialPointsDataFrame"
## attr(,"package")
## [1] "sp"
# Get the projection from 'ind_state' shapefile
>proj4string(ind_state)
## [1] "+proj=longlat +datum=WGS84 +no_defs +ellps=WGS84 +towgs84=0,0,0"
# Assign the projection of 'ind_state' shapefile to the spatial object
>proj4string(pop_data) =CRS(proj4string(ind_state))
>proj4string(pop_data) #check projection
## [1] "+proj=longlat +datum=WGS84 +no_defs +ellps=WGS84 +towgs84=0,0,0"
#plot spatial object or point vector file based on population values
>options(scipen =999) #disable scientific notation
>spplot(pop_data, "Population", colorkey =TRUE)
```

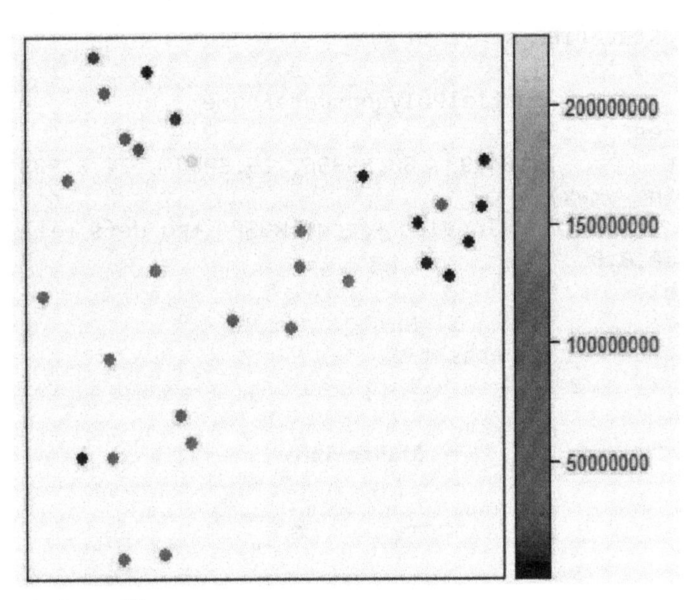

5.5. Subset shapefile

Subset of a shapefile can be created based on attribute type by selecting and saving it. In the following example, shapefile of the state 'Maharshtra' is created by subsetting it from the India shapefile imported in the section 5.1.

```
>ind_state$ST_NM#view attribute of shapefile
##  [1] Andaman &-Nicobar Island Arunanchal Pradesh
##  [3] Assam                    Bihar
##  [5] Chandigarh               Chhattisgarh
```

```
##  [7] Dadara& Nagar Havelli   Daman & Diu
##  [9] Goa                      Gujarat
## [11] Haryana                  Himachal Pradesh
## [13] Jammu & Kashmir          Jharkhand
## [15] Karnataka                Kerala
## [17] Lakshadweep              Madhya Pradesh
## [19] Maharashtra              Manipur
## [21] Meghalaya                Mizoram
## [23] Nagaland                 NCT of Delhi
## [25] Puducherry               Punjab
## [27] Rajasthan                Sikkim
## [29] Tamil Nadu               Telangana
## [31] Tripura                  Uttar Pradesh
## [33] Uttarakhand              West Bengal
## [35] Odisha                   Andhra Pradesh
## 36 Levels: Andaman & Nicobar Island Andhra Pradesh ... West
Bengal
# Creating subset for Maharashtra state
>MH =ind_state[ind_state$ST_NM== 'Maharashtra',]
>MH #View metadata of subset created
## class       : SpatialPolygonsDataFrame
## features    : 1
## extent      : 72.64293, 80.89956, 15.60679, 22.02903  (xmin,
xmax, ymin, ymax)
## crs        : +proj=longlat +datum=WGS84 +no_defs +ellps=WGS84
+towgs84=0,0,0
## variables   : 1
## names       :       ST_NM
## value       : Maharashtra
>plot(MH, main ="Maharashtra")#Plot the created subset
```

Maharashtra

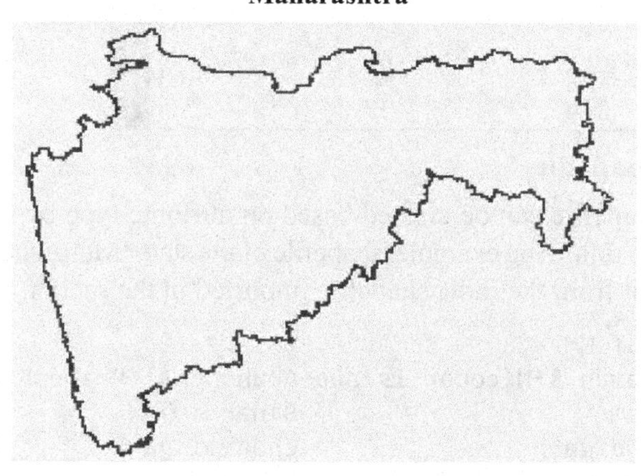

5.6. Exporting shapefile

A new shapefile can be created in the destination directory by using `writeOGR` function of `rgdal` package. In the following example, shapefile of Maharashtra state subsetted in the section 5.5 is exported in the target folder.

```
#Export or write shapefile in the disk
>writeOGR(obj = MH, # Vector data object or spatial file
dsn ='E:\\..........\\state_shpfile', # Destination folder
layer ="Maharashtra", # Name of shapefile to be created
   driver ="ESRI Shapefile") # Format
#Load the created shapefile
>Maharashtra
=readOGR('E:\\........\\state_shpfile','Maharashtra')
## OGR data source with driver: ESRI Shapefile
## Source: "E:\.........\state_shpfile", layer: "Maharashtra"
## with 1 features
## It has 1 fields
#Plot the created ESRI shapefile
>plot(Maharashtra, col ='blue', main ="Maharashtra shapefile")
```

Maharashtra shapefile

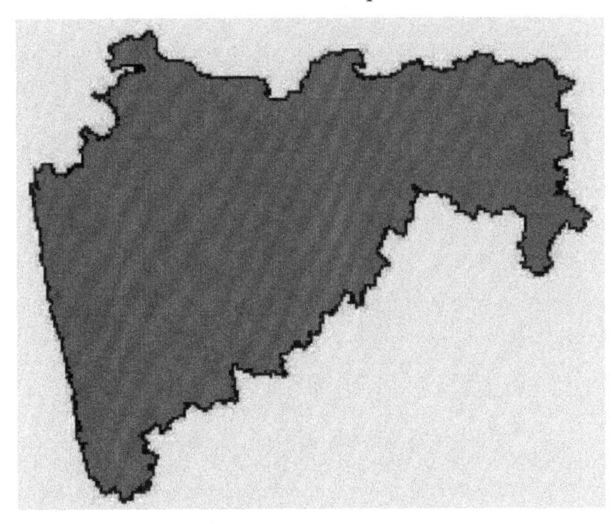

Chapter 6

Coordinate Reference Systems (CRS) in R

Coordinate Reference System (CRS) is a framework of coordinate based local, regional or global reference system used to represent the locations of geographic features. Coordinate Reference Systems (CRS) provide a standardized way to define the real-world geographic locations which further enable us to use common locations of geographic datasets for integration. The datasets from different sources and CRS are transformed to a common CRS in order to perform various integrated analytical operations like overlaying of data layers. Coordinate Reference Systems are mainly of two types:

1) **Geographic Coordinate System:** This is global or spherical coordinate system in which locations of features on earth are defined in terms of latitude and longitude values. Latitude and longitude are angles based on a point at the centre of the earth. In this system, locations are defined by degrees, therefore, distances cannot be measured accurately. The most common geographic coordinate system is the World Geodetic System 84 (WGS84).

2) **Projected Coordinate System:** In this system, locations of features are defined using cartesian x, y coordinates on a flat, two – dimensional surface. This system is based on sphere or ellipsoid projected on flat surface. Projected Coordinate System are generally referred as projection. It is required for preparing maps on paper or computer screen. Unlike Geographic Coordinate System, projected coordinate system has accurate measurements of distance, angles and areas across the two dimensions. However, there is alwaystrade-offs between area, direction, shape, and distance in this system due to distortions created while representing ellipsoid or spheroid on a flat map. So, the projection type is selected based on its application. Most common projections are Universal Transverse Mercator (UTM), Lambert Conformal Conic and Albers Equal Area.

For defining any CRS, the shape of the earth is determined first. The model describing shape of the earth is known as **ellipse**. Now a days, a global ellipsoid (flattened sphere) such as WGS84 is generally used ellipse for maintaining compatibility. CRS also include datum and projection (in case of projected coordinate system). The **datum** defines an origin point of the coordinate axes along with the direction of those axes. It consists of series of numbers describing shape and size of ellipsoid and its orientation in space. **Projection** is the mathematical transformation of points on the globe to the flat 2 – D surface. A particular CRS is mainly denoted by its **EPSGcode**which was originally compiled by European Petroleum Survey Group. The EPSG codes of CRS can be obtained from https://spatialreference.org/ and http://www.epsg-registry.org/ web portals.

6.1. About Coordinate Reference System (CRS) in R

In R software, three packages *i.e.* `rgdal`, `sp` and `raster` are required for retrieval, assigning and transformation of CRS of spatial data. These packages are imported as shown below.

```
#Import packages
>library(rgdal)
#Import packages
>library(sp)
>library(raster)
```

The CRS in R is represented by proj4string notation from the PROJ.4 library. The proj4string for a particular CRS in R can be obtained by specifying its EPSG code by using `CRS` function of `sp` package as given below.

```
#Obtain projection string of EPSG code: 4326
>CRS("+init=epsg:4326")
## CRS arguments:
##   +init=epsg:4326 +proj=longlat +datum=WGS84 +no_defs
+ellps=WGS84
## +towgs84=0,0,0
```

Proj4 consists of a list of parameters each prefixed by '+' character. Few parameters are given in the following table (Table 6.1).

Table 6.1: Parameters in Proj4

Parameter	Description
`+datum`	Datum name
`+ellps`	Ellipsoid name
`+proj`	Projection name
`+units`	Unit like meters, feet, etc.
`+vunits`	Vertical units
`+zone`	UTM zone
`+towgs84`	3 or 7 term datum transform parameters
`+no_defs`	No defaults read from default files
`+a`	Semimajor radius of the ellipsoid axis
`+b`	Semiminor radius of the ellipsoid axis
`+axis`	Axis rotation
`+lat_0`	Latitude of origin
`+lon_0`	Central meridian
`+lat_1`	Latitude of first standard parallel
`+lat_2`	Latitude of second standard parallel
`+lat_ts`	Latitude of true scale
`+x_0`	False easting
`+y_0`	False northing

A data frame of all the CRS available in R from the EPSG codes can be made by using following function.

```
#Make data frame of all EPSG codes
>EPSG =make_EPSG()
>head(EPSG)
##   code                                                    note
## 1 3819                                                # HD1909
## 2 3821                                                 # TWD67
## 3 3824                                                 # TWD97
## 4 3889                                                  # IGRS
## 5 3906                                              # MGI 1901
## 6 4001 # Unknown datum based upon the Airy 1830 ellipsoid
##
prj4
## 1 +proj=longlat +ellps=bessel +towgs84=595.48,121.69,515.35,4.115,-
2.9383,0.853,-3.408 +no_defs
## 2 +proj=longlat +ellps=aust_SA +no_defs
## 3 +proj=longlat +ellps=GRS80 +towgs84=0,0,0,0,0,0,0 +no_defs
## 4 +proj=longlat +ellps=GRS80 +towgs84=0,0,0,0,0,0,0 +no_defs
## 5 +proj=longlat +ellps=bessel +towgs84=682,-203,480,0,0,0,0
+no_defs

## 6 +proj=longlat +ellps=airy +no_defs
```

This created data frame has three columns:

code: EPSG code

note: notes as included in the file

prj4: proj4string of the corresponding EPSG code

This created EPSG data frame can be searched by using the following R code.

```
#Searching through EPSG data frame
>EPSG[grep("4131", EPSG$code),]
##    code          note
## 64 4131 # Indian 1960
##    prj4
## 64 +proj=longlat +a=6377276.345 +b=6356075.41314024
+towgs84=198,881,317,0,0,0,0 +no_defs
```

6.2. Retrieve CRS

Coordinate Reference System of both vector and raster data can be obtained by `proj4string` function of `sp` package whereas `projection` function from the `raster` package retrieves the CRS of raster data only. In the following example, both vector and raster spatial object were imported first, and then their CRS information was retrieved. The date used here are saved as:

Chapter5\data\India_state\Admin2.shp

Chapter6\data\ data_class.tif

Chapter6\data\ data_for_creating_vector.csv

```
#Import vector file
>x =readOGR('E:\\...........\\India_state', 'Admin2')
## OGR data source with driver: ESRI Shapefile
## Source: "E:\..............\India_state", layer: "Admin2"
## with 36 features
## It has 1 fields
#Import multi-Layered raster file
>y =brick('E:/........................./data_class.tif')
#Retrieve CRS of a spatial object
>proj4string(x) # CRS of vector Layer
## [1] "+proj=longlat +datum=WGS84 +no_defs +ellps=WGS84
+towgs84=0,0,0"
>proj4string(y) # CRS of raster Layer
## [1] "+proj=utm +zone=44 +datum=WGS84 +units=m +no_defs
+ellps=WGS84 +towgs84=0,0,0"
>projection(y) # Alternative function for getting CRS of raster
Layer
```

```
## [1] "+proj=utm +zone=44 +datum=WGS84 +units=m +no_defs
+ellps=WGS84 +towgs84=0,0,0"
```

6.3. Assign CRS

If a new raster or vector data is created in R, then a CRS is assigned to it in order to make it an accurate spatial object. In the following sections, CRS is assigned to the newly formed vector and raster data in R.

6.3.1. Assign CRS to vector data

CRS to vector data can be assigned by using CRS function. Both EPSG code or projection string can be used to assign it. In the following example, a vector data is created from a '.csv' file containing latitude and longitude information, and then CRS is assigned to it.

```
# Create point vector data from .csv file
>pop_data =read.csv('E:/……………………………………………/
data_for_creating_vector.csv', header =TRUE) # Read csv file
from the disk
>head(pop_data) # Display first few rows and columns of imported
data

##            State Latitude Longitude Population
## 1     Tamil Nadu  11.0598   78.3875   76481545
## 2      Telangana  17.1232   79.2088   38472769
## 3 Madhya Pradesh  23.4733   77.9480   82342793
## 4        Haryana  29.2385   76.4319   27388008
## 5   Chhattisgarh  21.2951   81.8282   28566990
## 6    Maharashtra  19.6012   75.5530  120837347

>pop_proj =CRS("+init=epsg:4326") #create CRS object
>pop_proj#display created CRS object

## CRS arguments:
##   +init=epsg:4326 +proj=longlat +datum=WGS84 +no_defs
+ellps=WGS84
## +towgs84=0,0,0

# Create point vector data from csv file with CRS assigned
>pop_pts =SpatialPointsDataFrame(pop_data[3:2], data =pop_data,
proj4string =pop_proj)
>pop_pts# Display the metadata of created point vector data

## class       : SpatialPointsDataFrame
## features    : 30
## extent     : 72.1362, 94.7278, 10.8505, 32.7266 (xmin, xmax, ymin,
ymax)
## crs         : +init=epsg:4326 +proj=longlat +datum=WGS84 +no_defs
+ellps=WGS84 +towgs84=0,0,0
```

```
## variables   : 4
## names       :           State, Latitude, Longitude, Population
## min values  : Andhra Pradesh,  10.8505,   72.1362,     671720
## max values  :    West Bengal,  32.7266,   94.7278,  228959599
```

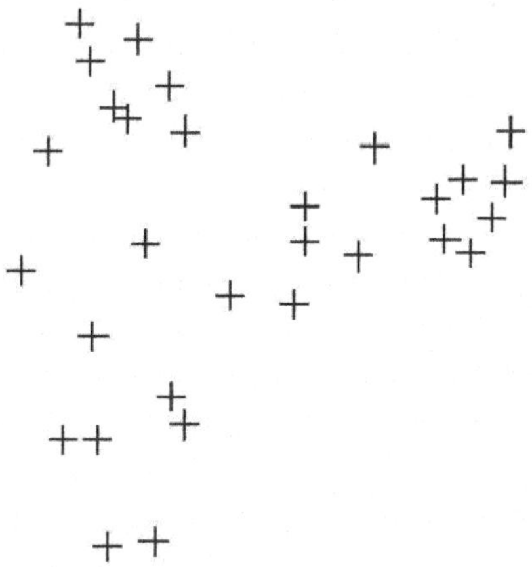

```
>plot(pop_pts) # Plot the created point vector data
# Assigning CRS using projection string
>proj1 =CRS('+proj=longlat +a=6377276.345 +b=6356075.41314024
+towgs84=198,881,317,0,0,0,0 +no_defs') #create CRS object
# Create point vector data from csv file with CRS assigned
>pop1 =SpatialPointsDataFrame(pop_data[3:2], data =pop_data,
proj4string = proj1)
>pop1 # Display the metadata of created point vector data
```

```
## class       : SpatialPointsDataFrame
## features    : 30
## extent      : 72.1362, 94.7278, 10.8505, 32.7266  (xmin, xmax, ymin,
ymax)
## crs           : +proj=longlat +a=6377276.345 +b=6356075.41314024
+towgs84=198,881,317,0,0,0,0 +no_defs
## variables   : 4
## names       :           State, Latitude, Longitude, Population
## min values  : Andhra Pradesh,  10.8505,   72.1362,     671720
## max values  :    West Bengal,  32.7266,   94.7278,  228959599
```

6.3.2. Assign CRS to raster data

CRS is assigned to raster data by using `CRS` and `projection` function. In the following example, a raster data is created from the matrix, and then extent and projection are assigned to it.

```
# Assigning CRS to raster data
>xy =matrix(rnorm(200), 50, 50) # create matrix
```

```
>xy[1:4, 1:4] # Display first four rows and columns of matrix
##               [,1]       [,2]      [,3]        [,4]
## [1,]  0.03347105 -1.0527627 1.7894638 -2.02957981
## [2,]  0.59065381  1.0024668 0.3213683  0.05169028
## [3,]  1.14829680 -0.2986094 1.1500742 -0.34850518
## [4,] -0.28916655 -1.1802231 1.2520187 -1.49037422
>rast =raster(xy) #create raster object from the matrix
>extent(rast) =c(78, 80, 10, 12) #define extent to created
raster
#assign projection or CRS to the created raster object
>projection(rast) =CRS("+init=epsg:4326")
>rast# Display attributes of created raster
## class      : RasterLayer
## dimensions : 50, 50, 2500  (nrow, ncol, ncell)
## resolution : 0.04, 0.04  (x, y)
## extent     : 78, 80, 10, 12  (xmin, xmax, ymin, ymax)
## crs         : +init=epsg:4326 +proj=longlat +datum=WGS84 +no_defs
+ellps=WGS84 +towgs84=0,0,0
## source     : memory
## names      : layer
## values     : -2.600655, 2.605163  (min, max)
>plot(rast) # Plot the created raster
```

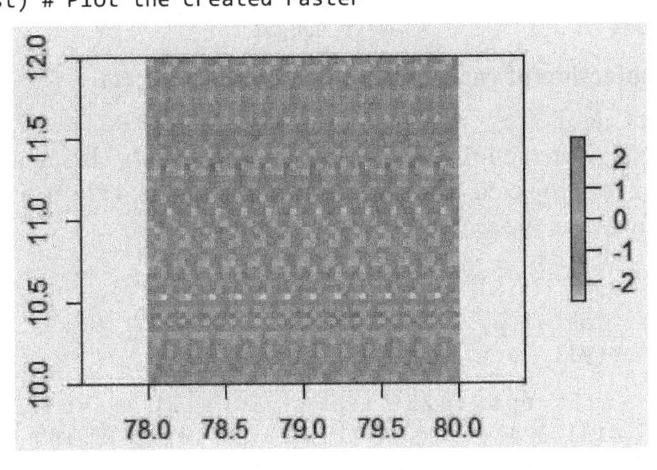

6.4. Transform CRS

Generally, there is need to transform the CRS or projection of vector and raster data to a common CRS system in order to overlay them for performing certain analytical operations. This transformation of CRS of a spatial object to another CRS is called as reprojection.

6.4.1. Reprojection of vector data through CRS object

A vector data can be re – projected by using `spTransform` function from `sp` package. In the following example, vector data '`x`' being imported in section

6.2 is re-projected from CRS with EPSG code 4326 to CRS with EPSG code 4239.

```
#Reprojection of vector data through CRS object
>reproj =CRS("+init=epsg:4239") # create CRS object
>reproj# Display projection string of CRS object

## CRS arguments:
##    +init=epsg:4239  +proj=longlat  +a=6377276.345
+b=6356075.41314024
## +towgs84=217,823,299,0,0,0,0 +no_defs

>Ind =spTransform(x, reproj) # Reprojection of vector data x
>Ind # Display metadata to check re-projection

## class       : SpatialPolygonsDataFrame
## features    : 36
## extent      : 68.18523, 97.41856, 6.753739, 37.07915  (xmin, xmax,
ymin, ymax)
## crs              : +init=epsg:4239 +proj=longlat +a=6377276.345
+b=6356075.41314024 +towgs84=217,823,299,0,0,0,0 +no_defs
## variables   : 1
## names       :                    ST_NM
## min values  : Andaman & Nicobar Island
## max values  :              West Bengal
```

6.4.2. Reprojection of raster data through CRS object

A raster data can be re – projected by using `projectRaster` function of `raster` package. In the following example, raster data '`y`' being imported in section 6.2 is re-projected from CRS with EPSG code 4326 to CRS object '`reproj`' created in the section 6.4.1.

```
#Reprojection of raster data through CRS object

>y1 =projectRaster(y, crs =reproj)
>proj4string(y1) #Display CRS to check reprojection

## [1] "+init=epsg:4239 +proj=longlat +a=6377276.345
+b=6356075.41314024 +towgs84=217,823,299,0,0,0,0 +no_defs"
```

6.4.3. Reprojection of vector data by taking CRS of raster data

Vector data can be re – projected by taking the CRS of raster data through applying `CRS(projection(raster layer name))` parameter in the place of CRS object in the `spTransform` function. The R – code example is given below.

```
# Reprojection of vector through raster data
>proj4string(pop1) # Display the CRS of vector layer to be re-
projected
```

```
## [1] "+proj=longlat +a=6377276.345 +b=6356075.41314024
+towgs84=198,881,317,0,0,0,0 +no_defs"

>proj4string(rast) # Display the CRS of raster Layer whose CRS
is taken

## [1] "+init=epsg:4326 +proj=longlat +datum=WGS84 +no_defs
+ellps=WGS84 +towgs84=0,0,0"

>pop2 =spTransform(pop1, CRS(projection(rast))) # Reprojection
>proj4string(pop2) # Display CRS of re-projected vector data

## [1] "+init=epsg:4326 +proj=longlat +datum=WGS84 +no_defs

+ellps=WGS84 +towgs84=0,0,0"
```

6.4.4. Reprojection of raster data by taking CRS of vector data

Raster data can be re – projected by using `proj4string(vector data name)` in the `crs` parameter of `projectRaster` function of `raster` package. The R – code example is given below.

```
#Reprojection of raster from vector data
>crs(rast) # Display the CRS of raster Layer to be re-projected

## CRS arguments:
##   +init=epsg:4326 +proj=longlat +datum=WGS84 +no_defs
+ellps=WGS84
## +towgs84=0,0,0

>crs(pop1) # Display the CRS of vector data whose CRS is taken

## CRS arguments:
##  +proj=longlat +a=6377276.345 +b=6356075.41314024
## +towgs84=198,881,317,0,0,0,0 +no_defs

>rast1 =projectRaster(rast, crs =proj4string(pop1)) # Reprojection
>crs(rast1) # Display CRS of re-projected raster data

## CRS arguments:
##  +proj=longlat +a=6377276.345 +b=6356075.41314024
## +towgs84=198,881,317,0,0,0,0 +no_defs
```

Chapter 7

Subset Raster

Raster sub-setting is done to select the desired area or bands of the raster image on which the user wants to perform analysis. Raster sub-setting is done in two ways:

Spectral subset: Spectral sub-setting is done to extract few desired bands from the multiband raster image which are required for analysis.

Spatial subset: Spatial sub-setting is done to minimize the area or extent of raster image to the desired area or extent for further analysis.

Raster sub-setting in R is done by functions available in the 'raster' package. The following packages are required in this chapter:

```
#Import packages
>library(raster)    #For analysis of raster data
>library(RStoolbox)     #For plotting raster image
>library(rgdal)         #For reading shapefile
```

In this chapter, the spectral and spatial sub-setting is done for the following multiband raster:

Chapter7\data\Landsat_Kanpur.tif

The input data contains four bands of Landsat 8 OLI data having band numbers 2 to 4 corresponding to Blue, Green, Red and NIR bands, respectively. First set the directory as the file location.

```
>setwd("E:/............/Chapter7/data")

#Import raster image
>img =brick('Landsat_Kanpur.tif')
>img#View metadata of raster image

## class      : RasterBrick
## dimensions : 3858, 2314, 8927412, 4  (nrow, ncol, ncell, nlayers)
## resolution : 30, 30  (x, y)
## extent     : 389775, 459195, 2867025, 2982765  (xmin, xmax,
```

```
ymin, ymax)
## crs        : +proj=utm +zone=44 +datum=WGS84 +units=m +no_defs
+ellps=WGS84 +towgs84=0,0,0
## source     : E:/......................./Chapter7/data/Landsat_Kanpur.tif
## names        : Landsat_Kanpur.1, Landsat_Kanpur.2,
Landsat_Kanpur.3, Landsat_Kanpur.4
## min values :          10130,          9161,          8322,
8694
## max values :          21049,         22192,         23430,
27690
```

```
>ggRGB(img, r =4, g =3, b =2, stretch ='lin') #Plot FCC of raster
```

7.1. Spectral subset

The spectral subset is created in R by using `subset` function of `raster` package as shown below. The created subset can be saved to the disk by specifying its name in the `filename` parameter.

```
# subset band number 2 and 3 from raster brick (spectral subset)
>subset1 =subset(img, 2:3, filename ='subset1.tif') #output raster
>subset1 #Display metadata of spectral subset image
```

```
## class      : RasterBrick
## dimensions : 3858, 2314, 8927412, 2  (nrow, ncol, ncell, nlayers)
## resolution : 30, 30  (x, y)
## extent     : 389775, 459195, 2867025, 2982765  (xmin, xmax, ymin, ymax)
## crs         : +proj=utm +zone=44 +datum=WGS84 +units=m +no_defs +ellps=WGS84 +towgs84=0,0,0
## source     : E:/......................../subset1.tif
## names      : subset1.1, subset1.2
## min values :      9161,      8322
## max values :     22192,     23430
```

In this example, spectral subset of two bands *i.e.* band number 2 and 3 is created and saved in the working directory with the name of *'subset1.tif'*.

7.2. Spatial subset

Spatial subset can be created by many methods as shown in the following sections.

7.2.1. Spatial subset by extent

Spatial subset can be done by defining the extent and cropping the input raster to that extent by using `crop` function as shown below.

```
#Spatial subset by extent
>extent(img) # Display original extent of input image 'img'
```

```
## class       : Extent
## xmin        : 389775
## xmax        : 459195
## ymin        : 2867025
## ymax        : 2982765
```

```
>e =extent(400000, 410000, 2900000, 2910000) # Define extent for subset
>subset2 =crop(img, e, filename ='subset2.tif') # Spatial sub-setting
>subset2 #Display metadata of spatial subset
```

```
## class      : RasterBrick
## dimensions : 334, 333, 111222, 4  (nrow, ncol, ncell, nlayers)
## resolution : 30, 30  (x, y)
## extent     : 400005, 409995, 2899995, 2910015  (xmin, xmax, ymin,
ymax)
## crs         : +proj=utm +zone=44 +datum=WGS84 +units=m +no_defs
+ellps=WGS84 +towgs84=0,0,0
## source     : E:/............................../subset2.tif
## names      : subset2.1, subset2.2, subset2.3, subset2.4
## min values :     10450,      9539,      9005,      9923
## max values :     14730,     15880,     17294,     22266
```

```
#Plot spatial subset
>ggRGB(subset2, r =4, g =3, b =2, stretch ='lin')
```

The extent can also be selected by drawing on the plotted map interactively by using `drawExtent` function. First you need to plot the map using `plot` or `plotRGB` function, then assign extent by drawing extent on the plot. This extent parameters can be passed to the crop function for spatial sub-setting.

```
#Spatial subset by selecting extent interactively by drawing on plot
>plotRGB(img[[4:2]], stretch ='lin', axes =TRUE) #Plot to draw extent
>ext =drawExtent(show =TRUE, col ='yellow') #Draw extent on map
>ext# Display the selected extent
```

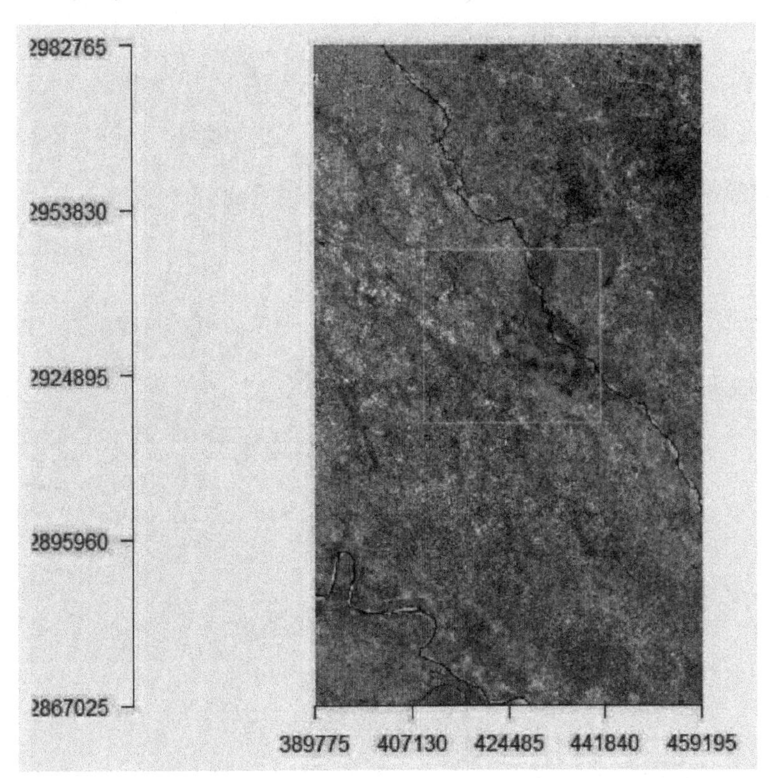


```
## class      : Extent
## xmin       : 409193
## xmax       : 441071
## ymin       : 2916377
## ymax       : 2946762
```

```
>subset3 =crop(img, ext, filename ='subset3.tif') # Spatial sub-setting
>subset3 #Display metadata of spatial subset
```

```
## class      : RasterBrick
## dimensions : 913, 1079, 985127, 4 (nrow, ncol, ncell, nlayers)
## resolution : 30, 30  (x, y)
## extent     : 409193, 441071, 2916377, 2946762 (xmin, xmax, ymin, ymax)
## crs        : +proj=utm +zone=44 +datum=WGS84 +units=m +no_defs +ellps=WGS84 +towgs84=0,0,0
## source     : E:/…………………/subset3.tif
```

```
## names      : subset3.1, subset3.2, subset3.3, subset3.4
## min values : 10342,     9264,      8629,      8829
## max values : 21049,     22192,     23430,     26773
#Plot spatial subset
>ggRGB(subset3, r =4, g =3, b =2, stretch ='lin')
```

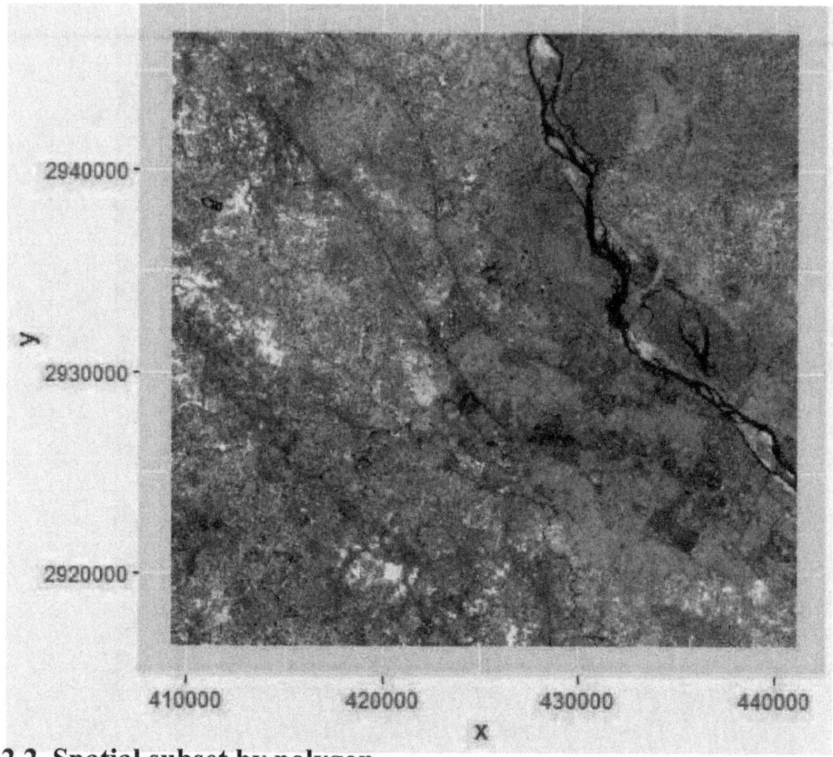

2940000
2930000
2920000
y
410000 420000 430000 440000
x

7.2.2. Spatial subset by polygon

For spatial subset by polygon, `crop` and `mask` functions are used. The polygon and raster must have same CRS before sub-setting the raster image. `crop` function doesn't subset the raster in accordance with the geometry of the polygon, but it just takes extent of that polygon and returns the spatial subset in rectangular form. The spatial subset of raster image with the same geometry or shape of the overlaying polygon can be created using `mask` function. In the following example, spatial subset by both these methods is done. The shapefile used for this purpose is saved as: Chapter7\data\polygon_test\ polygon_test.shp.

```
##Steps for spatial subset by polygon without maintaining geometry
# Import polygon
>polygon_test =readOGR('E:\\............\\polygon_test',
'polygon_test')
```

```
## OGR data source with driver: ESRI Shapefile
## Source: "E:\...............\polygon_test", layer:
"polygon_test"
## with 1 features
## It has 1 fields
## Integer64 fields read as strings:  id
```

```
# Compare CRS of both raster and polygon to check similarity
>compareCRS(polygon_test, img)
```

```
## [1] FALSE
```

As the result of `compareCRS` is false, it shows that the CRS of polygon and raster image are different. Therefore, we need to re-project the polygon in order to match its CRS with that of raster image by using `spTransform` function as shown below.

```
#Reproject polygon to the CRS of raster image
>poly_CRS =spTransform(polygon_test, CRS(projection(img)))
#Create a plot of polygon over the raster image
>plotRGB(img[[4:2]], stretch ='lin', axes =TRUE) #Plot FCC of
raster
>plot(poly_CRS, border ='blue', lwd =3, add =TRUE) #Plot polygon
```

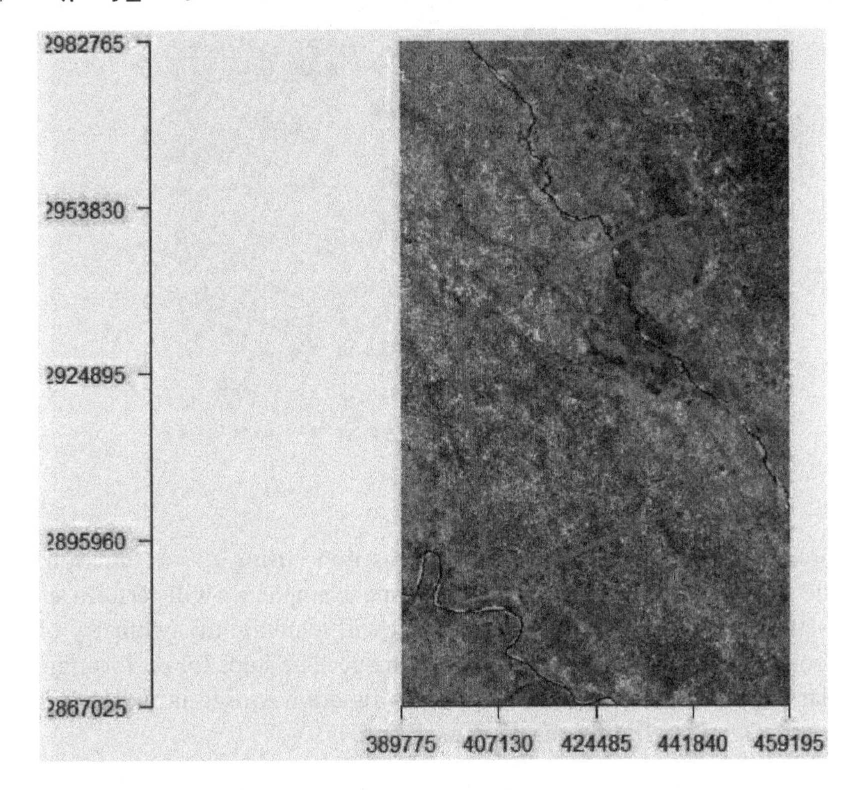

```
#Spatial subset by polygon without maintaining geometry
>subset4 =crop(img, poly_CRS, filename ='subset4.tif')
>subset4 #View metadata of spatial subset
```

```
## class      : RasterBrick
## dimensions : 2164, 1779, 3849756, 4  (nrow, ncol, ncell, nlayers)
## resolution : 30, 30  (x, y)
## extent     : 400005, 453375, 2887215, 2952135  (xmin, xmax, ymin,
ymax)
## crs          : +proj=utm +zone=44 +datum=WGS84 +units=m +no_defs
+ellps=WGS84 +towgs84=0,0,0
## source     : E:/............................/subset4.tif
## names      : subset4.1, subset4.2, subset4.3, subset4.4
## min values :     10266,      9264,      8629,      8829
## max values :     21049,     22192,     23430,     26773
```

```
>ggRGB(subset4, r =4, g =3, b =2, stretch ='lin') #Plot FCC of
subset
```

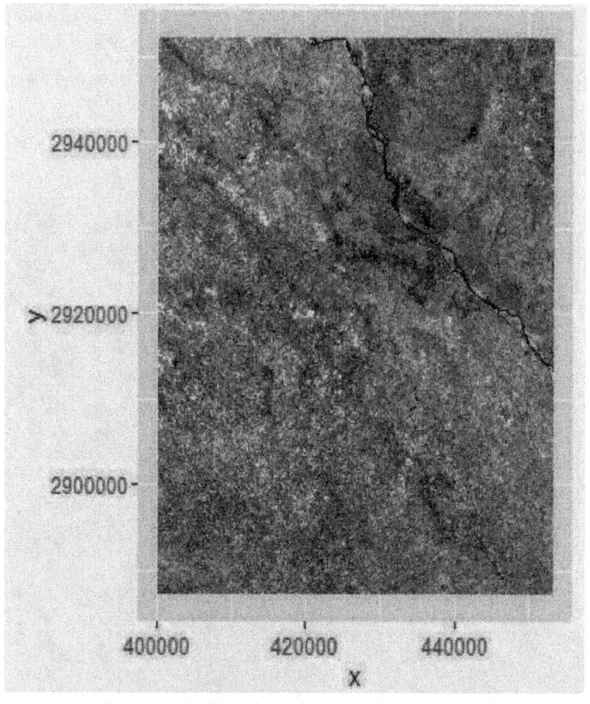

In the former example, spatial subset was done using `crop` function, so, geometry was not maintained. In the following example, we will perform spatial sub-setting by using `mask` function which will maintain the geometry of the polygon in the spatial subset of the raster image. The steps for performing it is similar to the above example except `mask` function is used in place of `crop` function. The shapefile for masking used here is:

Chapter7\data\Kanpur_UTM\Kanpur_UTM.shp

```
## Steps for spatial subset by polygon with maintained geometry
>Kanpur_shp =readOGR('Kanpur_UTM', 'Kanpur_UTM') #import polygon

## OGR data source with driver: ESRI Shapefile
## Source: "E:\..........\Chapter7\data\Kanpur_UTM", layer:
"Kanpur_UTM"
## with 1 features
## It has 5 fields

>compareCRS(Kanpur_shp, img) #compare CRS

## [1] TRUE

#Plot overlaying of polygon over raster
>plotRGB(img[[4:2]], stretch ='lin', axes =TRUE) #Plot FCC of raster
>plot(Kanpur_shp, border ='blue', lwd =4, add =TRUE) #Plot polygon
```

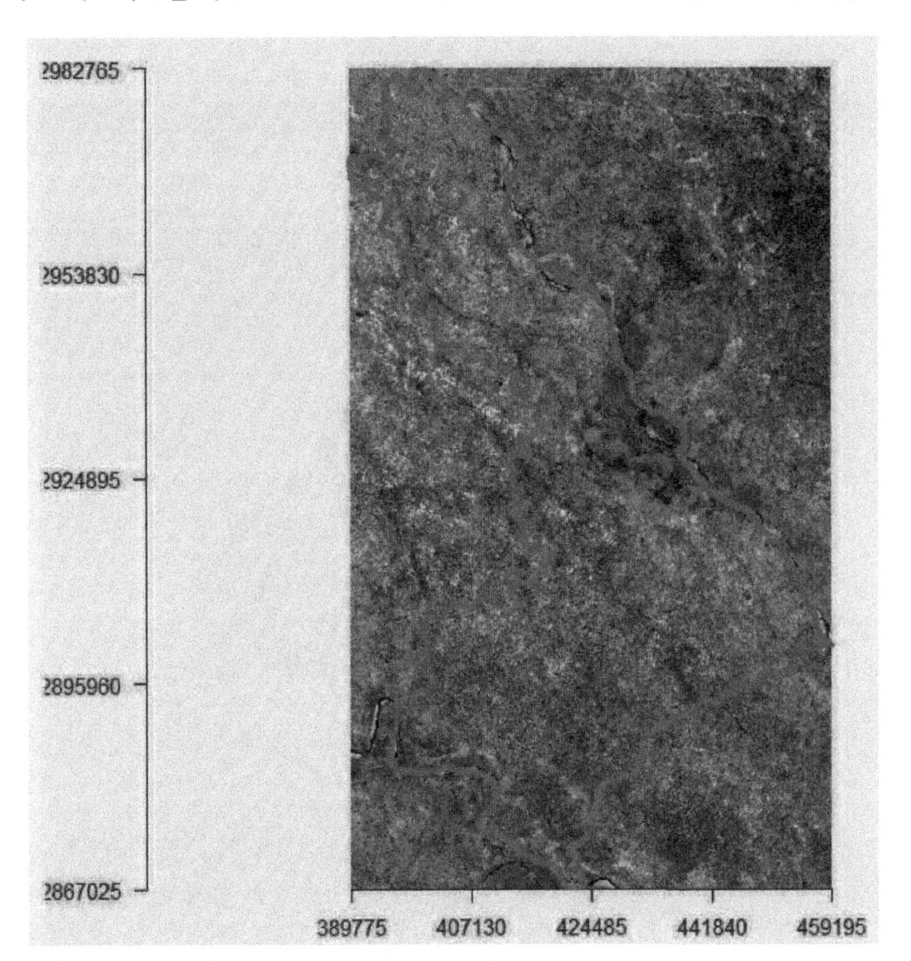

```
#Spatial subset by polygon with maintained geometry
>subset5 =mask(img, Kanpur_shp, filename ='subset5.tif')
>subset5 #View metadata of spatial subset
```

```
## class      : RasterBrick
## dimensions : 3858, 2314, 8927412, 4  (nrow, ncol, ncell, nlayers)
## resolution : 30, 30  (x, y)
## extent     : 389775, 459195, 2867025, 2982765  (xmin, xmax, ymin,
ymax)
## crs         : +proj=utm +zone=44 +datum=WGS84 +units=m +no_defs
+ellps=WGS84 +towgs84=0,0,0
## source     : E:/............................../subset5.tif
## names      : subset5.1, subset5.2, subset5.3, subset5.4
## min values :     10161,      9320,      8419,      8697
## max values :     21049,     22192,     23430,     27690
```

```
>ggRGB(subset5, r =4, g =3, b =2, stretch ='lin') #Plot FCC of
subset
```

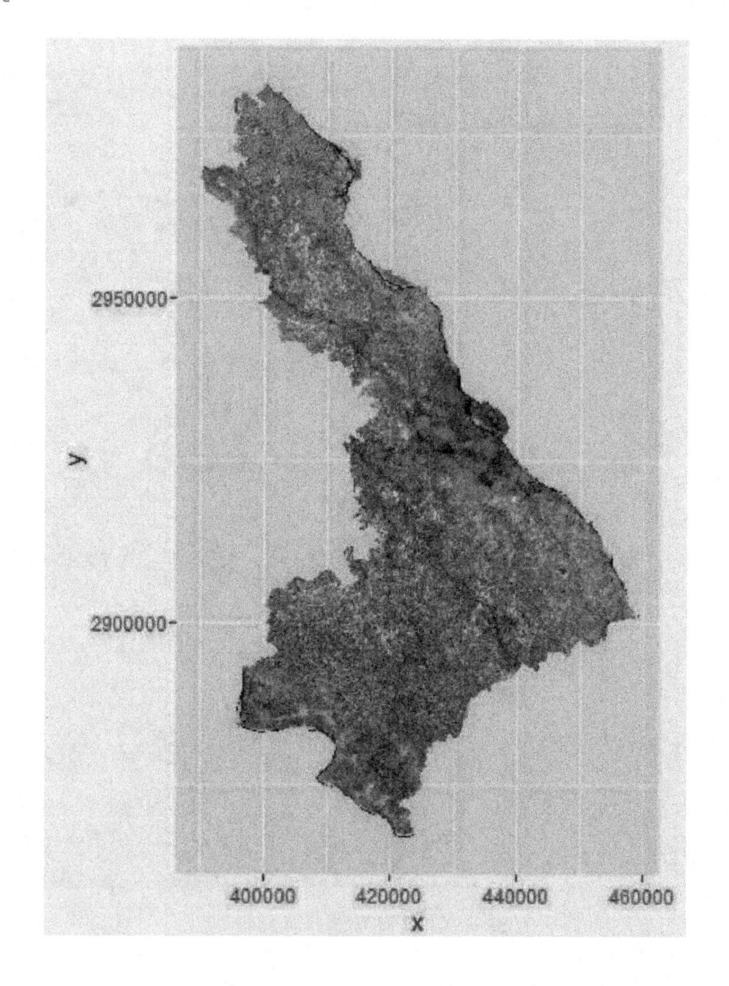

Thus, you can see that by using mask function, the raster was subset by maintaining the exact shape of the polygon. This function is useful if you want to perform raster analysis for a particular area and the shapefile of that area is available with you.

7.2.3. Spatial subset to retain area outside the polygon

If you want to get the raster image in which area inside the polygon is masked out and only the area outside the polygon is retained in the spatial subset, then inverse =TRUE parameter is used in the mask function as depicted in the following example.

```
# Spatial subset to retain area outside the polygon
>subset6 =mask(img, Kanpur_shp, inverse =TRUE, #Mask out inside
area
filename ='subset6.tif')
>subset6 #View metadata of spatial subset

## class      : RasterBrick
## dimensions : 3858, 2314, 8927412, 4  (nrow, ncol, ncell, nlayers)
## resolution : 30, 30  (x, y)
## extent     : 389775, 459195, 2867025, 2982765  (xmin, xmax, ymin,
ymax)
## crs         : +proj=utm +zone=44 +datum=WGS84 +units=m +no_defs
+ellps=WGS84 +towgs84=0,0,0
## source     : E:/................................/subset6.tif
## names      : subset6.1, subset6.2, subset6.3, subset6.4
## min values :     10130,      9161,      8322,      8694
## max values :     20201,     21356,     22195,     26556

# Plot FCC of the spatial subset
>ggRGB(subset6, r =4, g =3, b =2, stretch ='lin')
```

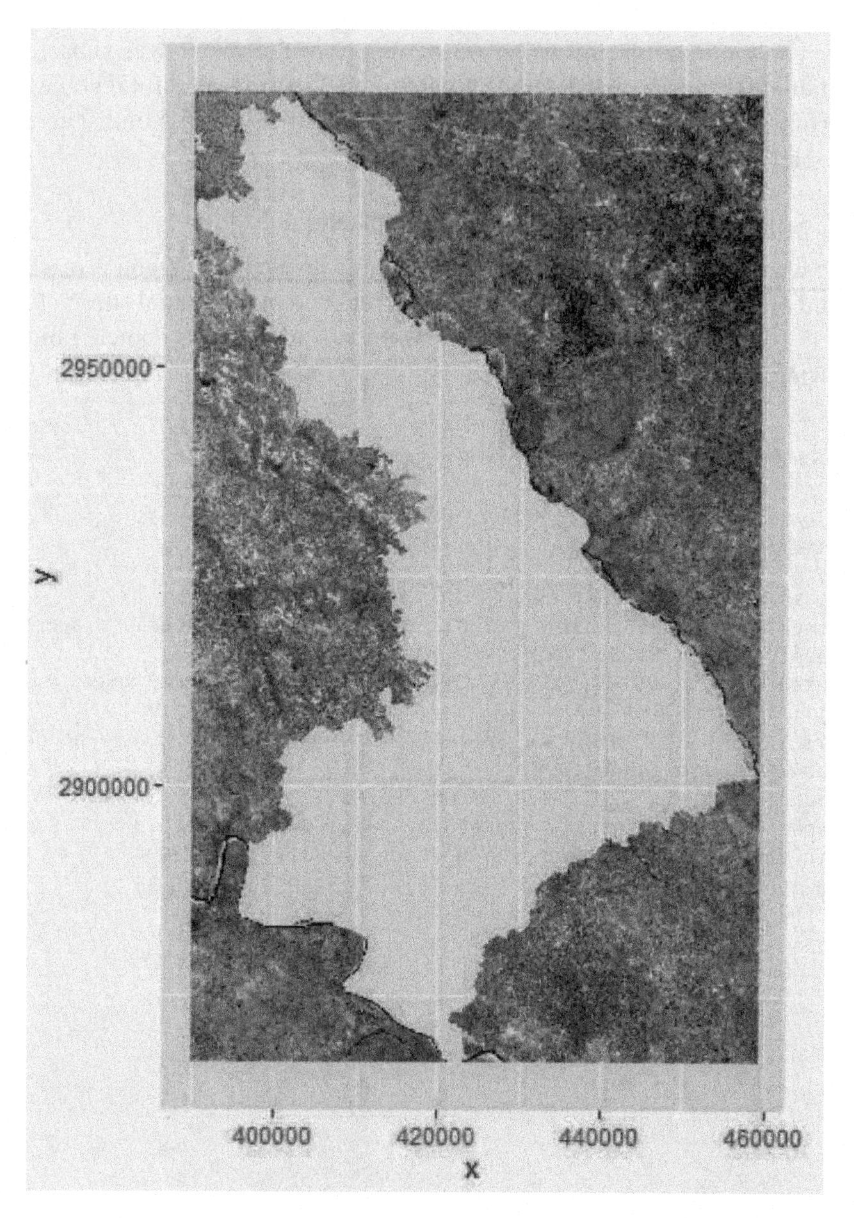

7.2.4. Spatial subset by raster

If you want to create a spatial subset of a raster from the extent of another raster image, then `crop` function is used. In the following example, another raster is imported from the disk whose extent we want to use for creating the spatial subset of our desired raster denoted as '`img`'. Both the raster should have same projection so that both can overlay. The R-code is given below. The raster used here for cropping the previous raster can be located as:

Chapter7\data\crop.tif

```
# Steps for creating spatial subset by raster
>rast =brick('crop.tif') #Import raster for getting extent
>rast#View metadata of imported raster
```

```
## class      : RasterBrick
## dimensions : 334, 333, 111222, 7  (nrow, ncol, ncell, nlayers)
## resolution : 30, 30  (x, y)
## extent     : 400005, 409995, 2899995, 2910015  (xmin, xmax, ymin, ymax)
## crs        : +proj=utm +zone=44 +datum=WGS84 +units=m +no_defs +ellps=WGS84 +towgs84=0,0,0
## source     : E:/................................/crop.tif
## names      : crop.1, crop.2, crop.3, crop.4, crop.5, crop.6, crop.7
## min values :  10797,  10021,   9101,   8084,  10345,   5992,   5330
## max values :  11635,  11082,  10817,  11050,  18488,  16498,  16603
```

```
# plot imported raster over target raster
>plotRGB(img[[4:2]], stretch ='lin', # Plot raster to be sub-setted
alpha =100, # Parameter to set transparency
axes =TRUE)
>plotRGB(rast[[5:3]], stretch ='lin', add =TRUE) # Plot imported raster
```

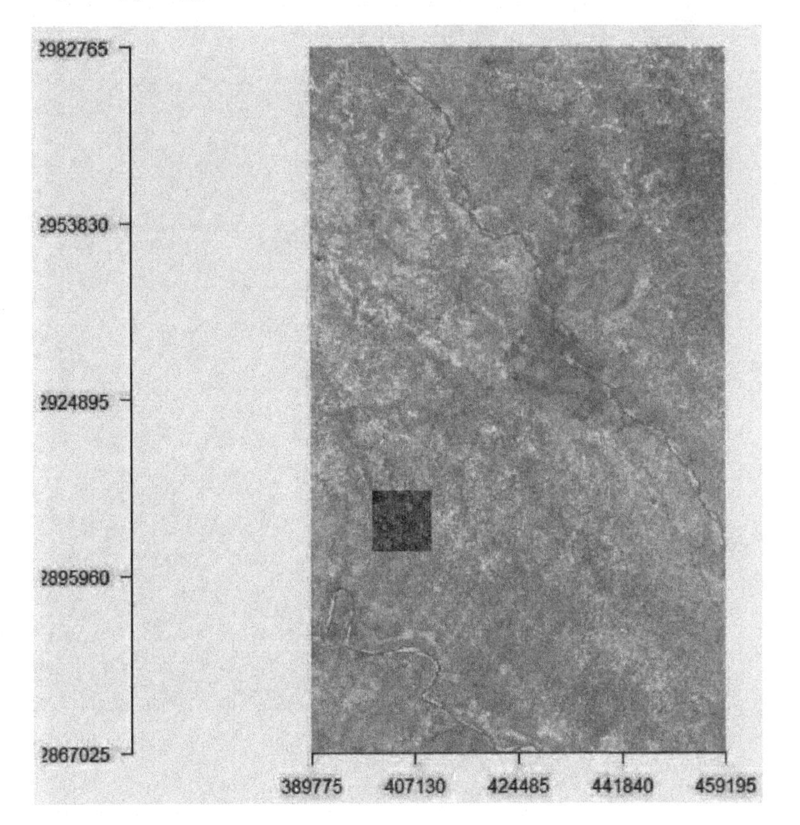

This plot shows the extent of area of target raster which will be subsetted by another imported raster image. This spatial subset is created as follow.

```
# Spatial subset of target raster by another imported raster
>subset7 =crop(img, rast, filename ='subset7.tif')
>subset7 #View metadata of spatial subset
```

```
## class      : RasterBrick
## dimensions : 334, 333, 111222, 4  (nrow, ncol, ncell, nlayers)
## resolution : 30, 30  (x, y)
## extent     : 400005, 409995, 2899995, 2910015  (xmin, xmax, ymin,
ymax)
## crs         : +proj=utm +zone=44 +datum=WGS84 +units=m +no_defs
+ellps=WGS84 +towgs84=0,0,0
## source     : E:/................................/subset7.tif
## names      : subset7.1, subset7.2, subset7.3, subset7.4
## min values :     10450,      9539,      9005,      9923
## max values :     14730,     15880,     17294,     22266
```

```
>ggRGB(subset7, r =4, g =3, b =2, stretch ='lin') #Plot FCC of
subset
```

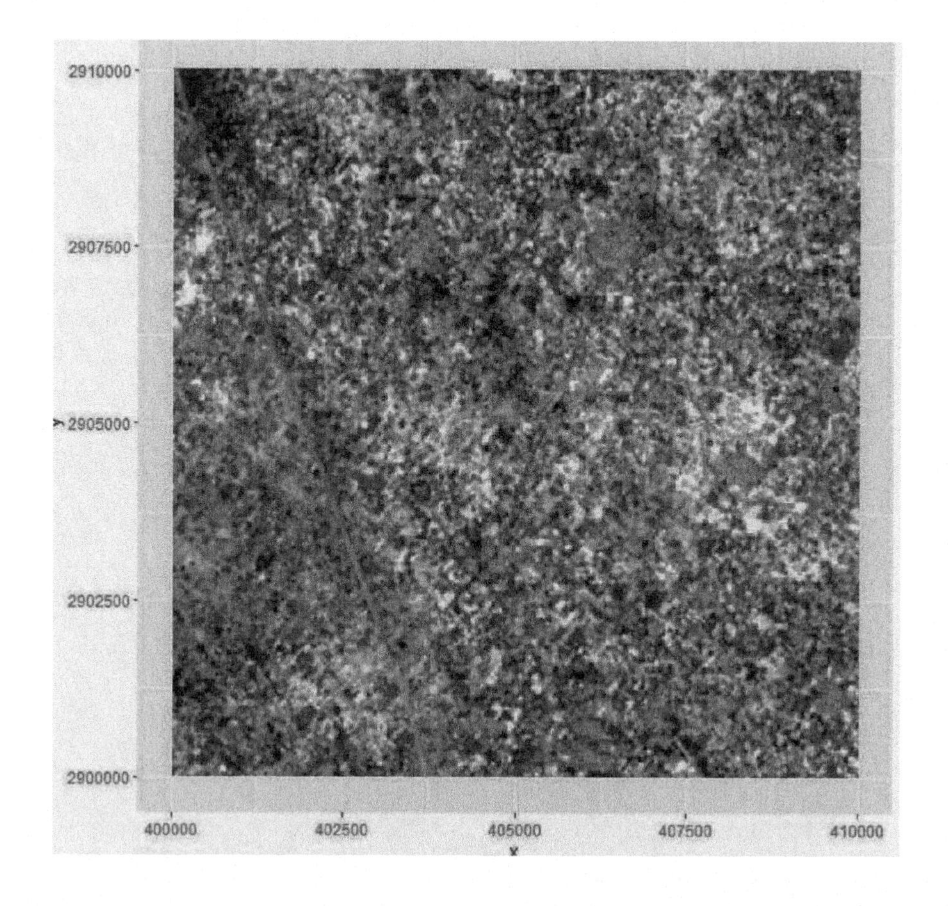

7.2.5. Spatial subset by cell numbers

If you want to create spatial subset by defining the range of cell numbers, then `rasterFromCells` command is used. This command works only for single raster layer unlike above discussed other functions. The example of R-code for this function is given further.

```
# Spatial subset raster by cell numbers
>subset8 =rasterFromCells(img[[5]], #5th band of 'img' image is used
     cells =c(3000:900000)) #range of cell numbers for sub-setting
>subset8 #View metadata of spatial subset
```

```
## class      : RasterLayer
## dimensions : 388, 2314, 897832  (nrow, ncol, ncell)
## resolution : 30, 30  (x, y)
## extent     : 389775, 459195, 2971095, 2982735  (xmin, xmax, ymin, ymax)
## crs         : +proj=utm +zone=44 +datum=WGS84 +units=m +no_defs +ellps=WGS84 +towgs84=0,0,0
## source     : memory
## names      : layer
## values     : 2315, 900146  (min, max)
```

```
>ggR(subset8) # Plot spatial subset
```

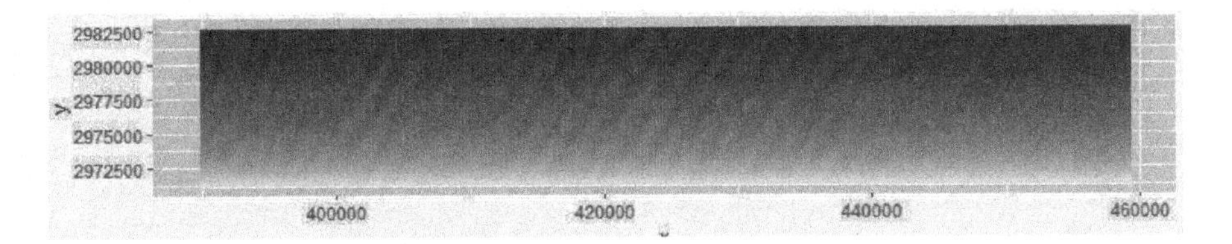

Chapter 8

Vector Data Analysis

Vector data analysis is performed to convert raw input vector data to output that provides new and meaningful information which ultimately supports decision making and disclose hidden patterns. There are various techniques to analyse vector data like buffering, intersect, union, and so on. Some of the most commonly used techniques with their R-code examples are explained in this chapter.

First of all, you need to import all the packages which are required for performing this exercise. These packages include `rgdal` and `sp` for importing and re-projecting vector data as well as `raster` and `rgeos` for performing vector analysis operations.

```
#Import packages
>library(rgdal)     #For importing vector data
>library(sp) #For re-projecting spatial data
>library(rgeos)     #For performing vector data analysis
>library(raster)    #For performing spatial analysis
```

Now, we will import vector data in the form of shapefiles from the disk which are required for performing vector analysis operations. In this exercise, we are using four shapefiles *i.e.* `'roads'`, `'forest'`, `'Maharashtra'` and `'Admin2'` containing spatial data of roads, forest, Maharashtra state and Indian states, respectively. We have used Indian state shapefile *i.e.* `'Admin2'` for attribute table joining and dissolve operations. Buffer analysis is performed on `'roads'` shapefile. Cropping the shapefile with respect to the pre-defined desired extent is done using `'forest'` shapefile. Both `'forest'` and `'Maharashtra'` shapefiles are used for performing intersect, union and erase operations. For importing these shapefiles in R, `readOGR` function from the `rgdal` package is used as shown below.

```
#set the working directory
>setwd("E:/....../Chapter8/data")
# Import shapefiles to perform vector analysis
>road =readOGR('Roads', 'roads') #Import roads shapefile
located in the Roads subfolder within the set directory
```

```
## OGR data source with driver: ESRI Shapefile
## Source: "E:\..............\Roads", layer: "roads"
## with 80 features
## It has 5 fields
```

```
>road #display metadata
```

```
## class       : SpatialLinesDataFrame
## features    : 80
## extent      : 73.30756, 74.85013, 16.31692, 16.85865  (xmin, xmax, ymin, ymax)
## crs          : +proj=longlat +datum=WGS84 +no_defs +ellps=WGS84 +towgs84=0,0,0
## variables   : 5
## names        :      MED_DESCRI,     RTT_DESCRI, F_CODE_DES, ISO, ISOCOUNTRY
## min values  :      Unknown, Primary Route,      Road, IND,     INDIA
## max values  : Without Median,      Unknown,      Road, IND,     INDIA
```

```
>plot(road) #plot roads shapefile
```

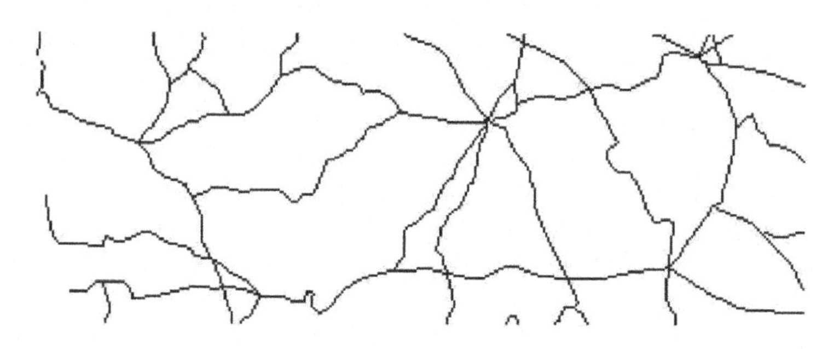

```
>forest =readOGR('Forest', 'forest') #Import forest shapefile
```

```
## OGR data source with driver: ESRI Shapefile
## Source: "E:\................\Forest", layer: "forest"
## with 1105 features
## It has 3 fields
```

```
>forest #display metadata
```

```
## class       : SpatialPolygonsDataFrame
## features    : 1105
## extent      : 72.68288, 82.98134, 15.4021, 21.96016  (xmin, xmax, ymin, ymax)
## crs          : +proj=longlat +datum=WGS84 +no_defs +ellps=WGS84 +towgs84=0,0,0
## variables   : 3
## names       :    osm_id,                 name,   type
## min values  : 101360250,  Ananthagiri hills, forest
## max values  :  99877536, Woodland and brush, forest
```

```
>plot(forest) #plot forest shapefile
```

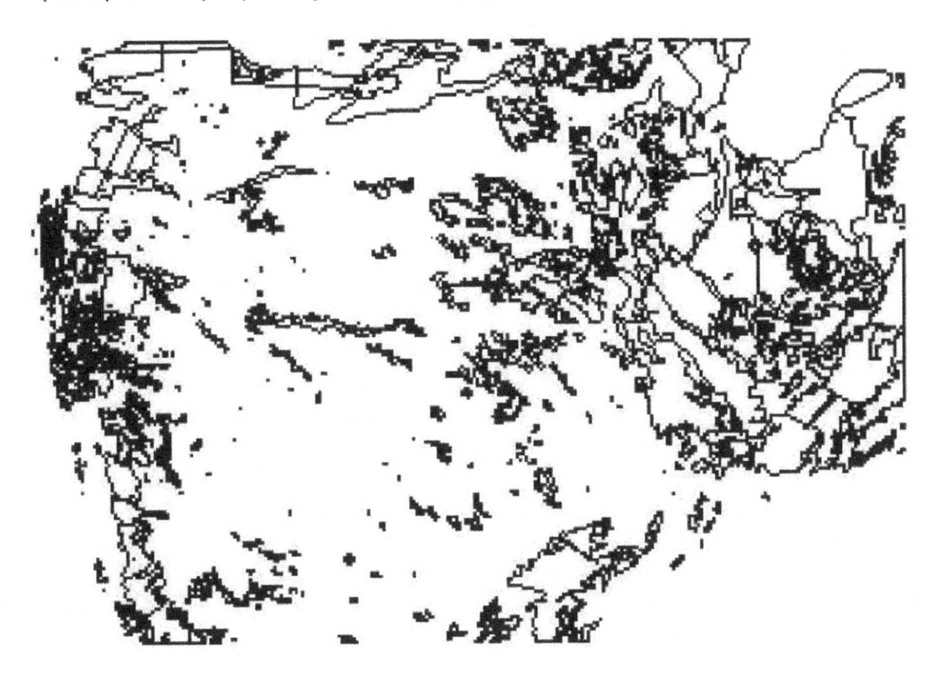

```
>MH =readOGR('Maharashtra','Maharashtra') #Import Maharashtra
shapefile located in Maharashtra subfolder within the set
directory

## OGR data source with driver: ESRI Shapefile
## Source: "E:\...............\Maharashtra", layer: "Maharashtra"
## with 1 features
## It has 1 fields

>MH #display metadata

## class       : SpatialPolygonsDataFrame
## features    : 1
## extent      : 72.64293, 80.89956, 15.60679, 22.02903  (xmin, xmax,
ymin, ymax)
## crs          : +proj=longlat +datum=WGS84 +no_defs +ellps=WGS84
+towgs84=0,0,0
## variables   : 1
## names       :       ST_NM
## value       : Maharashtra
```

```
>plot(MH) #plot Maharshtra state shapefile
```

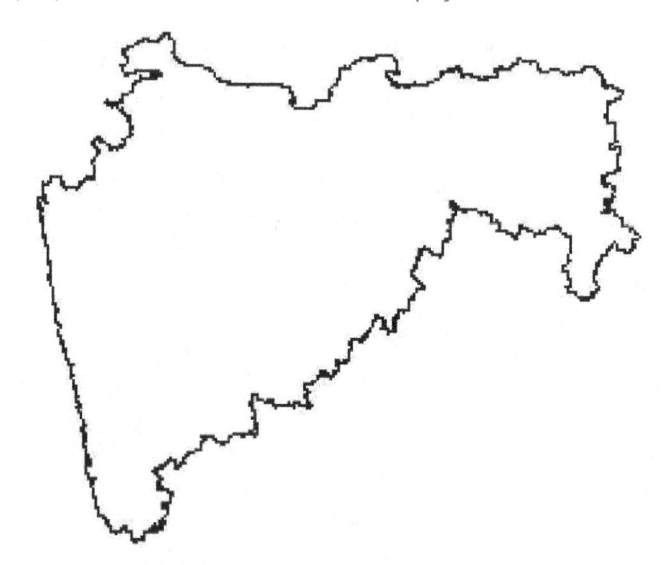

```
#Import India states shapefile
>India_state =readOGR('E:\\...........\\India_state',
'Admin2')

## OGR data source with driver: ESRI Shapefile
## Source: "E:\..............\India_state", layer: "Admin2"
## with 36 features
## It has 1 fields

>India_state #display metadata

## class       : SpatialPolygonsDataFrame
## features    : 36
## extent      : 68.18625, 97.41529, 6.755953, 37.07827  (xmin, xmax,
ymin, ymax)
## crs          : +proj=longlat +datum=WGS84 +no_defs +ellps=WGS84
+towgs84=0,0,0
## variables   : 1
## names       :                     ST_NM
## min values  : Andaman & Nicobar Island
## max values  :               West Bengal
```

```
>plot(India_state) #plot shapefile
```

8.1. Join attribute table

If you want to add data from a table to the attributes of the shapefile, then join attribute table operation is done. For joining the tables, one column of the attribute table should match to one column of the table which you wish to join. Thus, a table join appends all the columns from one table into another table based on unique ID column which is common in both tables. In R, this operation is performed by `merge` function of the `raster` package. In the following example, we are joining the state-wise population data present in a '.csv' file to the Indian states shapefile in order to create a map of state-wise distribution of population. So, first we need to import that '.csv' file, then find out the unique ID column which is common in both attribute table of shapefile and '.csv' file.

```
#Read '.csv' file containing population data
>pop_data =read.csv('data_for_creating_vector.csv',
header =TRUE)
>names(India_state) #display attribute name of shapefile
```

```
## [1] "ST_NM"
>names(pop_data) #display column names of '.csv' file
## [1] "State"       "Latitude"   "Longitude"  "Population"
```

The 'ST_NM' and 'State' columns of both tables contain names of Indian states, therefore, these columns act as unique ID column which is being used for joining these two tables as shown in the following R – code example.

```
#Joining shapefile attributes with table
>India_pop =merge(x = India_state, y = pop_data, #join data
by.x ='ST_NM', #name of attribute to be joined
by.y ='State', #column name identical with attribute
all.x =TRUE) #retain all rows of attribute table


#Display attribute table of shapefile to check joining
>data.frame(India_pop)
```

```
##                         ST_NM Latitude Longitude Population
## 1   Andaman & Nicobar Island       NA        NA         NA
## 3         Arunanchal Pradesh       NA        NA         NA
## 4                      Assam  26.2442   92.5378   34586234
## 5                      Bihar  25.0961   85.3131  119461013
## 6                 Chandigarh       NA        NA         NA
## 7               Chhattisgarh  21.2951   81.8282   28566990
## 8     Dadara & Nagar Havelli       NA        NA         NA
## 9                Daman & Diu       NA        NA         NA
## 10                       Goa  15.2993   74.1240    1542750
## 11                   Gujarat  22.3094   72.1362   63907200
## 12                   Haryana  29.2385   76.4319   27388008
## 13          Himachal Pradesh  32.0842   77.5712    7316708
## 14           Jammu & Kashmir  32.7266   74.8570   13635010
## 15                 Jharkhand  23.6102   85.2799   37329128
## 16                 Karnataka  15.3173   75.7139   66165886
## 17                    Kerala  10.8505   76.2711   35330888
## 18               Lakshadweep       NA        NA         NA
## 19            Madhya Pradesh  23.4733   77.9480   82342793
## 20               Maharashtra  19.6012   75.5530  120837347
## 21                   Manipur  24.6637   93.9063    3008546
## 22                 Meghalaya  25.4670   91.3662    3276323
## 23                   Mizoram  23.1645   92.9376    1205974
## 24                  Nagaland  26.1584   94.5624    2189297
## 25             NCT of Delhi       NA        NA         NA
## 27               Puducherry       NA        NA         NA
## 28                    Punjab  31.1471   75.3412   29611935
## 29                 Rajasthan  27.3913   73.4326   78230816
## 30                    Sikkim  27.5330   88.5122     671720
## 31                Tamil Nadu  11.0598   78.3875   76481545
## 32                 Telangana  17.1232   79.2088   38472769
## 33                   Tripura  23.7451   91.7468    4057847
## 34             Uttar Pradesh  28.2076   79.8267  228959599
```

```
## 35          Uttarakhand  30.0668   79.0193   11090425
## 36          West Bengal  22.9786   87.7478   97694960
## 26               Odisha  20.9409   84.8035   45429399
## 2        Andhra Pradesh  15.9129   79.7400   52883163
```

Now we will plot the map containing population data joined with Indian states shapefile so that we can display the state-wise distribution of population. For plotting this map, `spplot` command is used. A package containing pre-defined colour palettes `colorspace` is also used for assigning colours to the states based on their population intensity.

```
#Plot Indian states shapefile displaying distribution of
population with colour intensity
>library(colorspace) #Import package containing colour palettes
>options(scipen =999) #to disable scientific notation in plot
#Plot map
>spplot(India_pop, zcol ='Population', #data to be plotted
col.regions =rev(sequential_hcl(25))) #assign colour palette
```

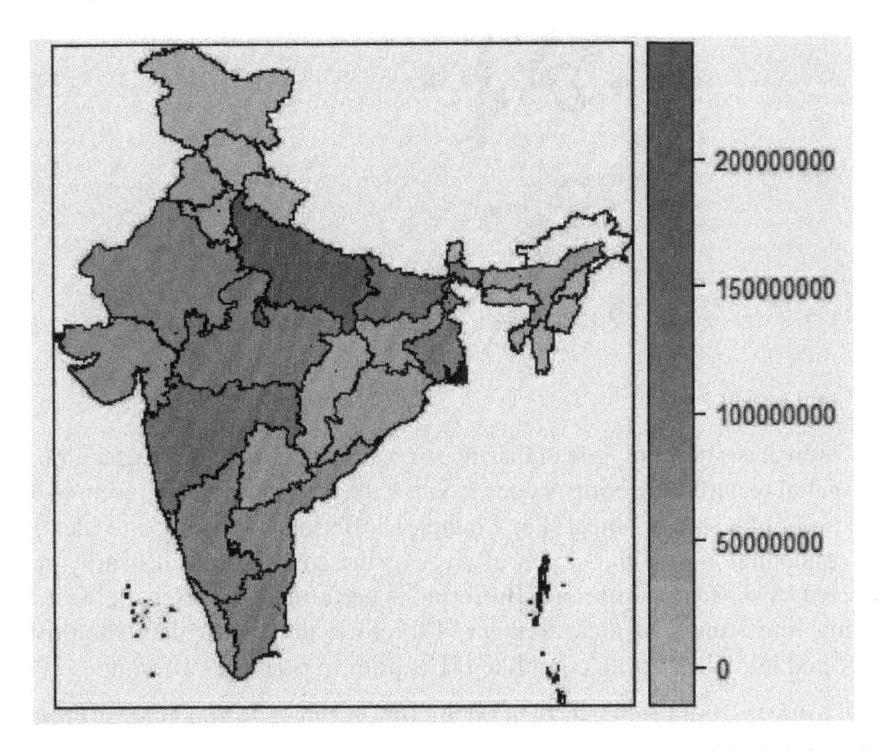

Thus, the map of state-wise population distribution is created by joining the population data contained in '.csv' file with the Indian states' shapefile. In this way, join attribute table operation is very useful to display spatial distribution of any data.

8.2. Crop by extent

Crop function is performed to minimize the area of shapefile or vector data to the desired extent. A vector file can be cropped by the extent provided by the user by using `crop` function of the raster data. In the following example, 'forest' shapefile is cropped to the desired extent.

```
#Crop by extent
>ext =c(73.0, 76.0, 16.0, 19.0) #define desired extent
>forest_crop =crop(forest, ext) #crop shapefile
>plot(forest_crop) #plot cropped shapefile
```

8.3. Buffer

Buffer creates a new polygon of distance or width as defined by the user around the spatial feature. Buffering is done to know the features or area present in the proximity of a spatial object. For example, buffering can be done to identify the residential area within 1 km of river or the composition of nearby area affected by traffic load of road. Buffering is performed on the spatial feature having linear units of measurement. Therefore, the vector data should be projected into planar projection like UTM prior to perform buffering.

In R software, buffering can be done by using `gBuffer` function of `rgeos` package is used. We are creating a vector containing buffer area of 1000 m around the roads in 'roads' shapefile in the following R-code example. As the 'roads' shapefile is in geographic coordinate system on which buffering is difficult to perform, therefore, it is first converted to UTM projection and then buffer function is applied.

```
#Reproject road shapefile to UTM having linear units of measurement
>utm =CRS("+init=epsg:32643") #get CRS of UTM zone 43
>utm #Display CRS

## CRS arguments:
##  +init=epsg:32643 +proj=utm +zone=43 +datum=WGS84 +units=m +no_defs
## +ellps=WGS84 +towgs84=0,0,0

>road_utm =spTransform(road, utm) #reproject road shapefile
>road_utm #display metadata

## class       : SpatialLinesDataFrame
## features    : 80
## extent      : 319654, 484027.8, 1804036, 1864684  (xmin, xmax, ymin,
ymax)
## crs         : +init=epsg:32643 +proj=utm +zone=43 +datum=WGS84
+units=m +no_defs +ellps=WGS84 +towgs84=0,0,0
## variables   : 5
## names       :     MED_DESCRI,     RTT_DESCRI, F_CODE_DES, ISO,
ISOCOUNTRY
## min values  :     Unknown, Primary Route,      Road, IND,     INDIA
## max values  : Without Median,      Unknown,      Road, IND,     INDIA

#Buffer applied to sub-geometries of the shapefile
>buffer1 =gBuffer(road_utm, width =1000, #distance of buffer
byid =TRUE) #buffer to sub-geometries
>buffer1 #display metadata

## class       : SpatialPolygonsDataFrame
## features    : 80
## extent      : 318654, 485027.8, 1803044, 1865680  (xmin, xmax, ymin,
ymax)
## crs         : +init=epsg:32643 +proj=utm +zone=43 +datum=WGS84
+units=m +no_defs +ellps=WGS84 +towgs84=0,0,0
## variables   : 5
## names       :     MED_DESCRI,     RTT_DESCRI, F_CODE_DES, ISO,
ISOCOUNTRY
## min values  :     Unknown, Primary Route,      Road, IND,     INDIA
## max values  : Without Median,      Unknown,      Road, IND,     INDIA


#Plot buffer over line shapefile
>plot(road_utm, col ='red')            #Plot roads
>plot(buffer1, add =TRUE)        #Plot buffer1
```

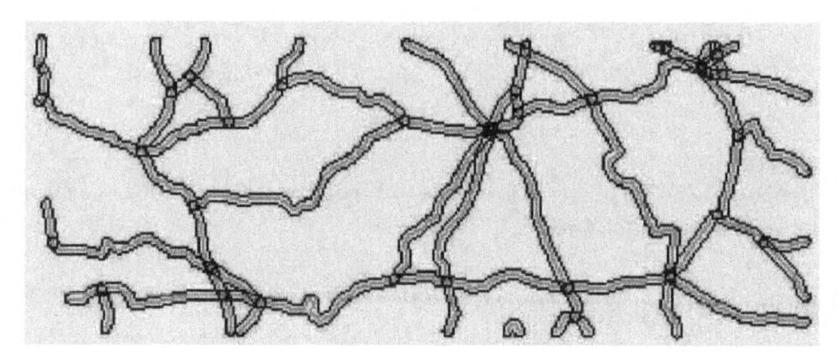

In `gBuffer` function, specify the distance of buffer from the input spatial object in the `width` argument. If you put `byid` argument as `TRUE`, then buffering is applied to the sub-geometries also specifying the point of intersection, but if this argument is `FALSE`, then buffering is applied on the entire geometry as shown below.

```
#Buffer applied to entire geometry
>buffer2 =gBuffer(road_utm, width =1000, byid =FALSE)
>buffer2 #display metadata

## class       : SpatialPolygons
## features    : 1
## extent      : 318654, 485027.8, 1803044, 1865680  (xmin, xmax, ymin, ymax)
## crs         : +init=epsg:32643 +proj=utm +zone=43 +datum=WGS84 +units=m +no_defs +ellps=WGS84 +towgs84=0,0,0

#Plot buffer over line shapefile
>plot(road_utm, col ='brown')     #Plot roads
>plot(buffer2, add =TRUE)         #Plot buffer2
```

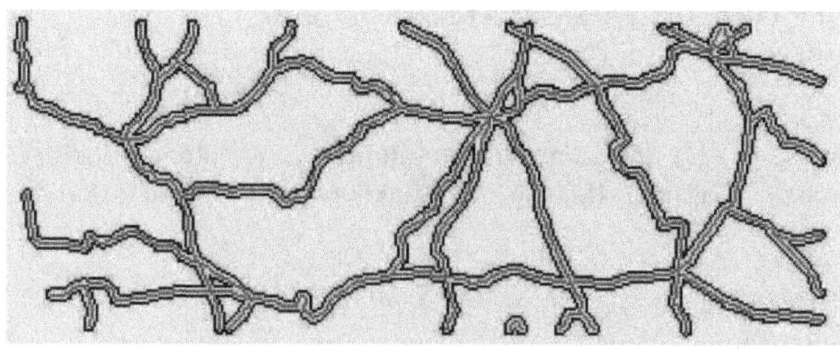

`gBuffer` command also provides some functions to customize the join style and end cap style of the buffer as shown in the following R-code example.

```
#Buffer with flat cap and bevel join style
>buffer3 =gBuffer(road_utm, width =1000, byid =FALSE, capStyle ='FLAT', joinStyle ='BEVEL')
>buffer3 #display metadata
```

```
## class        : SpatialPolygons
## features     : 1
## extent        : 318701.8, 484799.8, 1803385, 1865014  (xmin, xmax,
ymin, ymax)
## crs            : +init=epsg:32643 +proj=utm +zone=43 +datum=WGS84
+units=m +no_defs +ellps=WGS84 +towgs84=0,0,0
```

```
#Plot buffer over line shapefile
>plot(road_utm, col ='blue')      #Plot roads
>plot(buffer3, add =TRUE)         #Plot buffer3
```

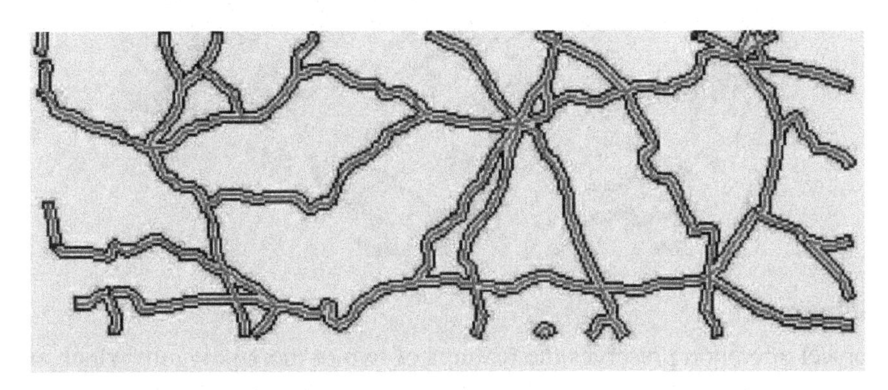

8.4. Union

In union, two or more polygons are combined into a single vector file including both overlapping and non-overlapping features. Union is performed in R by using **union** function of rgeos package. In the following R – code example, a union of forest and Maharashtra state shapefiles is created.

```
#Union of forest and Maharashtra shapefiles into one shapefile
>MH_forest_union =union(MH, forest)
>MH_forest_union #display metadata
```

```
## class        : SpatialPolygonsDataFrame
## features     : 1178
## extent       : 72.64293, 82.98134, 15.4021, 22.02903  (xmin, xmax,
ymin, ymax)
## crs            : +proj=longlat +datum=WGS84 +no_defs +ellps=WGS84
+towgs84=0,0,0
## variables    : 4
## names        :        ST_NM,    osm_id,                 name,   type
## min values   : Maharashtra, 101360250,  Ananthagiri hills, forest
## max values   : Maharashtra,  99877536, Woodland and brush, forest
```

```
>plot(MH_forest_union) #plot union
```

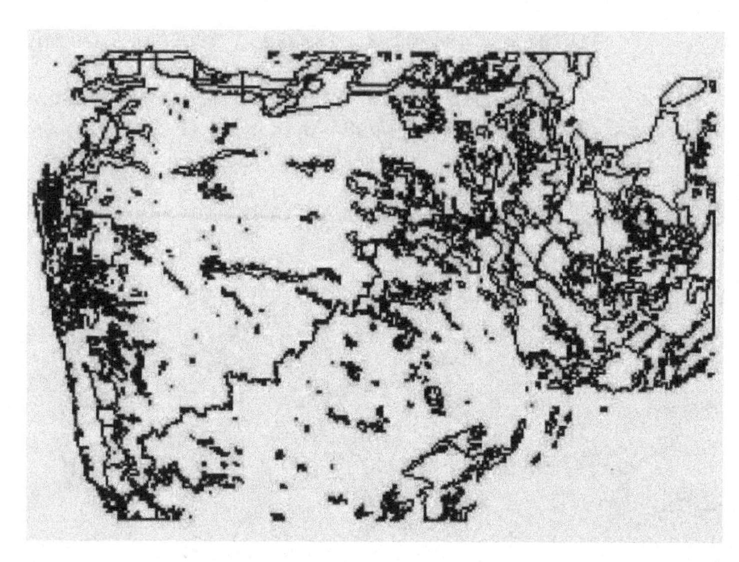

8.5. Intersect

Intersect operation preserves the features of two or more polygons which are intersecting with each other in the output vector. In the following R – code example, we are creating an intersect of forest and Maharashtra state shapefiles in order to create an output shapefile containing forest features of Maharashtra state only.

```
#Intersect to get forest inside Maharashtra state
>MH_forest_intersect =intersect(MH, forest)
>MH_forest_intersect #display metadata

## class       : SpatialPolygonsDataFrame
## features    : 566
## extent      : 72.68288, 80.89956, 15.61086, 21.96016  (xmin, xmax, ymin, ymax)
## crs           : +proj=longlat +datum=WGS84 +no_defs +ellps=WGS84 +towgs84=0,0,0
## variables   : 4
## names       :       ST_NM,    osm_id,                    name,   type
## min values  : Maharashtra, 101962412, Area of Parsi Dairy, forest
## max values  : Maharashtra,  99877536,  Woodland and brush, forest

>plot(MH_forest_intersect) #Plot intersect
```

8.6. Erase

Erase operation is used to create an output vector which erase the overlapping features of erase vector from the input vector. This operation is performed in R by using **erase** function of `raster` package. In the following R – code example, we want to get output shapefile having forest area outside the Maharashtra state. Therefore, 'forest' shapefile is used as input vector in the first argument and 'Maharashtra' shapefile act as erase vector in the second argument of **erase** function.

```
#Erase to get forest outside Maharashtra state
>Forest_not_MH =erase(forest, MH)
>Forest_not_MH #display metadata
```

```
## class       : SpatialPolygonsDataFrame
## features    : 611
## extent      : 72.70689, 82.98134, 15.4021, 21.96016  (xmin, xmax,
ymin, ymax)
## crs          : +proj=longlat +datum=WGS84 +no_defs +ellps=WGS84
+towgs84=0,0,0
## variables   : 3
## names       :    osm_id,                    name,    type
## min values  : 101360250,     Ananthagiri hills, forest
## max values  :  91667754, Vansda National Park, forest
```

```
>plot(Forest_not_MH) #Plot erased shapefile
```

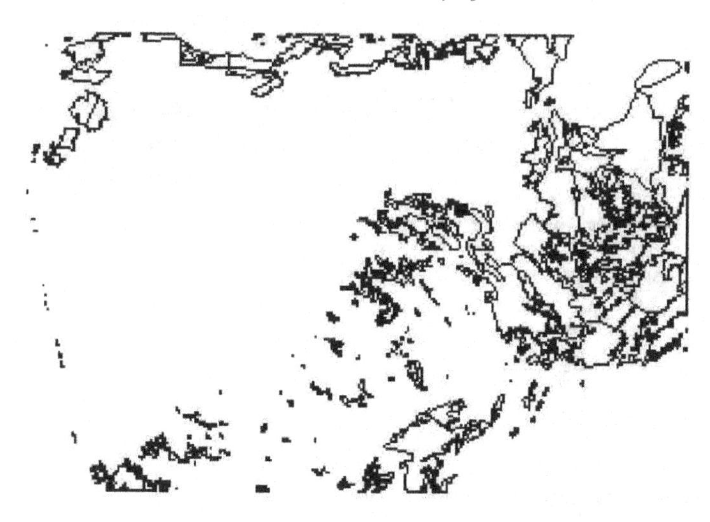

8.7. Dissolve

Dissolve operation creates a new vector by dissolving or merging the adjacent polygons which share common values for a specified attribute. In R, dissolve operation is performed by using `aggregate` function of `raster` package. In the following R – code example, dissolve operation is practiced on Indian states shapefile which created an output shapefile containing international boundaries of India only after dissolving the boundaries of all the states sharing common boundaries with other states as shown below.

```
#Dissolve the boundaries of states in India shapefile

>India =aggregate(India_state)

>India #display metadata

## class       : SpatialPolygons

## features    : 1

## extent      : 68.18625, 97.41529, 6.755953, 37.07827
(xmin, xmax, ymin, ymax)

## crs         : +proj=longlat +datum=WGS84 +no_defs
+ellps=WGS84 +towgs84=0,0,0

>plot(India) #Plot dissolved shapefile
```

Chapter 9

Mosaic Raster Images

Mosaic is the combination of two or more raster images into single image. Sometimes, single image captured by remote sensing satellite doesn't completely covers the desired study area mostly. Therefore, users resort for mosaicking two or more images captured by satellite sensors from different paths so that the whole study area is merged into a single raster image to facilitate the image processing and analysis. The pixel size of the raster images should be same for mosaicking.

In most of the cases, there is spatial overlapping among the raster images from which we want to create mosaic. These overlapping areas are handled in multiple ways which is being discussed further using R – code example. For mosaicking raster images in R, `mosaic` function of the `raster` package is used.

Landsat 8 OLI images are mosaicked in the R – code example of this chapter. The set of raster data used in this exercise has only three bands corresponding to green, red and NIR wavelengths, respectively. First of all, we will import the given raster data using the following R – code.

```
#Import individual images to be mosaicked
>setwd("E:/........../Chapter9/data")
>img1 =brick('raster_1.tif') #Import 1st image
>img1 #Display metadata of raster_1
```

```
## class      : RasterBrick
## dimensions : 7801, 7651, 59685451, 3  (nrow, ncol, ncell, nlayers)
## resolution : 30, 30  (x, y)
## extent     : 345585, 575115, 2916885, 3150915  (xmin, xmax, ymin, ymax)
## crs        : +proj=utm +zone=44 +datum=WGS84 +units=m +no_defs +ellps=WGS84 +towgs84=0,0,0
## source     : E:/............................/Chapter9/data/Raster_1.tif
## names      : Raster_1.1, Raster_1.2, Raster_1.3
## min values :          0,          0,          0
## max values :      25678,      27333,      31521
```

```
>NAvalue(img1) =0#Set no data value
>plotRGB(img1, r=3, g=2, b=1, stretch ='lin', axes =TRUE,
main ='1st Image') #Plot FCC
```

The no data value is '0' in the given set of raster data. This no data value is ignored by assigning the same in `NAvalue(raster data)` function as depicted above. Similarly, other three raster images of different paths are imported and standard False Colour Composite (FCC) plots are created. In the *'data'* folder of *'Chapter9'* directory of the provided DVD, four images are present which are used in this chapter.

The FCC plots of all four images which is being mosaicked in the example are shown below.

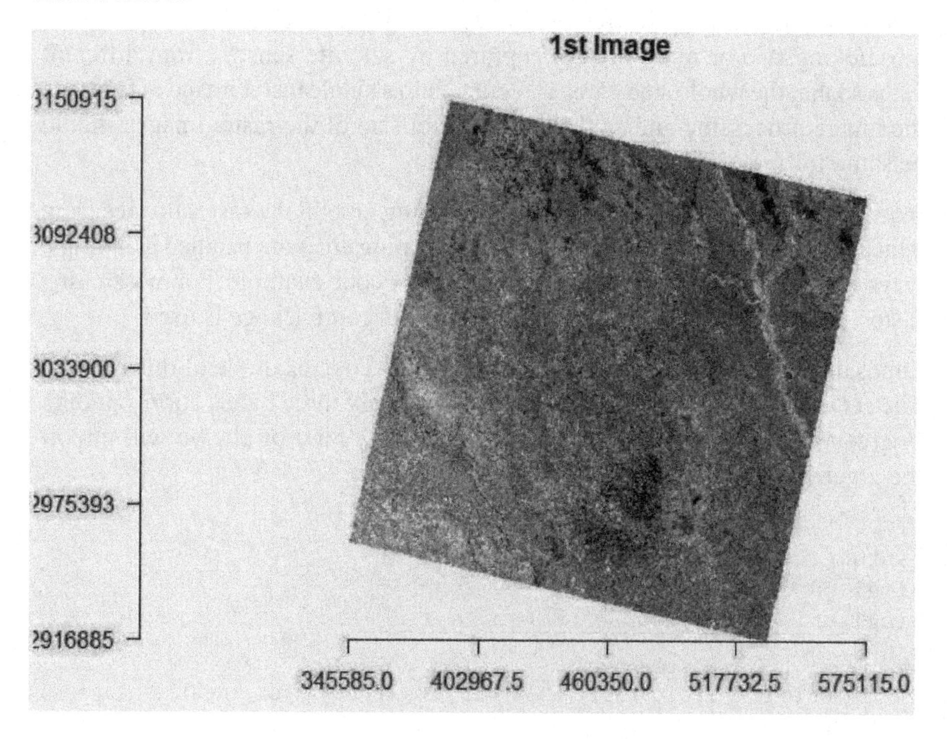

2nd Image
2991615
2933108
2874600
2816093
2757585
308985.0
366442.5
423900.0
481357.5
538815.0

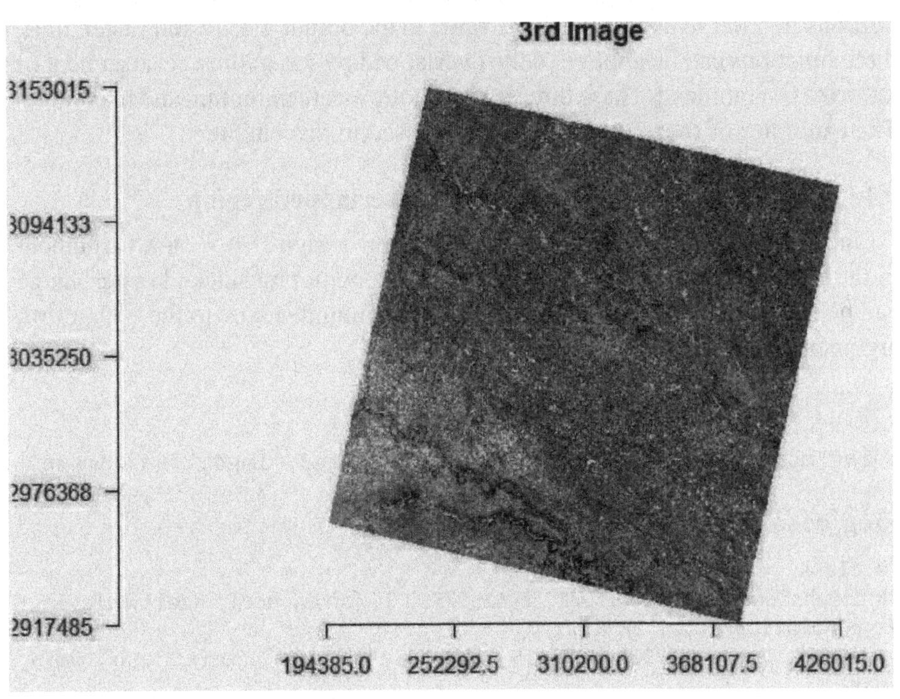
3rd Image
3153015
3094133
3035250
2976368
2917485
194385.0
252292.5
310200.0
368107.5
426015.0

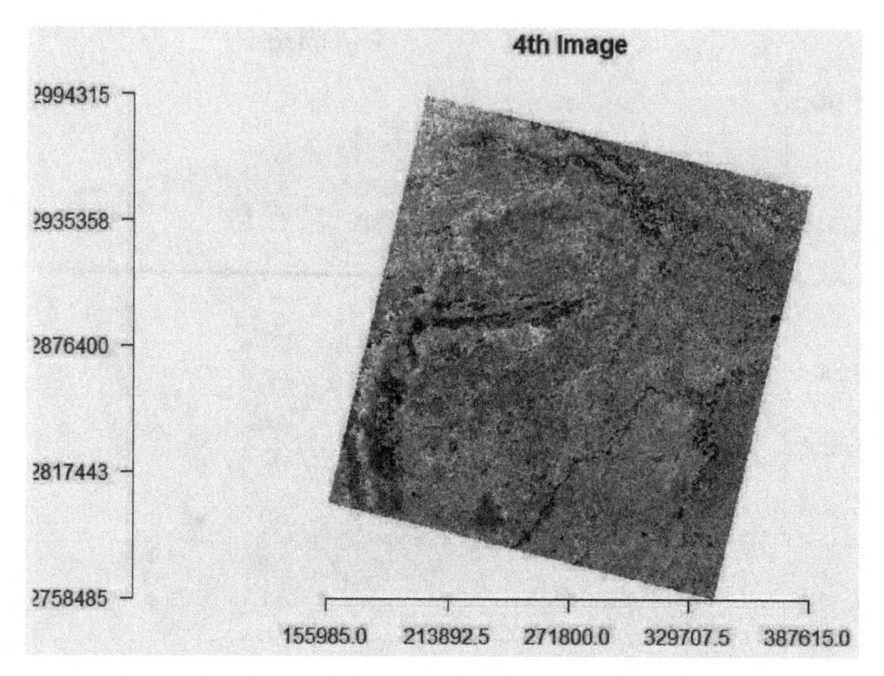

While mosaicking the raster images, edge portion of the images of different paths overlap with each other. So, the function is applied to the overlapped portions in order to assign the pixel value to the output mosaicked raster. In R, three functions can be applied to the overlapped region in the `fun` argument of the **`mosaic`** command. These three functions are mean, minimum and maximum. The examples of these functions are discussed in this chapter.

9.1. Mosaic with mean function to the overlapped region

To apply mean function to the overlapped region, assign `fun = mean` argument in the **`mosaic`** command as shown below. The output mosaicked raster image can be saved in the working directory by assigning its name to the `filename` argument.

```
# Mosaic these 4 raster bricks imported above with mean
function
> img_mosaic_mean =mosaic(img1, img2, img3, img4, fun = mean,
                    filename ='img_mosaic_mean.tif')
>img_mosaic_mean #Display attributes of mosaicked raster
```

```
## class      : RasterBrick
## dimensions : 13181, 13971, 184151751, 3  (nrow, ncol, ncell, nlayers)
## resolution : 30, 30  (x, y)
## extent     : 155985, 575115, 2757585, 3153015  (xmin, xmax, ymin,
ymax)
## crs         : +proj=utm +zone=44 +datum=WGS84 +units=m +no_defs
```

```
+ellps=WGS84 +towgs84=0,0,0
## source     : E:/................................/Chapter9/data/img_mosaic_mean.tif
## names      : img_mosaic_mean.1, img_mosaic_mean.2, img_mosaic_mean.3
## min values :              7378,              6702,              5546
## max values :             30733,             33683,             41757
```

```
>plotRGB(img_mosaic_mean, r=3, g=2, b=1, stretch ='lin', axes =TRUE,
main ="Mosaic with mean overlapping") #Plot FCC
```

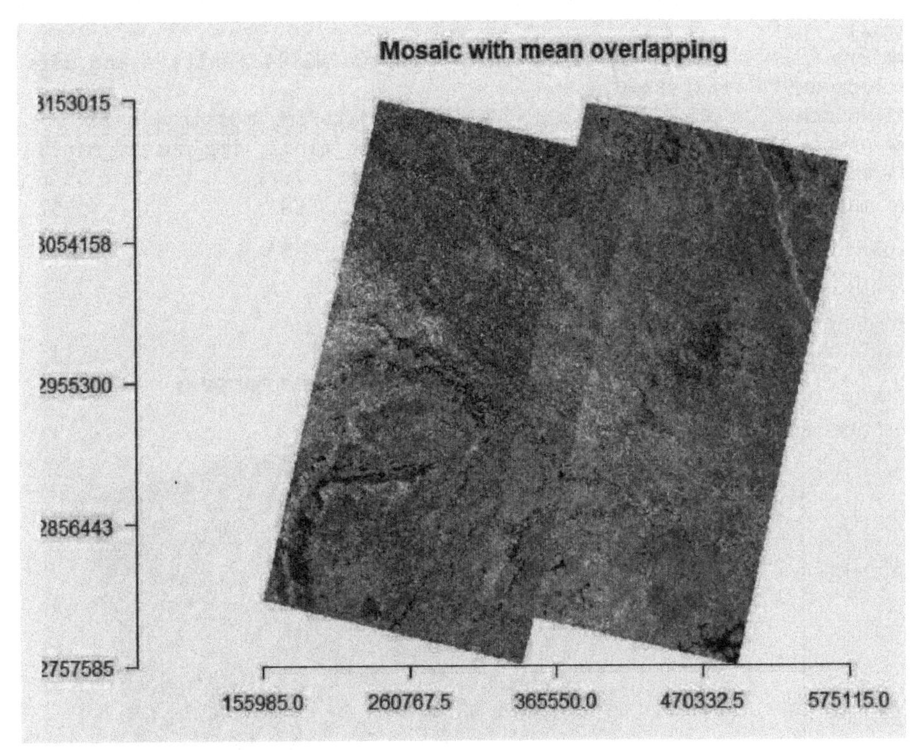

So, this mean function calculates the mean of the pixel values of the overlapping region and assign that mean value to the output raster as you can see in the plotted FCC.

9.2. Mosaic with minimum function to the overlapped region

In order to apply minimum function to the overlapped region, assign `fun = min` argument in the `mosaic` command as shown below. This function takes the minimum pixel value of the overlapping regions of the raster images to be mosaicked and assign it to the output as shown in the FCC plotted.

```
# Mosaic these 4 raster bricks imported above with minimum
function
> img_mosaic_min =mosaic(img1, img2, img3, img4, fun = min,
                        filename ='img_mosaic_min.tif')
> img_mosaic_min #Display attributes of mosaicked raster

## class      : RasterBrick
## dimensions : 13181, 13971, 184151751, 3  (nrow, ncol, ncell, nlayers)
## resolution : 30, 30  (x, y)
## extent     : 155985, 575115, 2757585, 3153015  (xmin, xmax, ymin,
ymax)
## crs         : +proj=utm +zone=44 +datum=WGS84 +units=m +no_defs
+ellps=WGS84 +towgs84=0,0,0
## source     : E:/................................/Chapter9/data/img_mosaic_min.tif
## names      : img_mosaic_min.1, img_mosaic_min.2, img_mosaic_min.3
## min values :             7378,             6702,             5546
## max values :            30733,            33683,            41757

>plotRGB(img_mosaic_min, r=3, g=2, b=1, stretch ='lin', axes
=TRUE,
main ="Mosaic with minimum overlapping") #Plot FCC
```

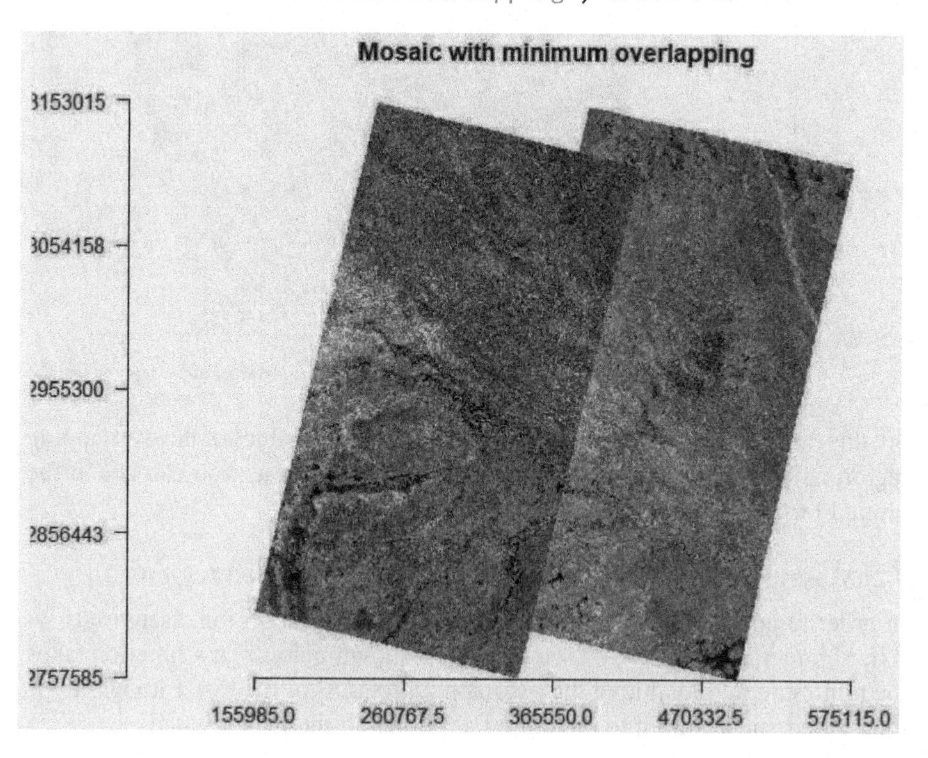

9.3. Mosaic with maximum function to the overlapped region

Maximum function assigns maximum pixel value of the overlapping regions to the output mosaicked raster. It can be applied by passing `fun = max` argument to the **`mosaic`** command. It creates best image without any loss of data.

```
# Mosaic these 4 raster bricks imported above with maximum
function
> img_mosaic_max =mosaic(img1, img2, img3, img4, fun = max,
                         filename ='img_mosaic_max.tif')
> img_mosaic_max #Display attributes of mosaicked raster

## class      : RasterBrick
## dimensions : 13181, 13971, 184151751, 3  (nrow, ncol, ncell, nlayers)
## resolution : 30, 30  (x, y)
## extent     : 155985, 575115, 2757585, 3153015  (xmin, xmax, ymin,
ymax)
## crs          : +proj=utm +zone=44 +datum=WGS84 +units=m +no_defs
+ellps=WGS84 +towgs84=0,0,0
## source    : E:/alka_PAT/Manual on R/Chapter9/data/img_mosaic_max.tif
## names      : img_mosaic_max.1, img_mosaic_max.2, img_mosaic_max.3
## min values :             7378,             6702,             5546
## max values :            30733,            33683,            41757

>plotRGB(img_mosaic_max, r=3, g=2, b=1, stretch ='lin', axes
=TRUE,
main ="Mosaic with maximum overlapping") #Plot FCC
```

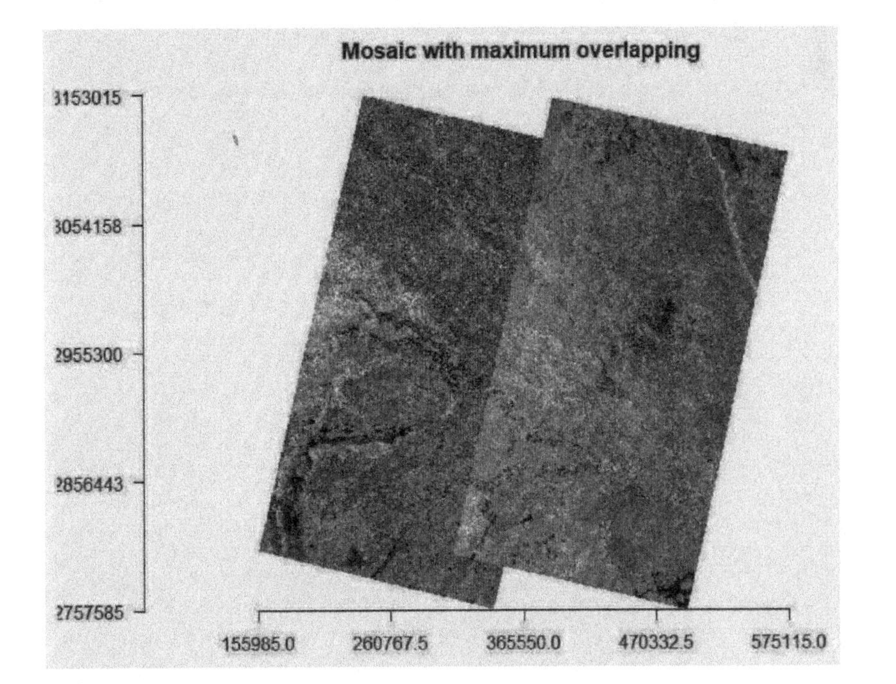

The function to be used for mosaicking depends upon the type of data and interest of user. Mostly maximum function is used for the mosaicking of the raster images.

Chapter 10

Resampling of Raster Images

Many times, users have to work with raster datasets of different spatial and radiometric resolutions. So, there is requirement of bringing their resolutions to the same resolution in order to facilitate their processing. **Resampling** is the process of deriving pixel values from the existing image to the new raster image of different pixel size. Resampling is most common pre-processing step in remote sensing. It is done both at spatial and radiometric scale. At spatial scale, resampling is done either to increase the spatial resolution *i.e.* decrease pixel size known as **Upsampling** or to decrease the spatial resolution *i.e.* increase pixel size known as **Downsampling**. **Radiometric resampling** is performed to increase or decrease the radiometric resolution of the image. In R, resampling of raster image is performed by `raster` package. The details of the resampling methods and their respective R – code with examples are discussed in this chapter.

First of all, we need to import the raster image on which we wish to perform resampling operations by using following R – code. In R, resampling is feasible on one band only. So, we will import single band raster image.

```
# Import single band raster image
>setwd("E:/........./Chapter10/data")
>img =raster('input_raster_30m.TIF')
>res(img) #Spatial resolution of imported raster image

## [1] 30 30

>ncell(img) #Number of cells in imported raster image

## [1] 59763461

>img #Display attributes of imported raster

## class      : RasterLayer
## dimensions : 7801, 7661, 59763461  (nrow, ncol, ncell)
## resolution : 30, 30  (x, y)
## extent     : 308985, 538815, 2757585, 2991615  (xmin, xmax, ymin,
ymax)
```

```
## crs          : +proj=utm +zone=44 +datum=WGS84 +units=m +no_defs
+ellps=WGS84 +towgs84=0,0,0
## source     : E:/......................................../Chapter10/data/input_raster_30m.TIF
## names      : input_raster_30m
## values     : 0, 65535  (min, max)
```

10.1. Spatial Upsampling

Spatial Upsampling is done to increase the spatial resolution or decrease the pixel size of the raster dataset. In R, upsampling is done by two methods *viz.,* nearest neighbour and bilinear interpolation. First of all, we need to create an empty raster of desired spatial resolution, say 10 m in this case. This new raster will act as output raster in which pixel values will be filled from the previously imported input raster through resampling methods. This empty raster can be created by using the given R – code.

```
#Create an empty raster layer of spatial resolution 10 m
>img10 =raster(crs =”+proj=utm +zone=44 +datum=WGS84 +units=m
+no_defs   +ellps=WGS84 +towgs84=0,0,0", #Define projection of
empty raster, same as input raster
resolution =10, #Define spatial resolution of empty raster
xmn =308985, xmx =538815, #define dimensions of new raster,
same as input raster
ymn =2757585, ymx =2991615) # same as input raster
```

10.1.1. Nearest neighbour

In the nearest neighbour method, pixel value of the output raster image is calculated from the nearest pixel of the input image. This method doesn't change the input cell value in the output image. It is very simple method. It is applicable to both continuous and discrete data. This produces blocky effect in upsampling.

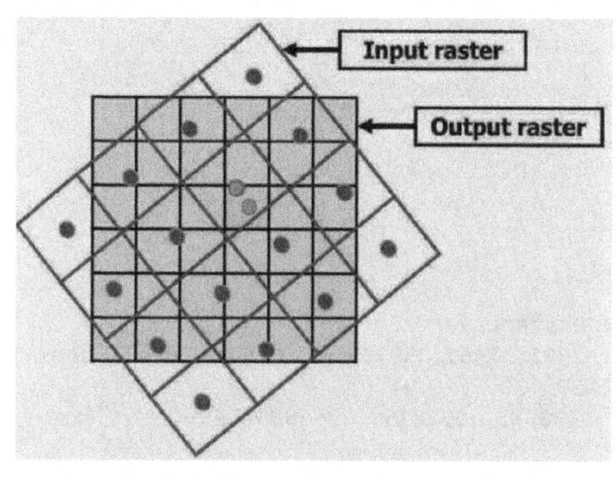

Fig. 10.1: Resampling by nearest neighbour method

In R, resampling by nearest neighbour method can be done by using `method = "ngb"` argument in the **`resample`** command of the **`raster`** package as shown below.

```
#Spatial Resampling to 10 m spatial resolution by nearest
neighbour method
> img10_ngb =resample(img, img10, method = "ngb")
>res(img10_ngb) #Spatial resolution of resampled image

## [1] 10 10

>ncell(img10_ngb) #Number of cells in resampled image

## [1] 537871149
```

The number of cells is nine times than that of the original image and spatial resolution is 10m.

10.1.2. Bilinear interpolation

In bilinear interpolation method of resampling, the new pixel value in output raster image is computed as distance weighted average of nearest four pixels. It is applicable to continuous data. It produces smoother appearance in the output raster image.

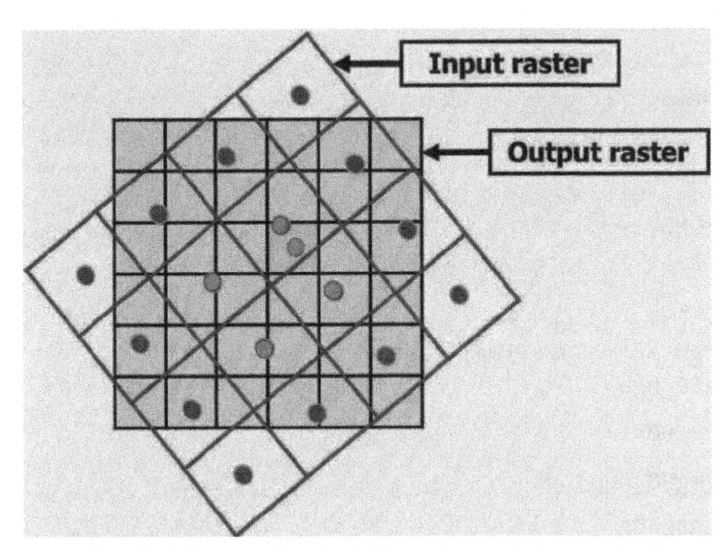

Fig. 10.2: Resampling by bilinear interpolation method

In R, resampling by bilinear interpolation can be done by using `method = "bilinear"` argument in the **`resample`** command of the **`raster`** package as shown below.

```
#Spatial resampling by bilinear interpolation to 10 m
resolution
> img10_bl =resample(img, img10, method =”bilinear”)
>res(img10_bl) #Spatial resolution of resampled image

## [1] 10 10

>ncell(img10_bl) #Number of cells in resampled image

## [1] 537871149
```

10.2. Spatial Downsampling

Spatial Downsampling is done to decrease the spatial resolution or increase the pixel size of the raster dataset. In R, downsampling is done by many methods likenearest neighbour, bilinear interpolation and spatial aggregation by mean, mode, minimum, maximum or whatever statistics user wish to apply.

10.2.1. Nearest neighbour and Bilinear Interpolation

Downsampling by nearest neighbour produces coarse and grainy effect in the output raster. For applying nearest neighbour and bilinear interpolation method for resampling in R, first we have to create an empty raster of lower spatial resolution, say 90 m in this case. Rest procedure is same as explained in the previous section.

```
#create an empty raster layer of spatial resolution 90 m
>img90 =raster(crs =”+proj=utm +zone=44 +datum=WGS84 +units=m
+no_defs +ellps=WGS84 +towgs84=0,0,0", #Define projection of
empty raster
resolution =90, #Define spatial resolution of empty raster
xmn =308985, xmx =538815, #define dimensions of new raster
ymn =2757585, ymx =2991615) # same as input raster

#Downsampling by nearest neighbour method to 90 m
>img90_ngb =resample(img, img90, method =”ngb”)
>res(img90_ngb) #Spatial resolution of resampled image

## [1] 90 90

>ncell(img90_ngb) #Number of cells in resampled image

## [1] 6640400

#Downsampling by bilinear interpolation method to 90 m
>img90_bl =resample(img, img90, method =”bilinear”)
>res(img90_bl) #Spatial resolution of resampled image

## [1] 90 90
```

```
>ncell(img90_bl) #Number of cells in resampled image
## [1] 6640400
```

10.2.2. Spatial aggregation

Spatial aggregation resamples input raster to the output raster of coarser resolution based on aggregation function specified by the user like mean, mode, minimum, maximum, etc. In R, spatial aggregation on the raster image is performed by `aggregate` function of the `raster` package. The statistics or function used to aggregate the value to the output raster is specified in `fun =` argument. The number of cells to aggregate in both horizontal and vertical direction is mentioned in aggregation factor `fact =` argument of the `aggregate` function which ultimately determine the final spatial resolution of the output raster. In the further sections, we will discuss the spatial aggregation to 90 m spatial resolution from the input raster of 30m resolution by four functions *i.e.* mean, mode, minimum and maximum. Users can specify the function with respect to their own interest.

10.2.2.1. Spatial aggregation by mean function

It computes mean of the pixel values of the number of pixels specified by the user in the input raster and specifies that mean value to the output raster.

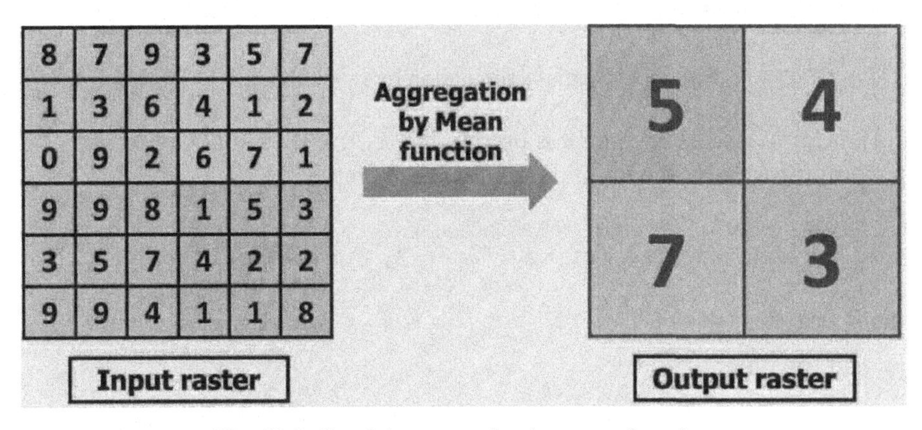

Fig. 10.3: Spatial aggregation by mean function

For performing this step in R, `fun = mean` argument is assigned. We are assigning `fact =3` argument in the R – code because it will create raster of pixel resolution 90 m by multiplying factor 3 with 30 m spatial resolution of input raster. If you want to remove the NA or missing data from the calculation for spatial aggregation, then `na.rm =TRUE` argument is used.

```
#Mean function for spatial aggregation
>img_90_mean =aggregate(img, fact =3, fun = mean, na.rm =TRUE)
>res(img_90_mean) #Spatial resolution of resampled raster image
## [1] 90 90
>ncell(img_90_mean) #Number of cells in resampled raster
## [1] 6642954
```

10.2.2.2. Spatial aggregation by mode function

The pixel values with higher frequency *i.e.* mode of pixel values in the input raster is assigned to the output raster as shown in the following figure.

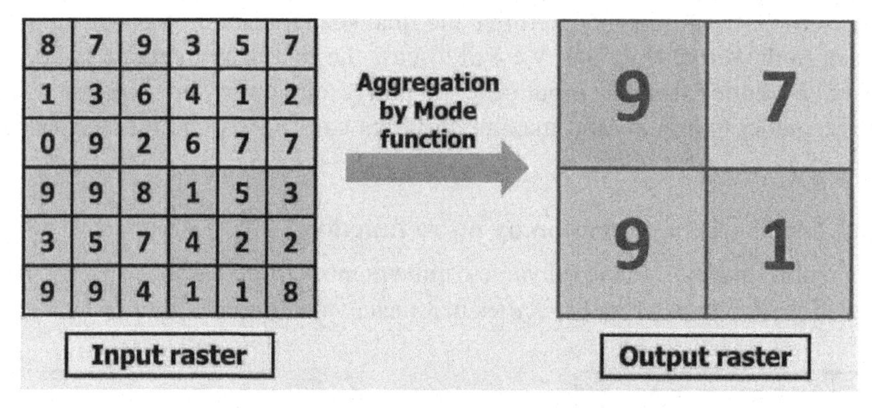

Fig. 10.4: Spatial aggregation by mode function

In R, `fun = modal` argument is used in the `aggregate` function to perform this step as shown in the following example.

```
#Mode function for downsampling
>img_90_mode =aggregate(img, fact =3, fun = modal, na.rm
=TRUE)
>res(img_90_mode) #Spatial resolution of resampled image
## [1] 90 90
>ncell(img_90_mode) #Number of cells in resampled raster
## [1] 6642954
```

10.2.2.3. Spatial aggregation by minimum function

The minimum pixel value among all the pixels from the input raster in the resolution window specified by the user is assigned as the pixel value in the output raster as shown in the figure.

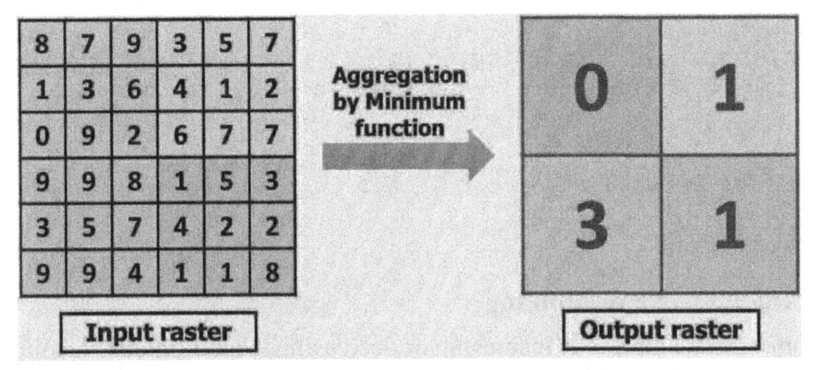

Fig. 10.5: Spatial aggregation by minimum function

Argument `fun = min` is used as shown in the following R – code for doing this step.

```
#Minimum function for downsampling
>img_90_min =aggregate(img, fact =3, fun = min, na.rm =TRUE)
>res(img_90_min) #Spatial resolution of resampled image

## [1] 90 90

>ncell(img_90_min) #Number of cells in resampled raster

## [1] 6642954
```

10.2.2.4. Spatial aggregation by maximum function

The maximum pixel value among all the pixels from the input raster in the resolution window specified by the user is assigned as the pixel value in the output raster as shown in the figure.

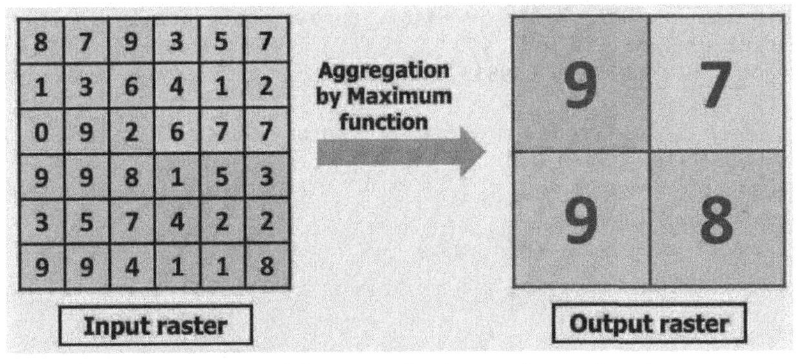

Fig. 10.6: Spatial aggregation by maximum function

Argument `fun = max` is used as shown in the following R – code for doing this step.

```
#Maximum function for downsampling
> img_90_max =aggregate(img, fact =3, fun = max, na.rm =TRUE)
>res(img_90_max) #Spatial resolution of resampled image

## [1] 90 90

>ncell(img_90_max) #Number of cells in resampled raster

## [1] 6642954
```

10.3. Radiometric resampling

Radiometric resampling is increasing or decreasing the radiometric resolution of the raster image. There is no particular function for performing this operation in R. Therefore, this operation is performed by applying mathematical formula. In the following R – code example, the image is converted from 16-bit size having radiometric values of pixels ranging from 0 to 65535 and INT2U data type into 8-bit size with pixel values ranging from 0 to 255 and INT1U data type.

```
#Radiometric resampling
>dataType(img) #Know the data type of input image

## [1] "INT2U"

#Apply formula for converting the pixel values to 8 bit (0-
255 range)
>img_8bit =calc(img, fun=function(x){((x -min(x)) *255)/
                                      (max(x)-min(x)) +0})
>img_8bit #Display the metadata of radiometric resampled
raster

## class      : RasterLayer
## dimensions : 7801, 7661, 59763461  (nrow, ncol, ncell)
## resolution : 30, 30  (x, y)
## extent     : 308985, 538815, 2757585, 2991615  (xmin, xmax, ymin,
ymax)
## crs         : +proj=utm +zone=44 +datum=WGS84 +units=m +no_defs
+ellps=WGS84 +towgs84=0,0,0
## source     : memory
## names      : layer
## values     : 0, 255  (min, max)

#Save the raster to working directory in 8-bit data type
>writeRaster(img_8bit, 'img_8bit.tif', datatype='INT1U')
>img_8bit =raster('img_8bit.tif') #Open the saved raster
>dataType(img_8bit) #Check the data type of saved raster

## [1] "INT1U"
```

Comparison of mean of raster images resampled with methods discussed above

Mean of each raster image resampled by the methods discussed above is computed using following formulae in R for the sake of comparison with the input raster data.

```
#Compute the mean of input raster
> a =cellStats(img, stat ='mean',
na.rm =TRUE) #Ignore no data value from calculation
#Compute the mean of upsampled raster to 10 m resolution by nearest neighbour method
> b =cellStats(img10_ngb, stat ='mean', na.rm =TRUE)
#Compute the mean of upsampled raster to 10 m resolution by bilinear interpolation method
> c =cellStats(img10_bl, stat ='mean', na.rm =TRUE)
#Compute the mean of downsampled raster to 90 m resolution by nearest neighbour method
> d =cellStats(img90_ngb, stat ='mean', na.rm =TRUE)
#Compute the mean of downsampled raster to 90 m resolution by bilinear interpolation method
> e =cellStats(img90_bl, stat ='mean', na.rm =TRUE)
#Compute the mean of downsampled raster to 90 m resolution by spatial aggregation by mean method
> f =cellStats(img_90_mean, stat ='mean', na.rm =TRUE)
#Compute the mean of downsampled raster to 90 m resolution by spatial aggregation by mode method
> g =cellStats(img_90_mode, stat ='mean', na.rm =TRUE)
#Compute the mean of downsampled raster to 90 m resolution by spatial aggregation by minimum method
> h =cellStats(img_90_min, stat ='mean', na.rm =TRUE)
#Compute the mean of downsampled raster to 90 m resolution by spatial aggregation by maximum method
> i =cellStats(img_90_max, stat ='mean', na.rm =TRUE)
#Compute the mean of radiometric resampled image to 8-bit
> j =cellStats(img_8bit, stat ='mean', na.rm =TRUE)
```

```
#Comparison of mean of all images

> mean_resample =c(a, b, c, d, e, f, g, h, i, j)
> mean_resample #Display mean

## [1] 11556.8749 11556.8749 11556.8749 11553.7912 11556.8827 11552.4048
##  [7] 11545.3234 11046.7103 12058.8787   106.3209

#Prepare data frame of mean value
> mean_resample_names =c('Input Raster', #List of row names
for data frame
                   'Upsampling by nearest neighbour',
'Upsampling by bilinear interpolation',
'Downsampling by nearest neighbour',
'Downsampling by bilinear interpolation',
'Aggregation by Mean', 'Aggregation by Mode',
'Aggregation by Minimum',
'Aggregation by Maximum',
'Radiometric resampling to 8 bit')

>data.frame(mean_resample, row.names = mean_resample_names)
#Data frame

##                                        mean_resample
## Input Raster                              11556.8749
## Upsampling by nearest neighbour           11556.8749
## Upsampling by bilinear interpolation      11556.8749
## Downsampling by nearest neighbour         11553.7912
## Downsampling by bilinear interpolation    11556.8827
## Aggregation by Mean                       11552.4048
## Aggregation by Mode                       11545.3234
## Aggregation by Minimum                    11046.7103
## Aggregation by Maximum                    12058.8787
## Radiometric resampling to 8 bit             106.3209
```

From the data frame, it is clear that the mean of the pixel values in upsampling is same to the input raster. But in downsampling, mean varies depending upon the method used.

Chapter 11

Raster Data Statistics

The calculation of raster data statistics is very essential before performing some geoprocessing operations. Raster data statistics involve knowing the distribution of pixel values in a band, comparing the pixel values of two bands, extracting values of all bands for a particular geographical location or point, and computing statistics like mean, median, standard deviation, etc. at pixel level across all bands of raster dataset.

In R, codes to perform raster data statistics are present in `raster` package. The visualisation of raster data statistics can be done by `rasterVis` package. So, we need to import these packages in R in order to perform this exercise.

```
#Import required packages
>library(raster)
>library(rasterVis)
```

Let's import the multiband raster image for which we want to compute statistics as shown below.

```
>setwd("E:/............/Chapter11/data")
#Import multiband raster image
>img =brick('Image_reflectance.tif')
>img#Display attributes of imported image
```

```
## class      : RasterBrick
## dimensions : 7801, 7661, 59763461, 4  (nrow, ncol, ncell, nlayers)
## resolution : 30, 30  (x, y)
## extent     : 308985, 538815, 2757585, 2991615  (xmin, xmax, ymin, ymax)
## crs        : +proj=utm +zone=44 +datum=WGS84 +units=m +no_defs +ellps=WGS84 +towgs84=0,0,0
## source     : E:/................................/Image_reflectance.tif
## names      : Image_reflectance.1, Image_reflectance.2, Image_reflectance.3, Image_reflectance.4
## min values : 0,          0,          0,          0
## max values : 0.3606562,  0.3919465,  0.4163525,  0.5113131
```

Now change the band names with respect to their wavelength for easy interpretation.

```
#Set the band names of multiband raster image
>names(img) =c('Blue', 'Green', 'Red', 'NIR')
>names(img) #Display the changed names
## [1] "Blue" "Green" "Red"   "NIR"
```

In this raster, the no data value is '0' which is considered in the calculation of statistics if not specified by the user. Hence, you need to specify it as no data value by using `NAvalue` R – code so that '0' is not taken into account while doing calculations. The conversion of numbers to scientific notations can be avoided by `options(scipen =999)` command.

```
#Set no data value
>NAvalue(img) =0
#Disable scientific notation
>options(scipen =999)
```

Plot the standard False Colour Composite (FCC) of the imported raster image to display the image.

```
#Plot FCC
>plotRGB(img, r=4, g=3, b=2, stretch ='lin',axes =TRUE)
```

11.1. Histogram

Histogram displays the frequency distribution of pixel values of the raster dataset. For plotting histogram of a single band in R, **histogram** function of **rasterVis** package is used as shown in the given example.

```
#Plot histogram of a band
>histogram(img[[4]], main ='Histogram of NIR band',
xlab ='Pixel value', ylab ='Frequency', col ='wheat')
```

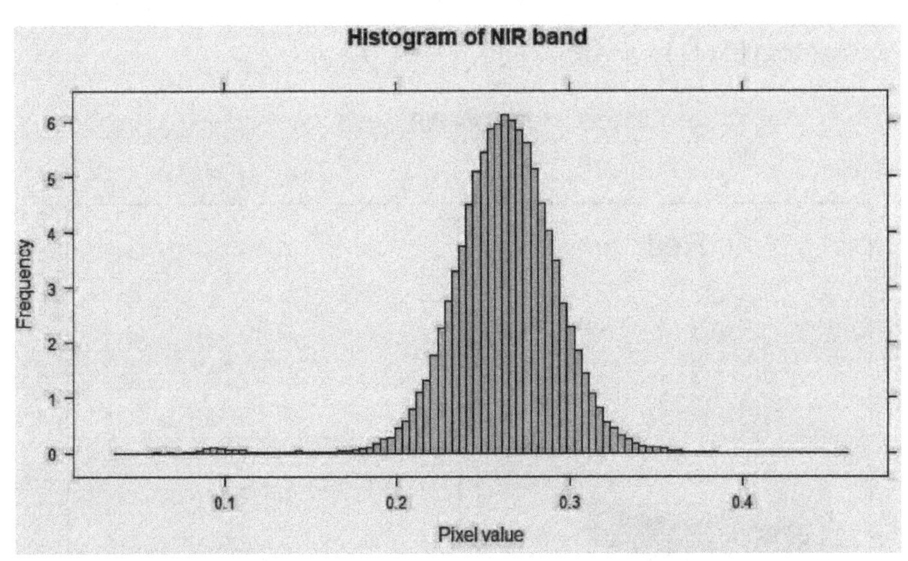

The histogram of all bands can also be plotted simultaneously by using same **histogram** function.

```
#Plot histograms of all the bands of raster brick
>histogram(img, main ='Histogram of all bands')
```

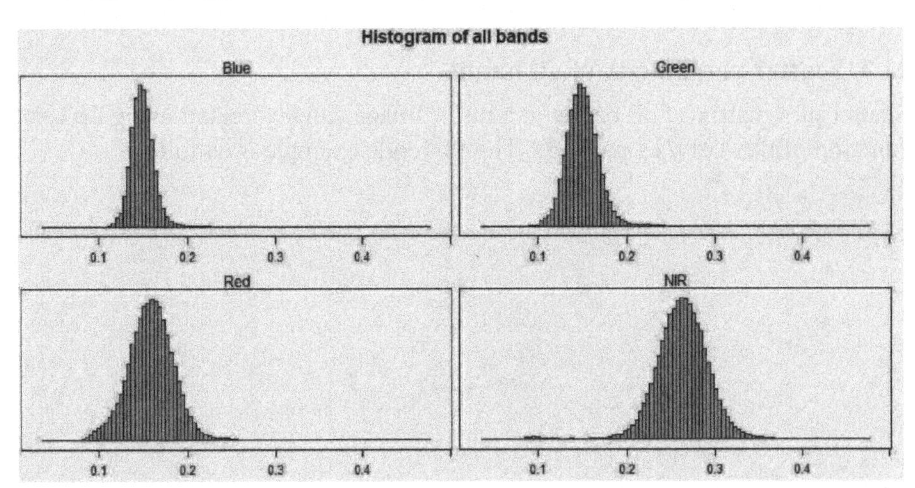

11.2. Spectra comparison between two bands

The comparison between spectra of two bands in a raster image can be done by using `pairs` function of `raster` package. This function returns the scatter plot between two specified bands, their histograms and correlation. Scatter plot tells about the value of a specific pixel in two bands and indicates the degree of relationship between these two bands. R – code example of this function is given below.

```
#Spectra comparison between 3rd and 4th bands
>pairs(img[[3:4]], main ="Red vs NIR")
```

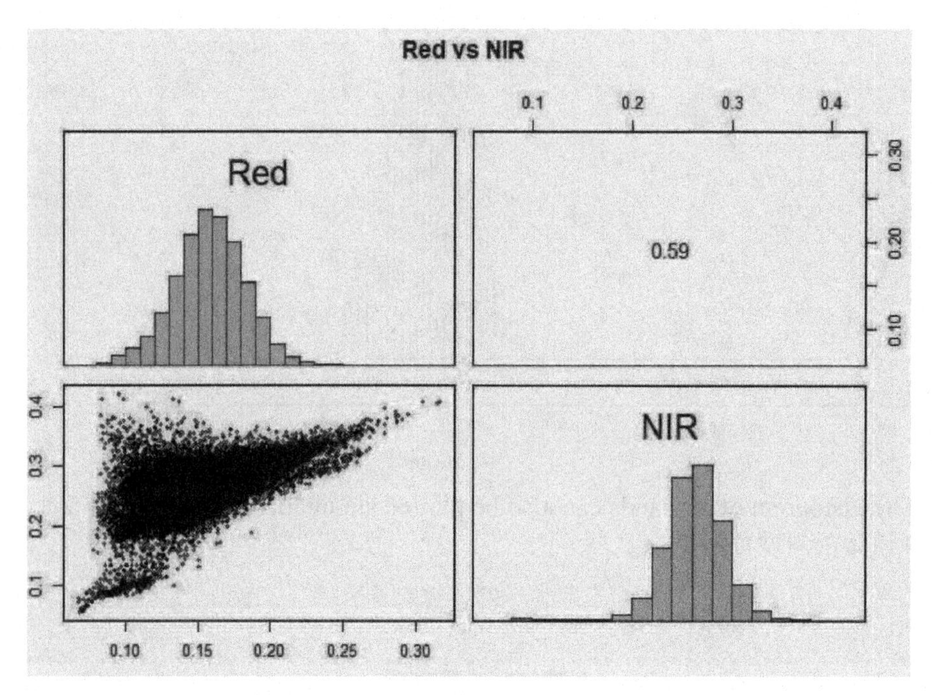

11.3. Scatter plot matrix of all bands

Scatter plot matrix of all bands in a raster image can be created using `splom` function of `rasterVis` package. The R – code example is as follow.

```
#Scatter plot matrix of all bands
>splom(img, main ='Scatter plot matrix of all bands')
```

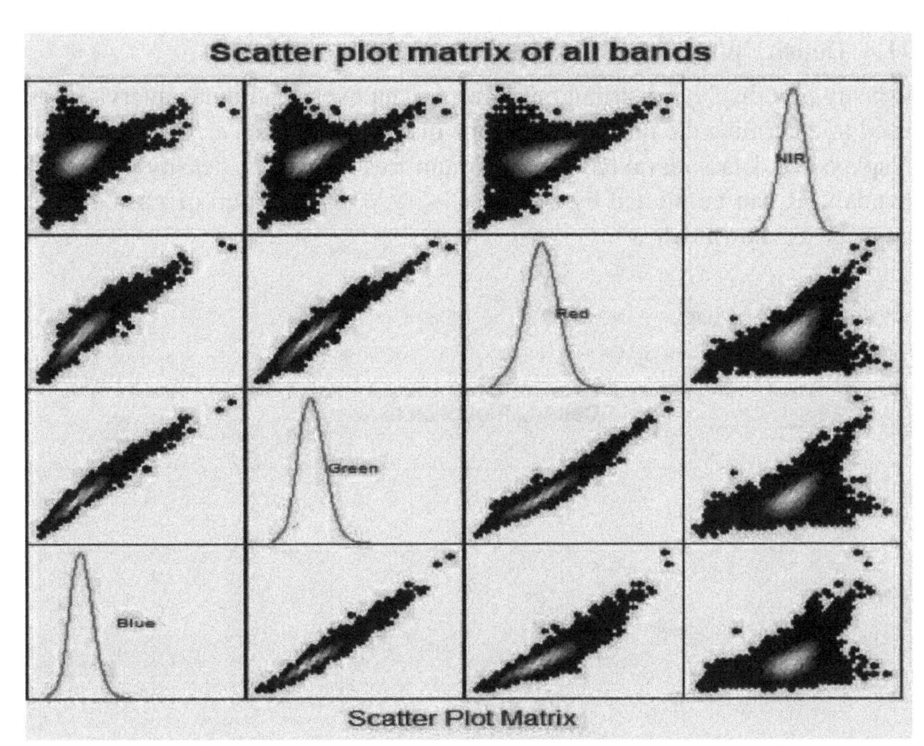

`pairs` function of `raster` package can also be used for this purpose which will give histograms and correlation along with scatter plot.

```
#Spectra comparison among all bands
>pairs(img, hist =TRUE, cor =TRUE,
main = "Scatter Plot and Spectra Comparison of all bands")
```

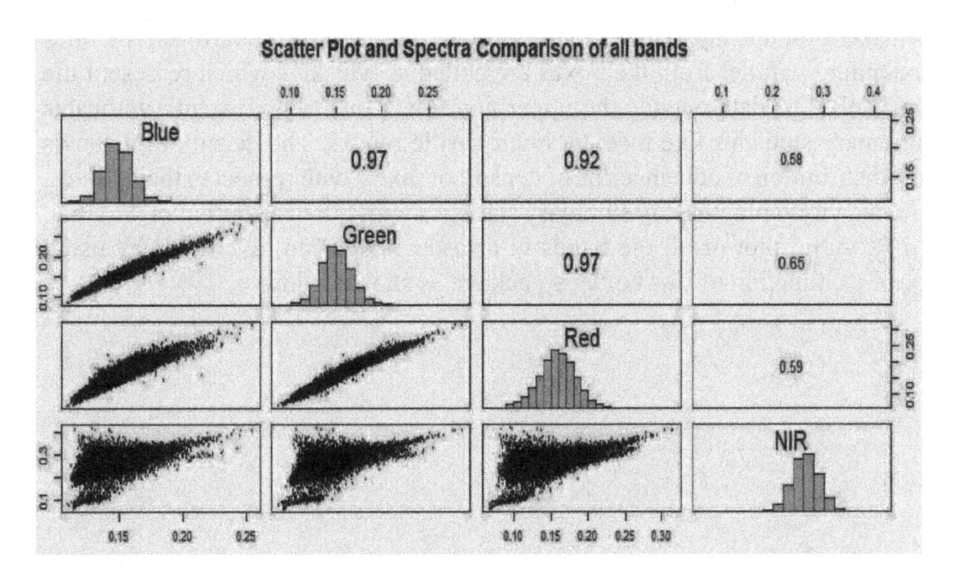

11.4. Density plot

Density plot displays distribution of raster data over continuous interval. It is used to determine the distribution shape of the data. The peak of density plot displays the data interval having maximum frequency. The density plot of all bands in R can be plotted by using `densityplot` function of `rasterVis` package as shown below.

```
#Density plot
>densityplot(img, xlab ='Reflectance values',
        main ='Density Plot ofall bands')
```

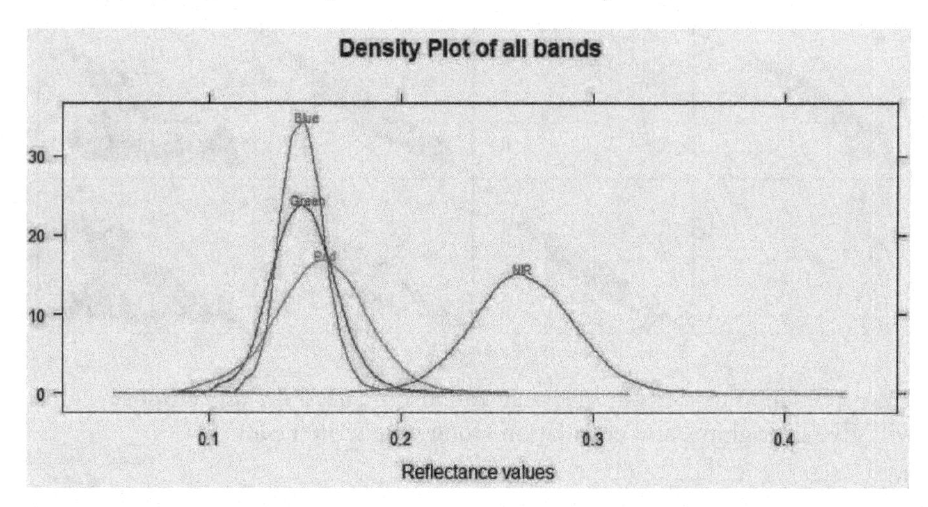

11.5. Violin plot

Violin plot is a combination of box plot and rotated density plot. Box and whisker plot display distribution of raster data in terms of quartiles. The lines extending parallel from the boxes are called as whiskers which represent the variability of data outside the upper and lower quartiles. Box plot indicates summary statistics like median, interquartile ranges. The density plot shows the distribution of data in terms of density of pixels with respect to their values. So, the violin plot display summary statistics with full distribution of variable. In R, violin plot of all the bands in a raster image can be created by using `bwplot` function of `rasterVis` package as shown below.

```
#Violin plot of all bands
>bwplot(img, main ='Violin Plot')
```

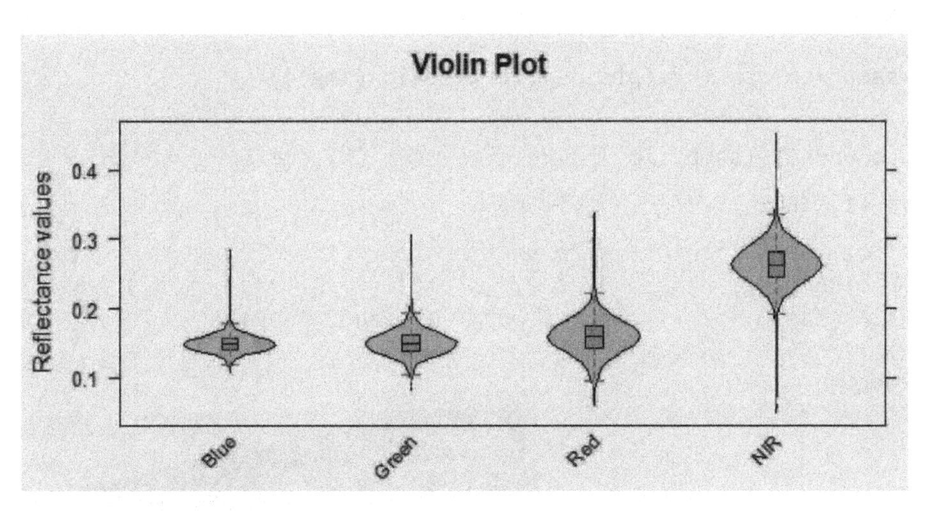

11.6. Extract pixel values of spatial points from raster

If you want to extract the pixel values for all the bands at a particular point, then 'extract' function of `raster` package is used. In the following R – code example, we are importing a vector file containing spatial points for which we want to extract pixel values using `rgdal` package.

```
#Import package for reading vector data
>library(rgdal)
#Read the shapefile containing spatial points from the directory
>pts =readOGR('spatial_pts_raster_extraction', #Data source name
'spatial_points_raster_extraction') #shapefile name
```

```
## OGR data source with driver: ESRI Shapefile
## Source: "E:\....................\spatial_pts_raster_extraction",
layer: "spatial_points_raster_extraction"
## with 20 features
## It has 1 fields
## Integer64 fields read as strings:  id
```

For extracting pixel values from spatial points, both vector and raster datasets should have same Coordinate Reference System (CRS). So, first check whether their CRS are same or not by using **compareCRS** function.

```
>compareCRS(pts, img) #Check whether CRS are same or not
```

```
## [1] FALSE
```

As we can see that the CRS of both raster and vector data are not same. Therefore, we will reproject the CRS of spatial points data to that of raster data by using **spTransform** function. After reprojection, the vector data can be plotted over the raster data to visualize the distribution of spatial points.

```
#Reproject the spatial point data to that of raster data
>samp =spTransform(pts, CRS(projection(img)))

>compareCRS(samp, img) #Check whether CRS are same or not
## [1] TRUE
#Plot spatial points data over raster data
>plotRGB(img, r=4, g=3, b=2, stretch ='lin',axes =TRUE)
>plot(samp, col ='yellow', pch = 19, add =TRUE)
```

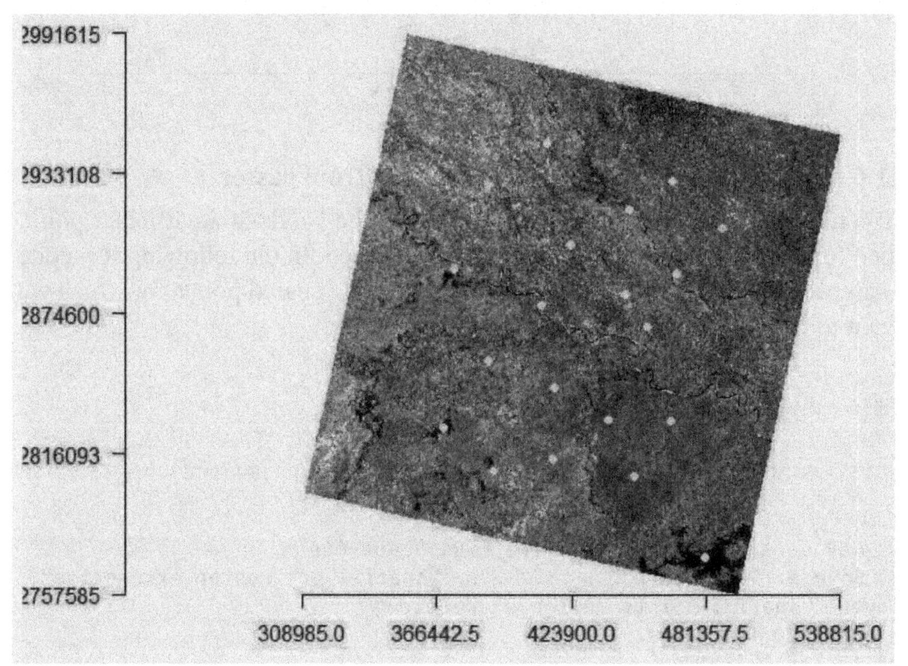

Now we can extract the data from all the raster bands at locations specified by spatial points through `extract` function of `raster` package as shown below.

```
#extract pixel values to points from all raster bands
>raster_pts =extract(img, samp, na.rm =TRUE)
>raster_pts#Display extracted data

##              Blue      Green        Red        NIR
##  [1,] 0.1611022 0.15698633 0.16093826 0.25905520
##  [2,] 0.1389445 0.13400555 0.13071147 0.26909330
##  [3,] 0.1417311 0.13979635 0.14413054 0.24258010
##  [4,] 0.1155958 0.09360681 0.07005453 0.05901339
##  [5,] 0.1459109 0.13922638 0.14609596 0.21658458
##  [6,] 0.1340006 0.12912670 0.13631403 0.22342668
##  [7,] 0.1457761 0.14654467 0.14989126 0.28057185
```

```
## [8,] 0.1465627 0.14818615 0.15824993 0.26153097
## [9,] 0.1183150 0.10450444 0.09700564 0.17096297
## [10,] 0.1519111 0.14809495 0.15958281 0.26391670
## [11,] 0.1551246 0.17023219 0.19416969 0.30760270
## [12,] 0.1640012 0.18042307 0.19233981 0.31124884
## [13,] 0.1392142 0.13701494 0.14830987 0.23814625
## [14,] 0.1608550 0.17059696 0.19143617 0.27616048
## [15,] 0.1485177 0.14743380 0.15592305 0.24768919
## [16,] 0.1489897 0.15224425 0.16186450 0.28408292
## [17,] 0.1415963 0.13943157 0.14632186 0.25835750
## [18,] 0.1582258 0.16391703 0.17998251 0.27573285
## [19,] 0.1500683 0.15260904 0.16509502 0.26967847
## [20,] 0.1408772 0.13988754 0.14790323 0.22797309
```

This extracted data can be attached with the spatial points to create a 'spatial points data frame' object as follow. This data can be saved to disk in '.csv' file.

```
#Attach extracted data with the spatial points vector data
>raster_pts_table =cbind(samp, raster_pts)
>raster_pts_table#Display the metadata of spatial object
created
```

```
## class       : SpatialPointsDataFrame
## features    : 20
## extent      : 368847, 488053.9, 2773023, 2945499  (xmin, xmax, ymin,
ymax)
## crs         : +proj=utm +zone=44 +datum=WGS84 +units=m +no_defs
+ellps=WGS84 +towgs84=0,0,0
## variables   : 5
## names       : id,    Blue,     Green,      Red,      NIR
## min values  :  1, 0.11559,  0.09360,  0.07005,  0.05901
## max values  :  9, 0.16400,  0.18042,  0.19416,  0.31124
```

```
#save table to disk in '.csv' format
>write.table(raster_pts_table, file="image_data.csv", append=FALSE,
sep=",", row.names =FALSE, col.names=TRUE)
```

This extracted data can be plotted to visualize the distribution of data at a particular pixel across all bands as shown in the following R – code example.

```
#Plot the data of extracted points across all bands
 #Plot 1st row
>plot(raster_pts[1,], type ='o', #for plotting both points and
lines
axes =FALSE,   #Disable default axis in the plot
ylim =c(0,0.40),  #Range of y-axis
ylab ='Reflectance', xlab ='Band Names', col ='blue',
main ='Spectral Plot of extracted points')
#Add customized axis to the existing plot
```

```
>axis(side =1, at =1:4, labels =colnames(raster_pts), tick =TRUE)
>axis(side =2, at =NULL, labels =TRUE, tick =TRUE)
 #Add other rows data to the existing plot
>lines(raster_pts[2,], type ='o', col ='red') #Plot 2nd row
>lines(raster_pts[4,], type ='o', col ='purple') #Plot 4th row
>lines(raster_pts[5,], type ='o', col ='green') #Plot 5th row
 #Add legends to the existing plot
>legend('topleft', legend =c(1,2,4,5), lty =1, bty ="n",
col =c('blue', 'red', 'purple', 'green'), y.intersp =0.5,
cex =0.7, title ='Spatial points ID')
```

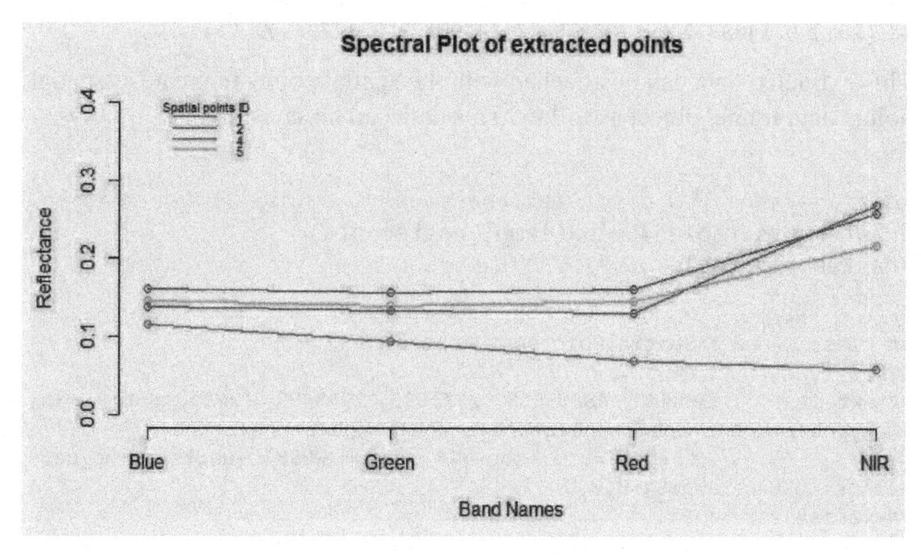

11.7. Zonal statistics with polygons

Zonal statistics with polygons compute the specified function *i.e.* mean, median, mode, etc. from the pixels of raster layer present within the zones or polygons of another vector dataset. In R, `extract` function from `raster` package is used for this purpose and the statistical functions like mean, median, mode, standard deviation etc. used to extract the pixel values within the zones are specified in `fun =` argument. Both raster and vector datasets should have same CRS before applying this function. In the following R – code example, 'zonal_stat_polygon.shp' shapefile containing five polygons are imported using `rgdal` package and mode of pixel values falling within the polygons is calculated from all bands of the raster image.

```
#Import shapefile containing polygons
> zones =readOGR('zonal_stat_polygon','zonal_stat_polygon')
```

```
## OGR data source with driver: ESRI Shapefile
## Source: "E:\..........\zonal_stat_polygon", layer:
"zonal_stat_polygon"
## with 5 features
## It has 1 fields
## Integer64 fields read as strings:  id

> zones  #Display metadata of imported shapefile containing
polygons

## class       : SpatialPolygonsDataFrame
## features    : 5
## extent      : 79.61956, 80.64933, 25.60427, 26.71429 (xmin, xmax,
ymin, ymax)
## crs         : +proj=longlat +datum=WGS84 +no_defs +ellps=WGS84
+towgs84=0,0,0
## variables   : 1
## names       : id
## min values  :  1
## max values  :  5

>compareCRS(zones, img) #check whether CRS same or not
## [1] FALSE
#Reproject the polygons to the CRS of raster
> zones =spTransform(zones, CRS(projection(img)))

>compareCRS(zones, img) #check whether CRS same or not

## [1] TRUE

#Plot polygons over raster layer
>plotRGB(img, r=4, g=3, b=2, stretch ='lin',axes =TRUE) #Plot
FCC
>plot(zones, col ='yellow',add =TRUE) #Plot polygons on FCC
```

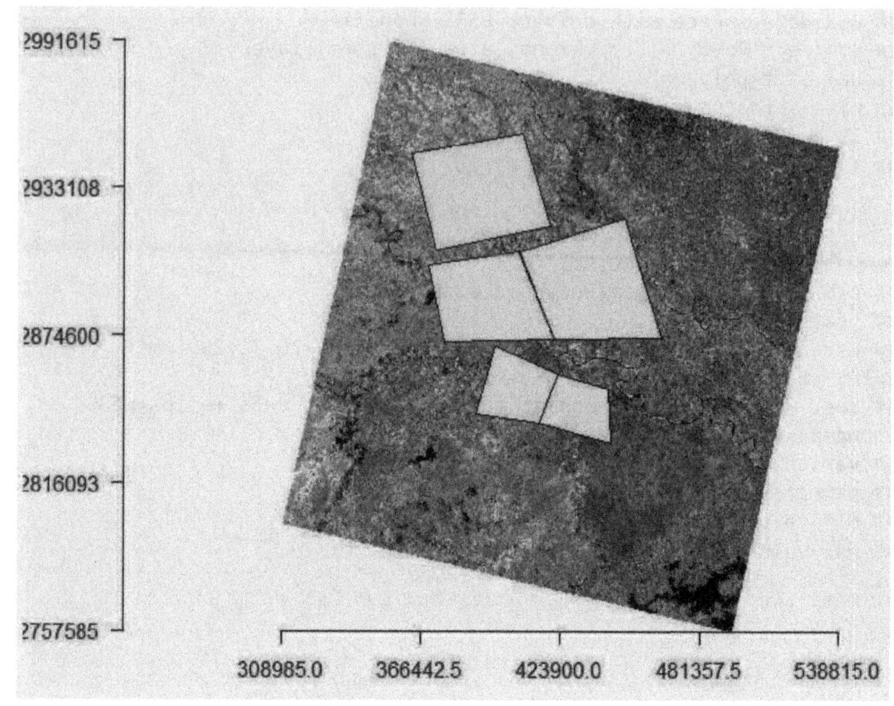


```
#Compute zonal statistics with polygons
>zonal_stat =extract(img, zones, fun ='modal', na.rm =TRUE)
>zonal_stat#Display mode of pixel values in zones for all bands

##           Blue     Green       Red       NIR
## [1,] 0.1491470 0.1499644 0.1654339 0.2607432
## [2,] 0.1408547 0.1371289 0.1488747 0.2529783
## [3,] 0.1486526 0.1537490 0.1598991 0.2743599
## [4,] 0.1484728 0.1536350 0.1641462 0.2679454
## [5,] 0.1582482 0.1630279 0.1764357 0.2882692
```

11.8. Band statistics

The statistics like mean, standard deviation, etc. of each band in a raster image can be derived by using `cellStats` function of the `raster` package as shown below.

```
#Compute Band statistics
>cellStats(img, stat ='sum') #Compute sum of all pixel values
in a band

##     Blue    Green      Red      NIR
##  6180323  6213589  6565129 10881884

>cellStats(img, stat ='mean') #Compute mean of all pixel values
```

```
##       Blue     Green       Red       NIR
## 0.1491504 0.1499495 0.1584340 0.2626073
```

```
>cellStats(img, stat ='sd') #Compute standard deviation of
all pixels
```

```
##        Blue       Green        Red         NIR
## 0.01253869 0.01793247 0.02537872 0.03112283
```

11.9. Cell statistics

The statistics can be computed at pixel level across all layers in a multiband raster image and a raster layer storing that pixel-wise data can be created by using `calc` function of the `raster` package. The statistical function like mean, median, mode, standard deviation or any other mathematical function defined by the user can be specified in `fun =` argument. The following R – code example depicts the calculation of mean of all the bands in entire raster brick at pixel level and the output raster layer of same dimensions is saved in the disk.

```
#Compute pixelwise mean of entire raster brick
>img_mean =calc(img, fun = mean,filename ='img_mean.tif',
na.rm =TRUE)
>img_mean#Display attributes of computed raster layer
```

```
## class      : RasterLayer
## dimensions : 7801, 7661, 59763461  (nrow, ncol, ncell)
## resolution : 30, 30  (x, y)
## extent     : 308985, 538815, 2757585, 2991615  (xmin, xmax, ymin,
ymax)
## crs         : +proj=utm +zone=44 +datum=WGS84 +units=m +no_defs
+ellps=WGS84 +towgs84=0,0,0
## source     : E:/................................/img_mean.tif
## names      : layer
## values     : 0.08185023, 0.4147161  (min, max)
```

```
#Plot computed raster band
>plot(img_mean, main ='Mean of all bands', axes =TRUE)
```

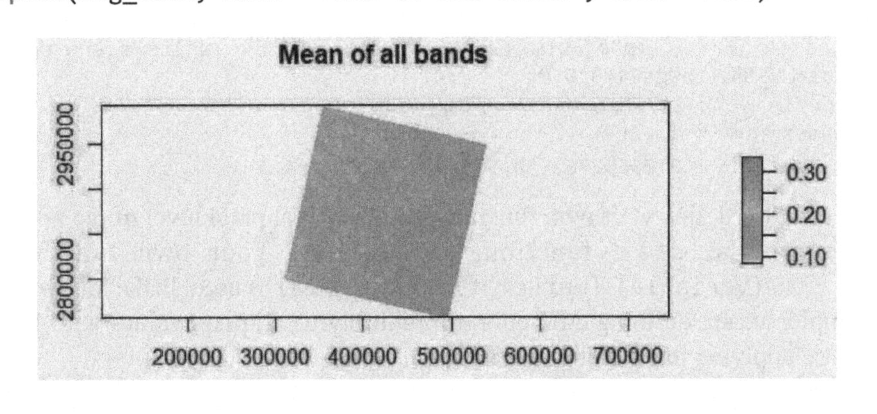

The distribution of pixel-wise mean values can be viewed by plotting the histogram of the computed raster layer.

```
#Plot histogram to display distribution of mean values
>histogram(img_mean, col ='wheat', main ='Histogram (Pixel-
wise mean of all bands)')
```

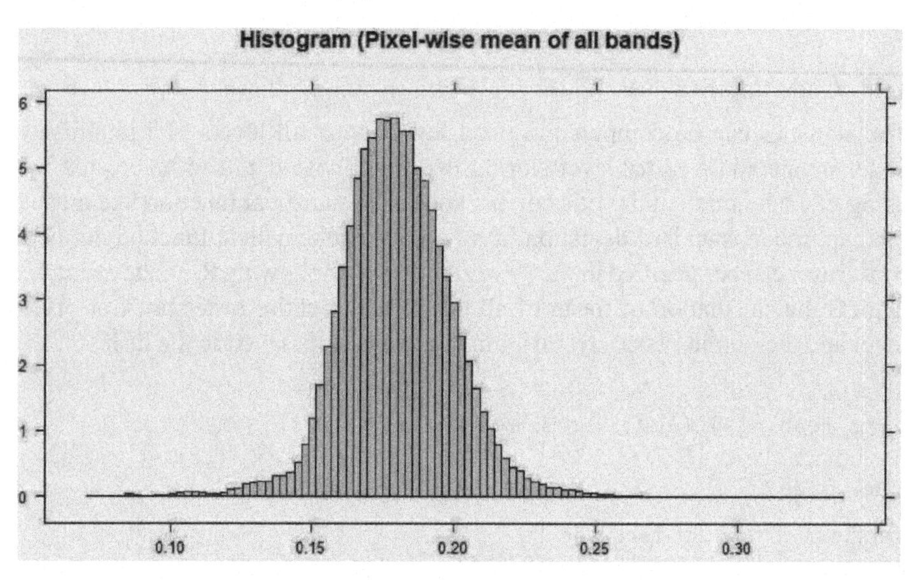

Similarly, mode can also be computed at pixel level as shown below.

```
#Compute pixelwise mode of entire raster brick
>img_mode=calc(img, fun = modal,filename ='img_mode.tif',
na.rm =TRUE)
>img_mode#Display attributes of computed raster layer

## class      : RasterLayer
## dimensions : 7801, 7661, 59763461  (nrow, ncol, ncell)
## resolution : 30, 30  (x, y)
## extent     : 308985, 538815, 2757585, 2991615  (xmin, xmax, ymin,
ymax)
## crs          : +proj=utm +zone=44 +datum=WGS84 +units=m +no_defs
+ellps=WGS84 +towgs84=0,0,0
## source     : E:/................................./img_mode.tif
## names      : layer
## values     : 0.05653762, 0.5090625  (min, max)
```

You can also define your own function and apply it at pixel level of the raster image in the `calc` function. For defining your own function, `function(variable) {mathematical formula}` is used. In the following example, we are defining a function for multiplying all pixel values with 100 and are applying this function to the raster image.

```
#Multiply all pixel values with 100
>func =function(x) {x*100} #Function defined by user
>img100 =calc(img, func) #Apply user-defined functions to all pixels
>img100 #Display attribute of output raster
```

```
## class      : RasterBrick
## dimensions : 7801, 7661, 59763461, 4  (nrow, ncol, ncell, nlayers)
## resolution : 30, 30  (x, y)
## extent     : 308985, 538815, 2757585, 2991615  (xmin, xmax, ymin, ymax)
## crs        : +proj=utm +zone=44 +datum=WGS84 +units=m +no_defs +ellps=WGS84 +towgs84=0,0,0
## source     : C:/......................./r_tmp_2019-07-11_111140_6640_18863.grd
## names      :   layer.1,   layer.2,   layer.3,   layer.4
## min values : 10.874178,  8.323355,  6.549114,  5.629004
## max values :  36.06562,  39.19465,  41.63525,  51.13131
```

So, the output raster shows that the reflectance values are multiplied by 100 as evident from the minimum and maximum values of pixels in the attributes of output raster. This function converts the reflectance values in the raster data in terms of percentage.

Chapter 12

Image Contrast Enhancement

The sensors present over remote sensing satellites are designed to collect energy or radiance from the varied surface features having wide range of radiance *i.e.* from features like water bodies having very low radiance to features like sand or snow having very high radiance. But, most of the time, the study area may not have all these features; therefore, radiance values may remain confined to a narrow range. This reduces the contrast of the image, thereby decreases its visual quality. In this context, image contrast enhancement techniques are used for effective visualization and display of remote sensing image on computer screens or hard copy so that maximum amount of information can be derived through its visual interpretation.

Contrast enhancement of a digital image is a technique to modify the original grey levels of input image to the entire available range of the grey levels. This increases the contrast between various features in the image which improves the visual quality of original image. Contrast enhancement can be done by both linear and non-linear methods. Some of the commonly used contrast enhancement techniques with their implementation in R software are discussed in this chapter.

We need following packages for performing this exercise in R. `raster` package is required for importing of raster image and for performing contrast enhancement. `RStoolbox` package is also required for contrast enhancement as well as for visualization of raster data with `ggplot2` package. `rasterVis` package is mainly used in this exercise for plotting histogram of the raster data.

```
#Import required packages
>library(raster)       #For importing and enhancement of raster
>library(rasterVis)    #For visualization of raster
>library(RStoolbox)  #For importing and visualization of raster
>library(ggplot2)  #For visualisation of raster with 'RStoolbox' package
```

Let's import the multiband raster data on which we want to perform contrast enhancement by using following R – code.

```
>setwd("E:/............/Chapter12/data") #Set working directory
#Import multiband raster image
>img =brick('img_enhancement.tif')
>names(img) #Display names of the bands
```

```
## [1] "img_enhancement.1" "img_enhancement.2" "img_enhancement.3"
## [4] "img_enhancement.4"
```

```
>names(img) =c('Blue', 'Green', 'Red', 'NIR') #Set the names
of bands
>img #Display attributes of imported image
```

```
## class      : RasterBrick
## dimensions : 3034, 3245, 9845330, 4  (nrow, ncol, ncell, nlayers)
## resolution : 30, 30  (x, y)
## extent     : 396255, 493605, 2830815, 2921835  (xmin, xmax, ymin,
ymax)
## crs         : +proj=utm +zone=44 +datum=WGS84 +units=m +no_defs
+ellps=WGS84 +towgs84=0,0,0
## source     : E:/......................................../Chapter12/data/img_enhancement.tif
## names      :  Blue, Green,   Red,   NIR
## min values : 10170,  9071,  8576,  8442
## max values : 19406, 20725, 21822, 25631
```

We can plot the histogram of the raster data to see the distribution of pixel values in the input image by using the following R – code.

```
#Plot histogram to show distribution of pixel values
>histogram(img, col ='wheat', main ='Histogram of input image',
xlab ='Pixel values')
```

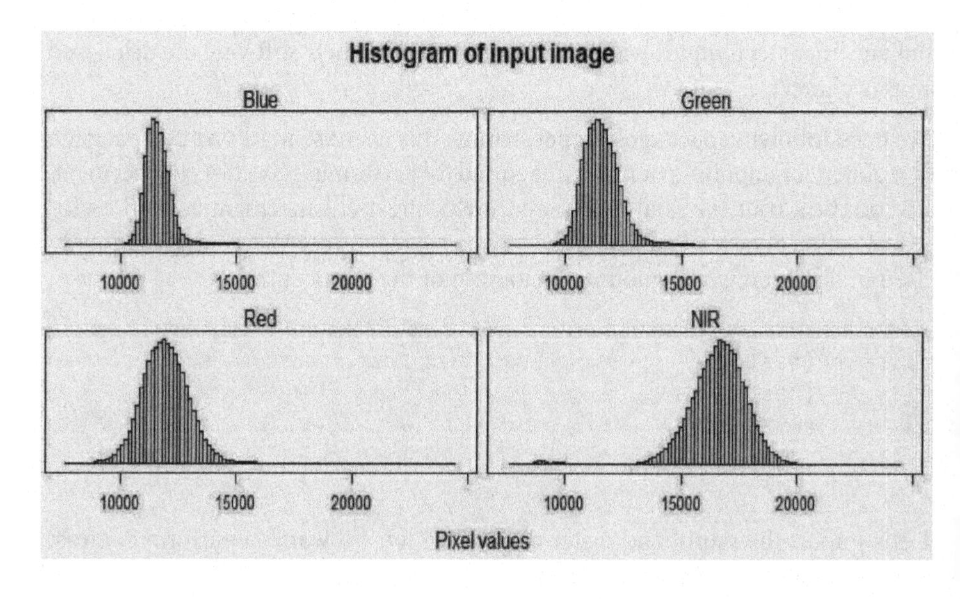

It is clear from the histograms of bands present in the raster data that the pixel values are confined only to a limited range. Let's plot the False Colour Composite (FCC) of the raster image to display its visual quality.

```
#Plot FCC without contrast enhancement
>ggRGB(img, r=4, g=3, b=2, stretch ='none') +#plot FCC
      ggtitle('No Stretch') +#plot title
      labs(x ='Longitude (m)', y ='Latitude (m)') +#axis labels
      theme(plot.title =element_text(hjust =0.5, #Align title
in centre
      size =30), #Adjust size of title
      axis.title =element_text(size =20)) #size of axis labels
```

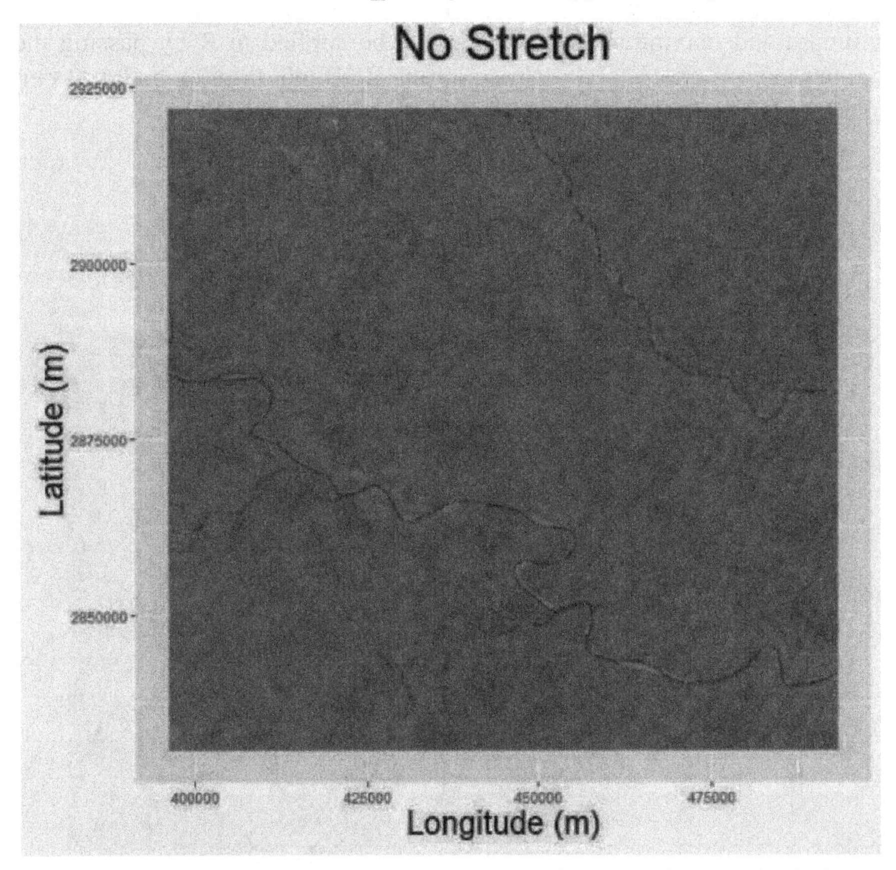

So, it is clear from the FCC plot that the contrast between features in the image is low. Therefore, we need to apply contrast enhancement techniques to improve the visual quality of the FCC image. R software provides us the functions to apply contrast enhancement while plotting the RGB plot of the multispectral image. So, you can use these pre-defined functions for contrast enhancement as discussed further in this chapter.

12.1. Linear Contrast Stretching

In linear contrast stretching, the digital numbers in the original image are linearly expanded to the entire range of the grey levels. This can be performed in R by two available methods *i.e.* Minimum-Maximum linear contrast stretch and percentage linear contrast stretch.

12.1.1. Minimum – Maximum linear contrast stretching

In minimum-maximum linear contrast stretching, the minimum and maximum pixel value of the original image is assigned to minimum and maximum values of the total available radiometric range of the output device, respectively and the in-between pixel values are linearly stretched between these assigned minimum and maximum values. This can be applied in R by passing the argument `stretch ="lin"` while plotting RGB plot through both `raster` and `RStoolbox` packages as shown below.

```
# Minimum-Maximum Linear Contrast Stretching by 'raster' package
>plotRGB(img, r=4, g=3, b=2, stretch ="lin",
main = "Minimum-Maximum Linear Contrast Stretch", axes =TRUE)
```

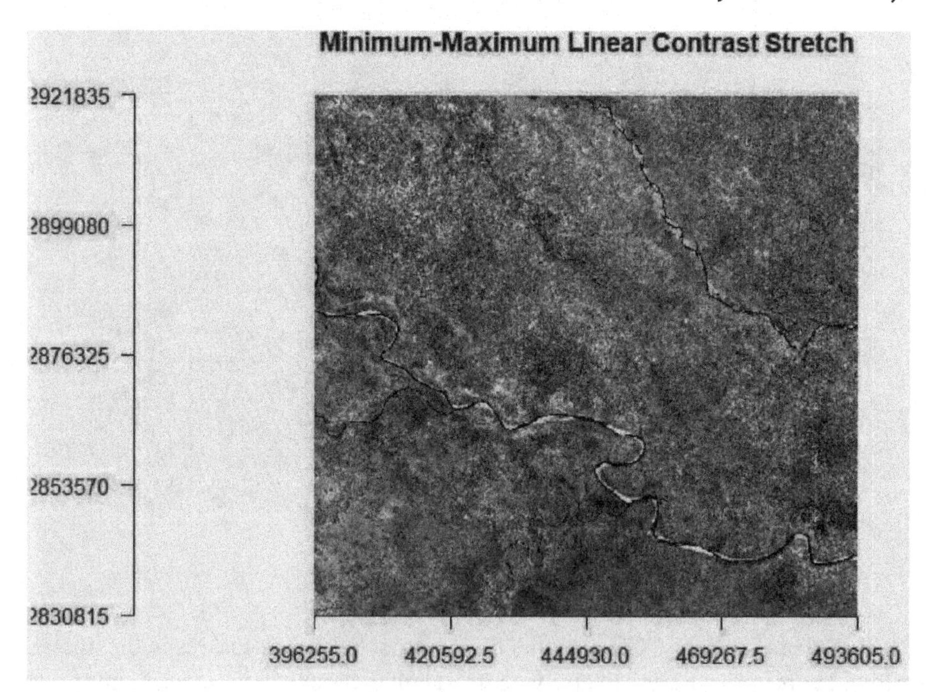


```
# Minimum-Maximum Linear Contrast Stretching by 'RStoolbox'
package
>ggRGB(img, r=4, g=3, b=2, stretch = 'lin') +
      ggtitle('Minimum-Maximum Linear Contrast Stretch') +#plot
title
```

```
    labs(x ='Longitude (m)', y ='Latitude (m)') +#axis labels
    theme(plot.title =element_text(hjust =0.5, #Align title
in centre
    size =30), #Adjust size of title
    axis.title =element_text(size =20)) #size of axis labels
```

Minimum-Maximum Linear Contrast Stretch

Latitude (m)

Longitude (m)

It is clear from the RGB plot of the linearly stretched image that the features are clearer, and colours are more visually appealing as compared to the unstretched image. This aptly depicts the importance of image contrast enhancement.

12.1.2. Percentage linear contrast stretching

Percentage linear contrast stretching is similar to the minimum-maximum linear contrast stretching except the minimum and maximum values used it this method represents certain percentage of pixels from the mean of the histogram. Suppose you want to apply 2% linear contrast stretching, then the lower 2% and upper 2% values in the histogram will be omitted while applying the linear stretch.

In the following R – code example, 2% linear contrast stretch is applied while plotting the FCC using `ggRGB` function of `RStoolbox` package. The lower and upper bounds are assigned in the `quantiles` argument as depicted below.

```
# Two percent Linear Contrast Stretching
>ggRGB(img, r=4, g=3, b=2, stretch ='lin', #plot FCC
quantiles =c(0.02, 0.98)) +#2% Linear Stretch
      ggtitle('2% Linear Contrast Stretch') +#plot title
      labs(x ='Longitude (m)', y ='Latitude (m)') +#axis labels
      theme(plot.title =element_text(hjust =0.5, #Align title
in centre
      size =30), #Adjust size of title
      axis.title =element_text(size =20)) #size of axis labels
```

2% Linear Contrast Stretch

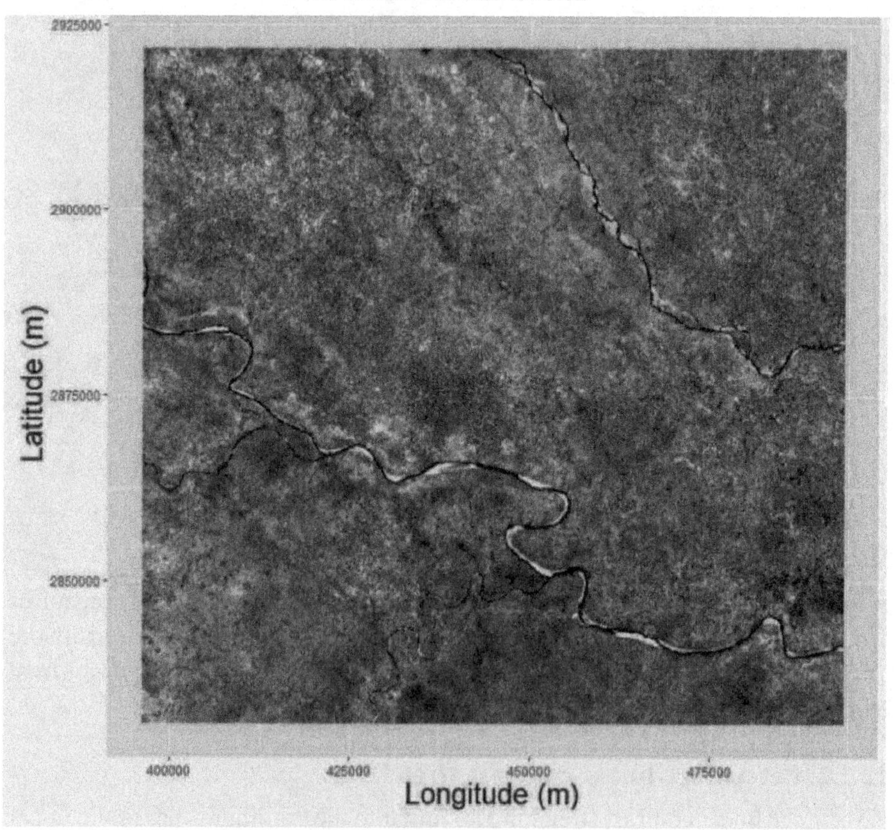

12.2. Non – linear Contrast Stretching

In non-linear contrast stretching, pixel values are not linearly stretched, but are stretched by applying some non-linear algorithm. Some of the methods for non-linear contrast stretching with their respective R – code examples are discussed below.

12.2.1. Histogram equalization

In this method, pixel values are stretched depending upon their frequency of occurrence. The pixel values which are occurring more frequently are assigned wider display range as compared to the less frequently occurring pixel values. In this way, this method provides better details through enhancement of higher frequency regions of the histogram. This can be performed in R by passing the argument `stretch ="hist"` while plotting RGB plot through both `raster` and `RStoolbox` packages as shown below.

```
# Histogram Equalisation Contrast Stretching by 'raster' package
>plotRGB(img, r=4, g=3, b=2, stretch ="hist",    #Plot FCC
main ="Histogram Stretch", axes =TRUE)
```

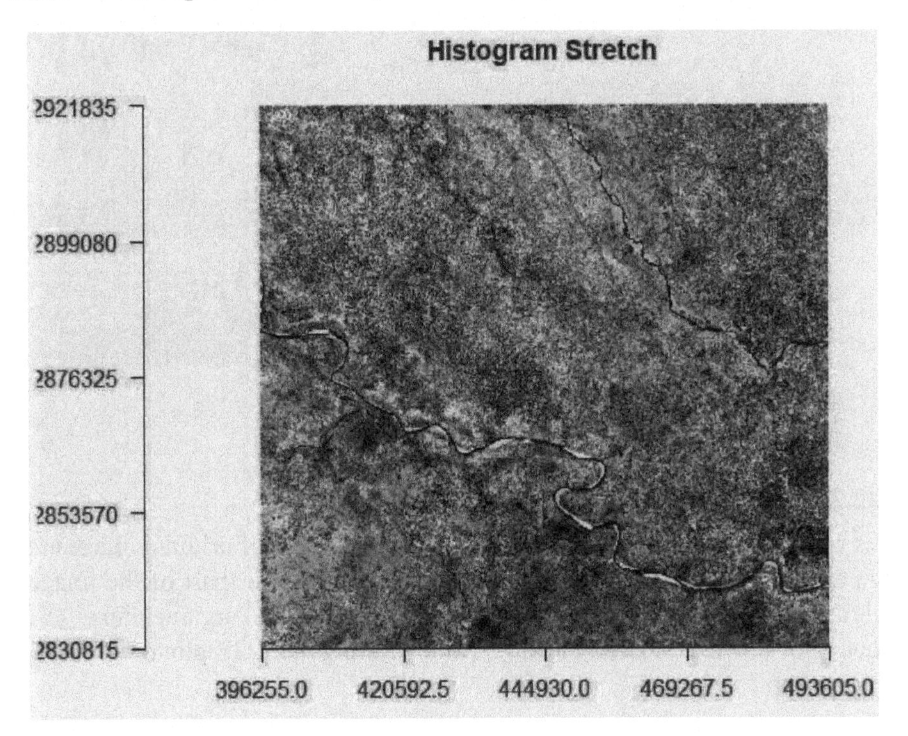


```
# Histogram Equalisation Contrast Stretching by 'RStoolbox'
package
>ggRGB(img, r=4, g=3, b=2, stretch ='hist') +#Plot FCC
      ggtitle('Histogram Stretch') +#plot title
      labs(x ='Longitude (m)', y ='Latitude (m)') +#axis labels
      theme(plot.title =element_text(hjust =0.5, #Align title
in centre
size =30), #Adjust size of title
      axis.title =element_text(size =20)) #size of axis labels
```

Histogram Stretch

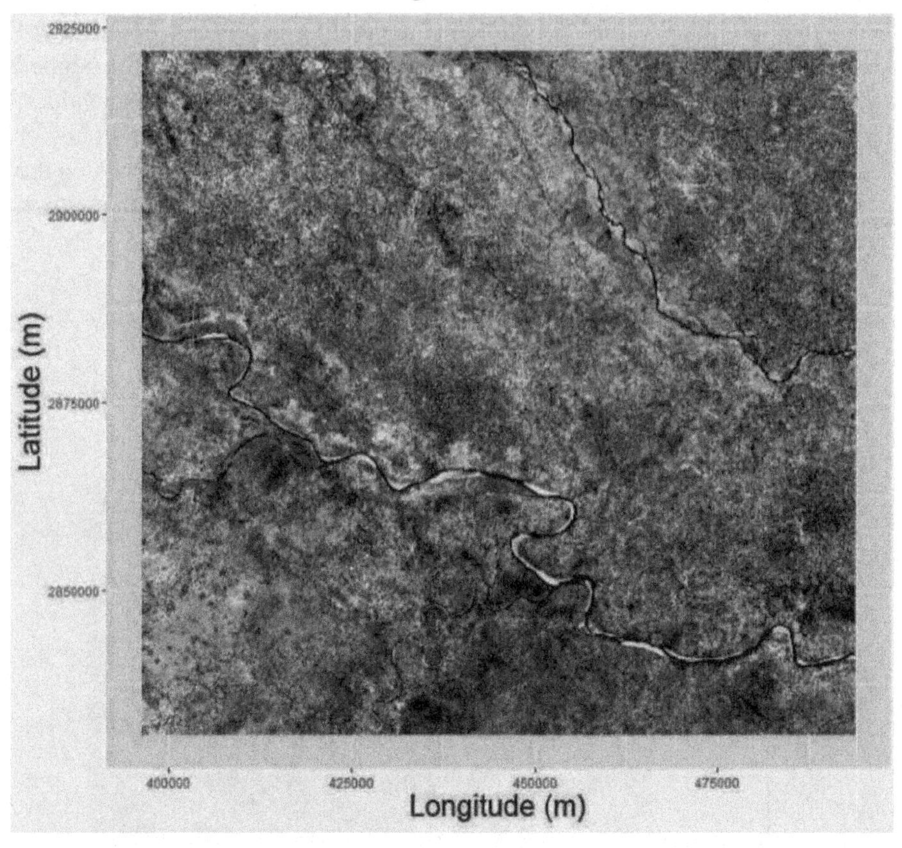

12.2.2. Square root contrast stretching

This method computes the square root of the histogram of original image and then applies linear stretch. It mainly enhances the darker part of the image. This can be performed in R by passing `stretch ='sqrt'` argument in **ggRGB** function of the `RStoolbox` package while plotting the RGB plot of the raster image.

```
# Square root contrast stretching
>ggRGB(img, r=4, g=3, b=2, stretch ='sqrt') +#Plot FCC
      ggtitle('Square-root Stretch') +#plot title
      labs(x ='Longitude (m)', y ='Latitude (m)') +#axis labels
      theme(plot.title =element_text(hjust =0.5, #Align title
in centre
      size =30), #Adjust size of title
      axis.title =element_text(size =20)) #size of axis labels
```

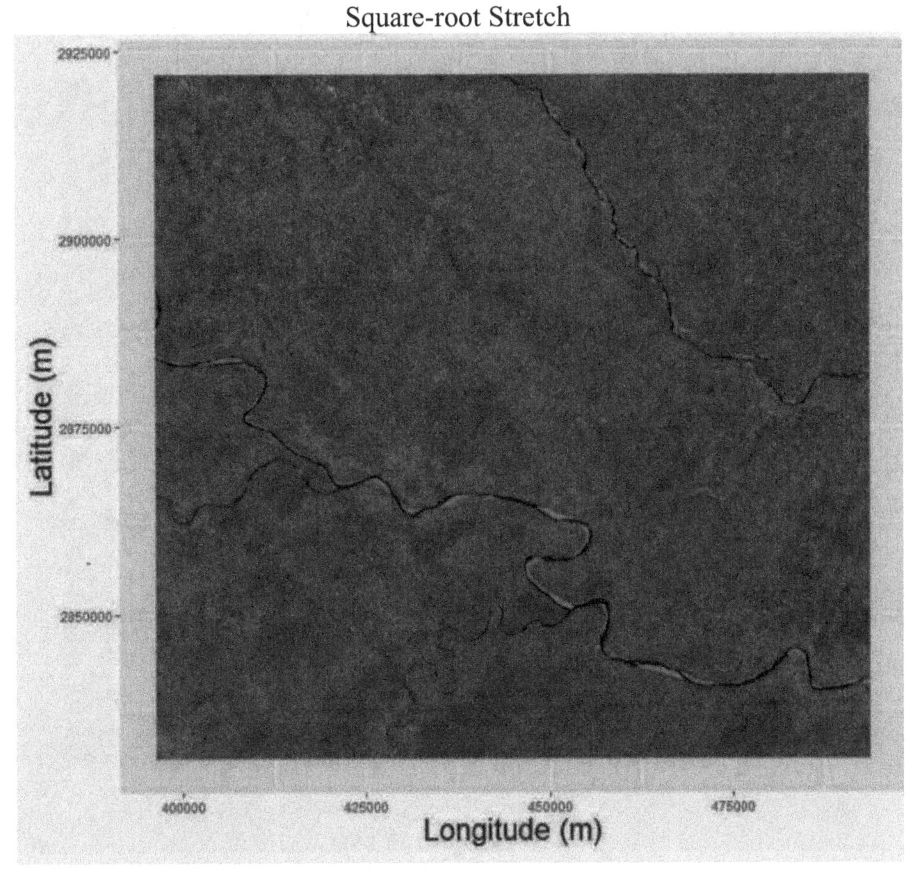

12.2.3. Logarithmic contrast stretching

Logarithmic contrast stretching rescales the input pixel values logarithmically. This method assigns the smaller pixel values to a wider range whereas higher pixel values have narrower output range. In this way, this method enhances greater details of darker features in the image. In R, this can be implemented by assigning `stretch ='log'` argument in **ggRGB** function of the `RStoolbox` package while plotting the RGB plot of the raster image as depicted in the following R – code example.

```
# Logarithmic contrast stretching
>ggRGB(img, r=4, g=3, b=2, stretch ='log')+#Plot FCC
      ggtitle('Logarithmic Contrast Stretch') +#plot title
      labs(x ='Longitude (m)', y ='Latitude (m)') +#axis labels
      theme(plot.title =element_text(hjust =0.5, #Align title
in centre
      size =30), #Adjust size of title
      axis.title =element_text(size =20)) #size of axis labels
```

Logarithmic Contrast Stretch

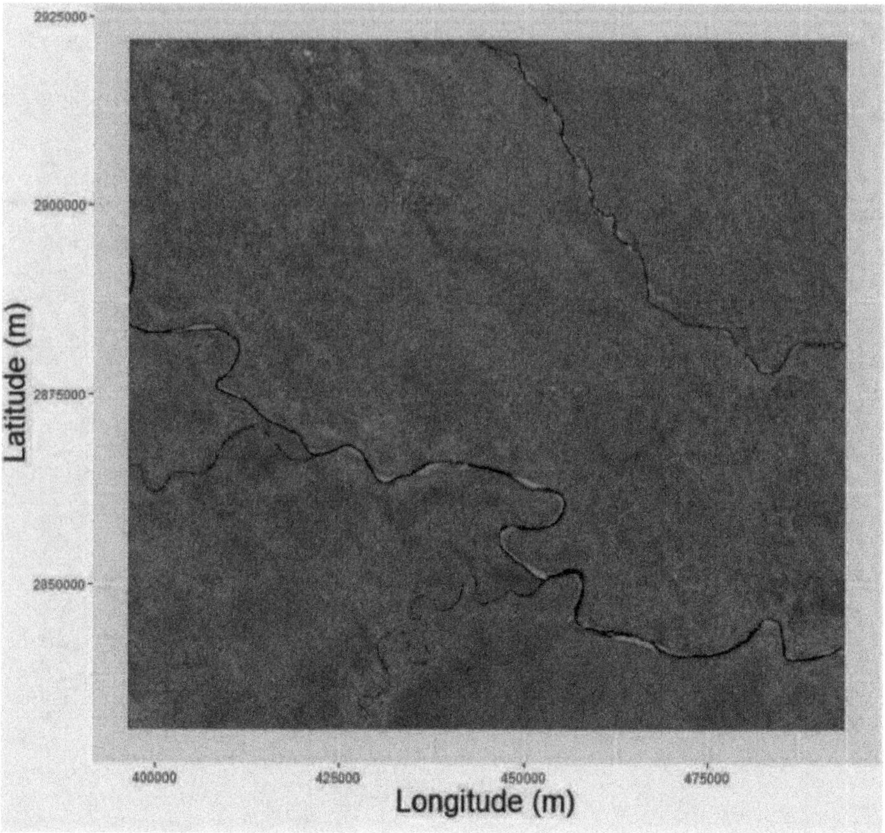

12.2.4. Gaussian stretching

In Gaussian stretching, histogram of pixel values of the raster image is modified to resemble a Gaussian or normal distribution. The raster image can be normalized by using `normImage` function of `RStoolbox` package as shown below.

```
# Gaussian contrast stretching
>img_norm =normImage(img, norm =TRUE) #Normalize Image
>img_norm #Display attributes of normalized stretched image

## class      : RasterBrick
## dimensions : 3034, 3245, 9845330, 4  (nrow, ncol, ncell, nlayers)
## resolution : 30, 30  (x, y)
## extent     : 396255, 493605, 2830815, 2921835  (xmin, xmax, ymin, ymax)
## crs         : +proj=utm +zone=44 +datum=WGS84 +units=m +no_defs +ellps=WGS84 +towgs84=0,0,0
## source      : C:/Users/Alka Rani/AppData/Local/Temp/RtmpKUK5lc/raster/r_tmp_2019-09-08_200312_5632_98635.grd
```

```
## names      :   layer.1,   layer.2,   layer.3,   layer.4
## min values : -2.745454, -3.323642, -3.274018, -5.978208
## max values : 15.145227, 12.304211,  9.554964,  6.703021

#Plot FCC of Gaussian stretched image
>ggRGB(img_norm, r=4, g=3, b=2)+#Plot FCC
      ggtitle('Gaussian Stretch') +#plot title
      labs(x ='Longitude (m)', y ='Latitude (m)') +#axis labels
      theme(plot.title =element_text(hjust =0.5, #Align title
in centre
      size =30), #Adjust size of title
      axis.title =element_text(size =20)) #size of axis labels
```

Gaussian Stretch

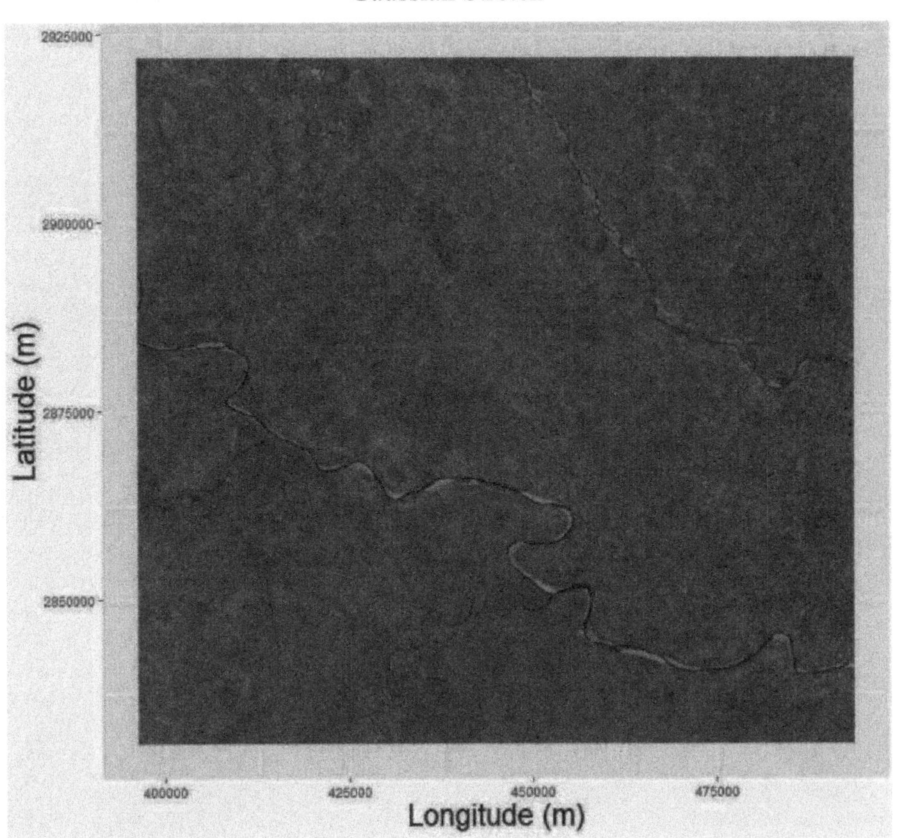

Chapter 13

Spatial Filters

Remote sensing images have an important characteristic known as spatial frequency. Spatial frequency is defined as the number of changes in pixel value per unit distance. Few changes in pixel or brightness value over an area in an image are known as low spatial frequency area like agricultural fields, forests, water bodies, etc. If the changes are large, then it is known as high frequency area like roads, built up areas, etc. Spatial filtering is an approach for extracting quantitative spatial information by taking into account the neighbouring pixel values along with the independent pixel value. Therefore, filtering is a technique for modifying or enhancing an image. Filtering is done to improve the interpretability of raster image or to extract certain information from them. It is a neighbourhood operation because the value of any pixel in the output image is determined by applying certain algorithm on the neighbourhood pixels of the corresponding input pixel. The nature of algorithm to be applied depends upon the information which user wants to derive from the raster image. Filtering operations are mainly done for smoothening, sharpening, noise removal and edge enhancement. A filter is a kernel of small array which is applied to input pixel and its neighbouring pixels in the raster. There are many types of filters which can be applied to a raster image in R. These types with their respective R – code examples are explained further.

All the functions required for performing filtering operations are available in the `raster` package. First of all, you need to import that package as well as the multiband input raster image on which you wish to do filtering as shown below.

```
#Import required package

>library(raster)
>setwd("E:/................/Chapter13/data") #set working directory
#Import multiband input raster image on which you wish to do
filtering
>img =brick('sentinel_filter.tif')
#Set the names of bands in the input raster
>names(img) =c('NIR', 'Red', 'Green')
#Display attributes of imported input raster image
```

```
>img
## class      : RasterBrick
## dimensions : 265, 273, 72345, 3  (nrow, ncol, ncell, nlayers)
## resolution : 10, 10  (x, y)
## extent     : 301340, 304070, 2919700, 2922350  (xmin, xmax, ymin, ymax)
## crs        : +proj=utm +zone=44 +datum=WGS84 +units=m +no_defs +ellps=WGS84 +towgs84=0,0,0
## source     : E:/......................................./sentinel_filter.tif
## names      :  NIR,  Red, Green
## min values :  992, 1127,  1317
## max values : 3682, 2575,  2101

#Plot standard False Colour Composite of input image
>plotRGB(img, stretch ='lin', main ='Original Image', axes =TRUE)
```

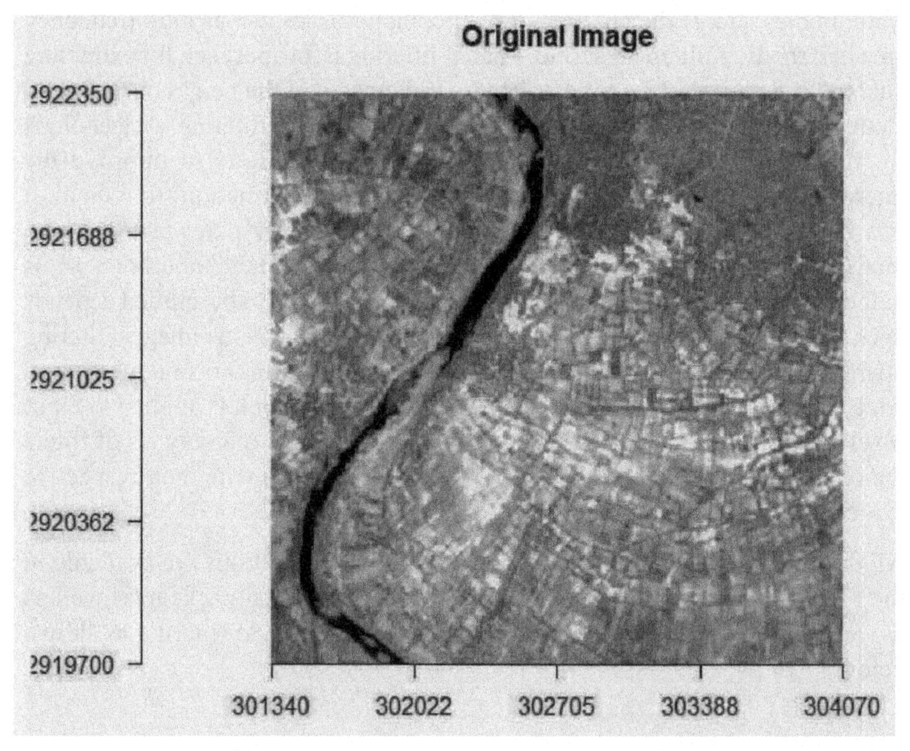

For applying any filter on raster image in R, a kernel of matrix with values depending upon the type of filter is created. This kernel is then applied to each pixel of the raster image by using `focal` function of the `raster` package. This `focal` function is applicable on single band only. Therefore, you need to apply that function on each band separately in a multiband raster image. In this exercise, we are going to use raster image of three bands as depictedabove in the attributes of the imported raster image. Each filter function will be applied to all the bands separately and a raster brick will be created from these filtered raster layers. This raster brick will be used for plotting standard False Colour Composite. The details of matrix for a particular filter along with its purpose and effect on raster image are discussed further.

13.1. Low pass filters

Low pass filters are used to retain low frequency information and reduce high frequency information in the raster image, thereby, producing smoothing effect. Therefore, low pass filters are also known as 'blurring' or 'smoothing' filter. This smoothing effect is created by computing mean, median, mode, etc. of the pixels falling into the given kernel size which in turn reduces the variation in brightness values from one pixel to the next. With increase in the size of kernel, more smoothing effect of raster image is produced. Most commonly used low pass filters with their respective R – code examples are discussed below.

13.1.1. Mean filter

Mean filter computes mean of all the pixels in the given kernel size in order to assign a value to the output pixel. It is used to reduce noise in the raster image. This filter smooths the image, therefore, blur the sharp edges. If a single pixel has unrepresentative value, then it can significantly affect the mean value which can further produce undesired results. This filter is simple and easy to implement. In the following R – code example, we are applying mean filter with three different kernel sizes *i.e.* 3x3, 5x5, and 25x25 in order to see the effect of kernel size on the output raster image. For applying this filter in R, matrix of desired kernel size is specified in its `w=` argument and mean function as `fun = mean` argument of the `focal` function.

```
# Applying 3x3 mean filter on each band of the input raster
> mean3_filter1 =focal(img[[1]], w=matrix(1,3,3), fun = mean)
#1st band
> mean3_filter2 =focal(img[[2]], w=matrix(1,3,3), fun = mean)
#2nd band
> mean3_filter3 =focal(img[[3]], w=matrix(1,3,3), fun = mean)
#3rd band
# Create brick of the filtered raster layers
> mean3_filter =brick(mean3_filter1, mean3_filter2,
mean3_filter3)

> mean3_filter #Display attributes of 3x3 mean filtered output
raster
```

```
## class      : RasterBrick
## dimensions : 265, 273, 72345, 3  (nrow, ncol, ncell, nlayers)
## resolution : 10, 10  (x, y)
## extent     : 301340, 304070, 2919700, 2922350  (xmin, xmax, ymin,
ymax)
## crs         : +proj=utm +zone=44 +datum=WGS84 +units=m +no_defs
+ellps=WGS84 +towgs84=0,0,0
## source     : memory
## names      :  layer.1,  layer.2,  layer.3
## min values : 1014.000, 1169.556, 1356.111
## max values : 3455.000, 2499.444, 2052.778
```

Similarly, apply mean filter of kernel size 5x5 and 25x25 on the input raster and create the brick of output raster layers.

```
# 5x5 mean filter

> mean5_filter1 =focal(img[[1]], w=matrix(1,5,5), fun = mean)
> mean5_filter2 =focal(img[[2]], w=matrix(1,5,5), fun = mean)
> mean5_filter3 =focal(img[[3]], w=matrix(1,5,5), fun = mean)
> mean5_filter =brick(mean5_filter1, mean5_filter2,
mean5_filter3)
> mean5_filter #Display attributes of 5x5 mean filtered output
raster

## class      : RasterBrick
## dimensions : 265, 273, 72345, 3  (nrow, ncol, ncell, nlayers)
## resolution : 10, 10  (x, y)
## extent     : 301340, 304070, 2919700, 2922350  (xmin, xmax, ymin,
ymax)
## crs         : +proj=utm +zone=44 +datum=WGS84 +units=m +no_defs
+ellps=WGS84 +towgs84=0,0,0
## source     : memory
## names      : layer.1, layer.2, layer.3
## min values : 1035.00, 1210.84, 1382.32
## max values : 3137.28, 2368.16, 1977.56

# 25x25 mean filter

> mean25_filter1 =focal(img[[1]], w=matrix(1,25,25), fun = mean)
> mean25_filter2 =focal(img[[2]], w=matrix(1,25,25), fun = mean)
> mean25_filter3 =focal(img[[3]], w=matrix(1,25,25), fun = mean)
> mean25_filter =brick(mean25_filter1, mean25_filter2,
mean25_filter3)
> mean25_filter #Display attributes of 25x25 mean filtered output
raster

## class      : RasterBrick
## dimensions : 265, 273, 72345, 3  (nrow, ncol, ncell, nlayers)
## resolution : 10, 10  (x, y)
```

```
## extent      : 301340, 304070, 2919700, 2922350  (xmin, xmax, ymin,
ymax)
## crs         : +proj=utm +zone=44 +datum=WGS84 +units=m +no_defs
+ellps=WGS84 +towgs84=0,0,0
## source      : memory
## names       :  layer.1,  layer.2,  layer.3
## min values : 1932.237, 1461.298, 1490.048
## max values : 2750.830, 2181.869, 1834.526
```

Now we will create the standard False Colour Composite (FCC) plot of filtered images to see the effect of kernel size.

```
# Plot FCC of 3x3 mean filtered raster image
>plotRGB(mean3_filter, stretch ='lin', axes =TRUE,
main ='3x3 Mean Filter')
# Plot FCC of 5x5 mean filtered raster image
>plotRGB(mean5_filter, stretch ='lin', axes =TRUE,
main ='5x5 Mean Filter')
# Plot FCC of 25x25 mean filtered raster image
>plotRGB(mean25_filter, stretch ='lin', axes =TRUE,
main ='25x25 Mean Filter')
```

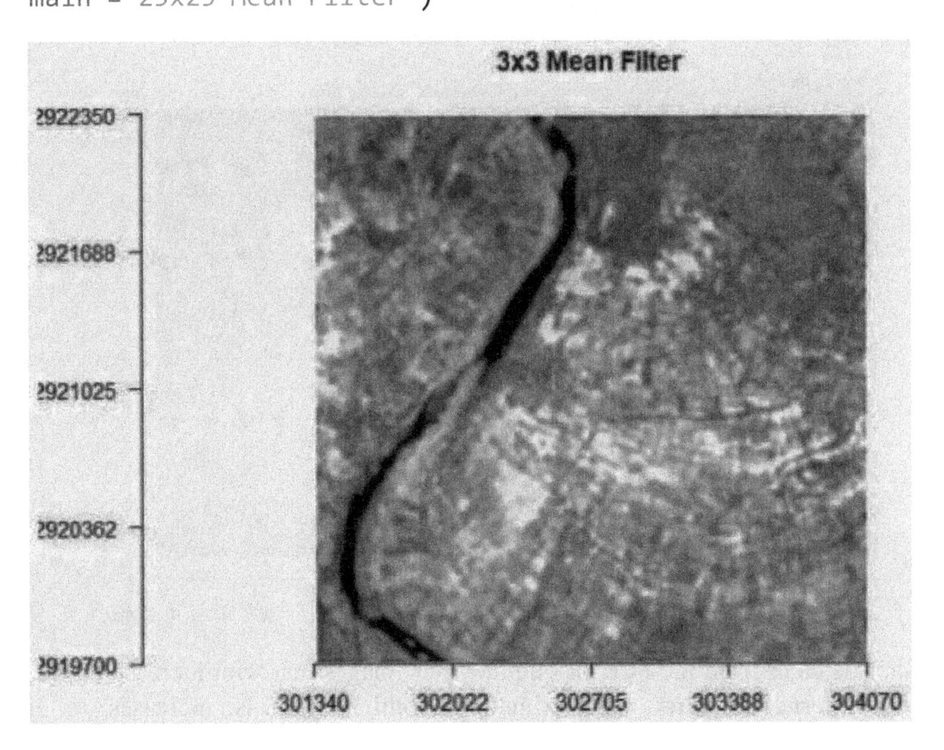

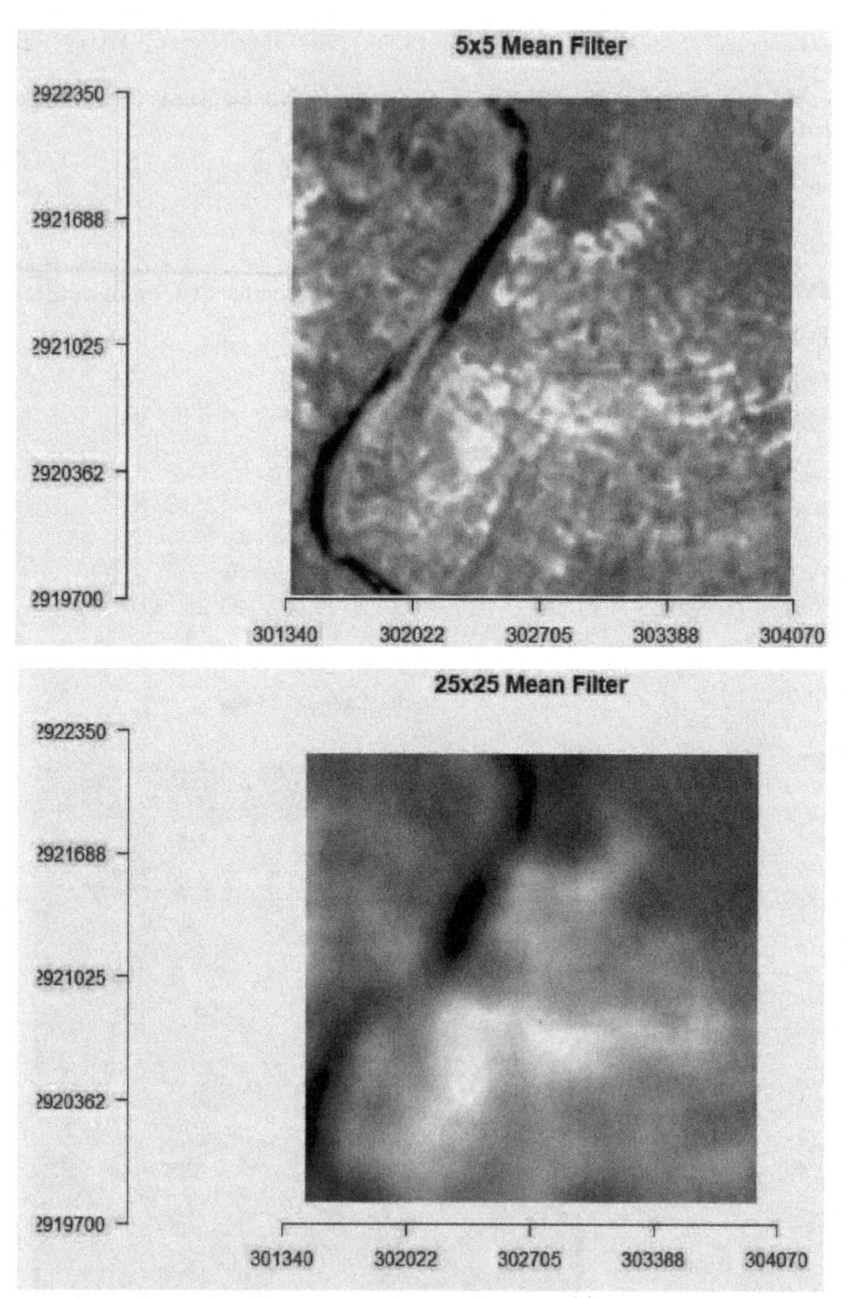

So, it is clear from the FCC of output raster images that with increase in the size of kernel or matrix, the blurring or smoothing effect also increases.

13.1.2. Median filter

Median filter replaces the pixel value in the raster image with the median of all the pixel values falling in the given kernel. Median is computed by arranging all the pixel values within kernel surrounding the input pixel in numerical order and assigning its middle value to the output pixel. As the value it takes is the original pixel value, therefore, it doesn't create unrealistic pixel values at edges unlike mean filter. In this way, it is better than mean filter in preserving edges and avoiding the effect of unrepresentative pixels. It is generally used to reduce noise, but it takes longer time in computation. In R, `fun = median` argument is used in the `focal` function to apply median filter as shown below.

```
# Applying 5x5 median filter on each band of the input raster
> median_filter1 =focal(img[[1]], w=matrix(1,5,5), fun = median)
#1st band
> median_filter2 =focal(img[[2]], w=matrix(1,5,5), fun = median)
#2nd band
> median_filter3 =focal(img[[3]], w=matrix(1,5,5), fun = median)
#3rd band

# Create brick of the filtered raster layers
>median_filter =brick(median_filter1, median_filter2,
median_filter3)
>median_filter#Display attributes of 5x5 median filtered output
raster

## class      : RasterBrick
## dimensions : 265, 273, 72345, 3  (nrow, ncol, ncell, nlayers)
## resolution : 10, 10  (x, y)
## extent     : 301340, 304070, 2919700, 2922350  (xmin, xmax, ymin,
ymax)
## crs        : +proj=utm +zone=44 +datum=WGS84 +units=m +no_defs
+ellps=WGS84 +towgs84=0,0,0
## source     : memory
## names      : layer.1, layer.2, layer.3
## min values :    1022,    1211,    1381
## max values :    3118,    2420,    1994

>plotRGB(median_filter, stretch ='lin', axes =TRUE,
main ='5x5 Median Filter') #Plot FCC of the median filtered image
```

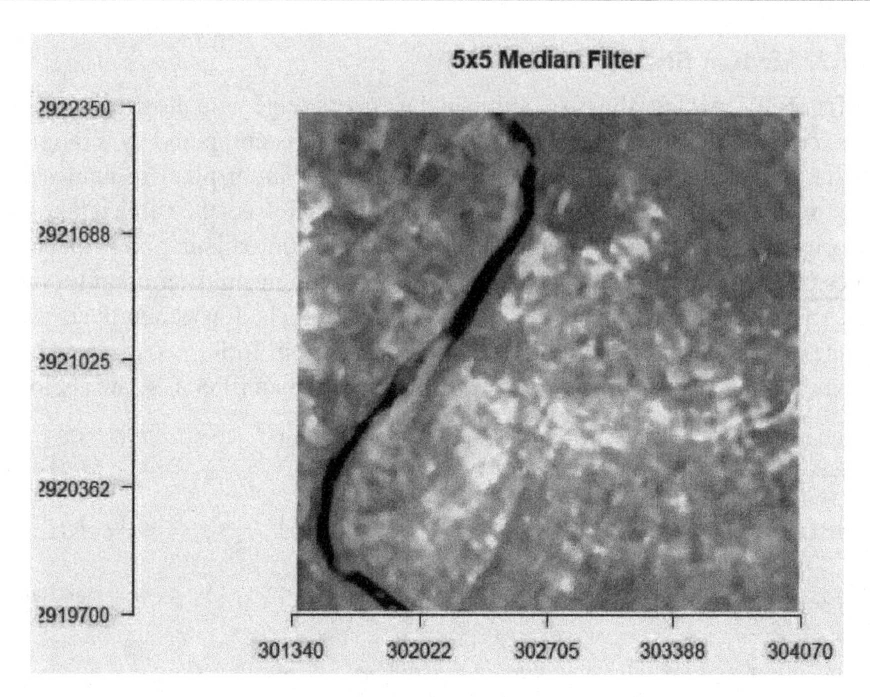

Thus, you can see that the edges are preserved in the output raster image along with smoothing.

13.1.3. Mode filter

Mode filter replaces the pixel value in the raster image with the most frequently occurring pixel value in the given kernel. This filter has the benefits similar to median filter.It reduces noise along with preserving edges. In R, `fun = modal` argument is used in the `focal` function to apply mode filter as shown below.

```
# Applying 5x5 mode filter on each band of the input raster
> mode_filter1 =focal(img[[1]], w=matrix(1,5,5), fun = modal) #1st band
> mode_filter2 =focal(img[[2]], w=matrix(1,5,5), fun = modal) #2nd band
> mode_filter3 =focal(img[[3]], w=matrix(1,5,5), fun = modal) #3rd band

# Create brick of the filtered raster layers
>mode_filter =brick(mode_filter1, mode_filter2, mode_filter3)
>mode_filter#Display attributes of 5x5 mode filtered output raster

## class      : RasterBrick
## dimensions : 265, 273, 72345, 3  (nrow, ncol, ncell, nlayers)
## resolution : 10, 10  (x, y)
## extent     : 301340, 304070, 2919700, 2922350  (xmin, xmax, ymin, ymax)
## crs         : +proj=utm +zone=44 +datum=WGS84 +units=m +no_defs +ellps=WGS84 +towgs84=0,0,0
```

```
## source      : memory
## names       : layer.1, layer.2, layer.3
## min values  :     998,    1148,    1317
## max values  :    3628,    2534,    2101
```

```
>plotRGB(mode_filter, stretch ='lin', axes =TRUE,
main ='5x5 Mode Filter') #Plot FCC of the mode filtered image
```

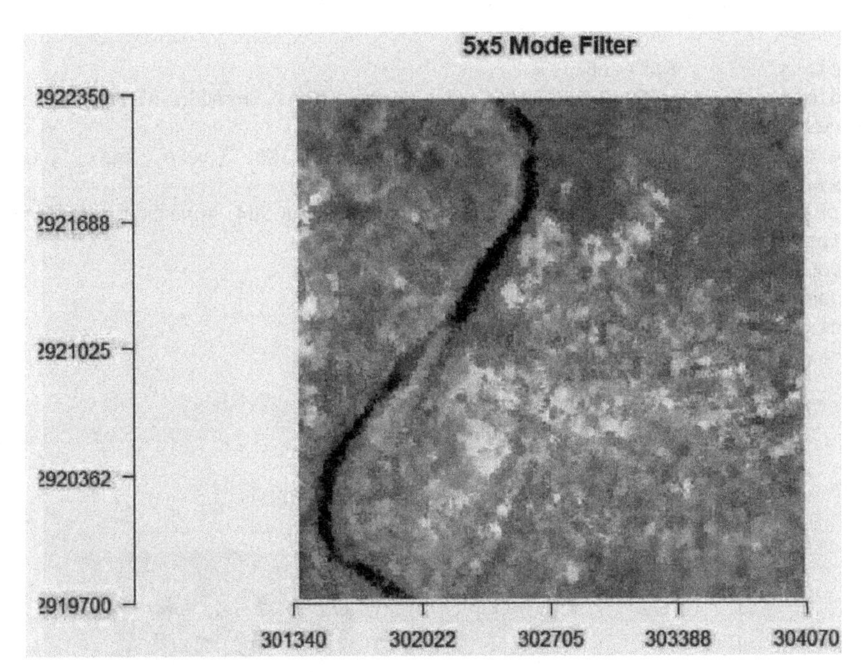

13.1.4. Gaussian filter

Gaussian filter is also used to remove noise from the image by smoothing it, but the kernel used in it has Gaussian or bell-shaped distribution. In R, you have to design the Gaussian kernel matrix of the desired size as there is no specific function for it unlike mean, median and mode. In the given R – code example, we are applying a 5x5 gaussian filter to the raster image.

```
# Create a 5x5 Gaussian filter matrix
>gaus =matrix(c(1,4,7,4,1,4,16,26,16,4,7,26,41,26,7,4,16,26,16,4,
1,4,7,4,1), nrow=5)/273
>gaus#Display Gaussian filter matrix
```

```
##             [,1]       [,2]       [,3]       [,4]        [,5]
## [1,] 0.003663004 0.01465201 0.02564103 0.01465201 0.003663004
## [2,] 0.014652015 0.05860806 0.09523810 0.05860806 0.014652015
## [3,] 0.025641026 0.09523810 0.15018315 0.09523810 0.025641026
## [4,] 0.014652015 0.05860806 0.09523810 0.05860806 0.014652015
## [5,] 0.003663004 0.01465201 0.02564103 0.01465201 0.003663004
```

```
# Apply the gaussian filter to each band of the input raster
> gaus_filter1 =focal(img[[1]], w=gaus) #1st band
> gaus_filter2 =focal(img[[2]], w=gaus) #2nd band
> gaus_filter3 =focal(img[[3]], w=gaus) #3rd band
```

```
# Create brick of the filtered raster layers
>gaus_filter =brick(gaus_filter1, gaus_filter2, gaus_filter3)
>gaus_filter#Display attributes of gaussian filtered output raster
```

```
## class      : RasterBrick
## dimensions : 265, 273, 72345, 3  (nrow, ncol, ncell, nlayers)
## resolution : 10, 10  (x, y)
## extent     : 301340, 304070, 2919700, 2922350  (xmin, xmax, ymin,
ymax)
## crs        : +proj=utm +zone=44 +datum=WGS84 +units=m +no_defs
+ellps=WGS84 +towgs84=0,0,0
## source     : memory
## names      :  layer.1,  layer.2,  layer.3
## min values : 1024.593, 1190.007, 1361.147
## max values : 3368.385, 2466.674, 2033.443
```

```
>plotRGB(gaus_filter, stretch ='lin', axes =TRUE,
main ='Gaussian filter') #Plot FCC of gaussian filtered image
```

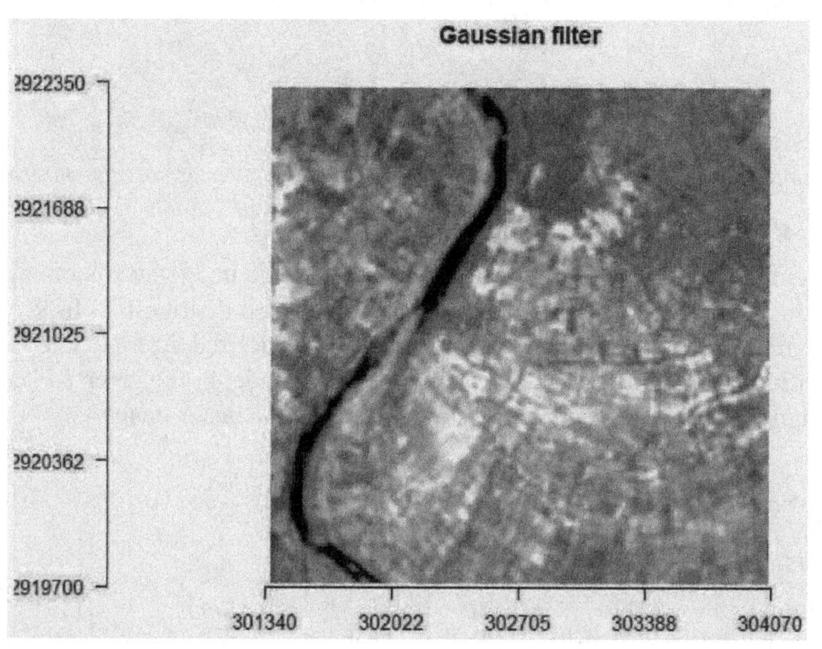

13.2. High pass filters

High pass filters are used for sharpening the raster image by enhancing the contrast with the adjoining areas. It retains high frequency information and reduces low frequency information in the raster image. Thus, it emphasizes the finer details in the raster image in contrary to the low pass filters. `raster` package of R doesn't have any specific function for applying this filter directly. Therefore, kernel matrix is designed to perform that task. The kernel matrix is designed in such a way that a single positive value is present in the centre surrounded by negative values. This increases the brightness value of center pixel relative to the surrounding pixels. In the following R – code example, a matrix of high pass filter is created and applied to the input raster image.

```
# Create a 3x3 High pass filter matrix
> hpf1 =matrix(c(-1,-1,-1,-1,9,-1,-1,-1,-1), nrow =3)
> hpf1 #Display High pass filter matrix

##      [,1] [,2] [,3]
## [1,]   -1   -1   -1
## [2,]   -1    9   -1
## [3,]   -1   -1   -1

# Apply high pass filter to each band of the input raster

> high_pass1 =focal(img[[1]], w = hpf1) #1st band
> high_pass2 =focal(img[[2]], w = hpf1) #2nd band
> high_pass3 =focal(img[[3]], w = hpf1) #3rd band

# Create brick of the filtered raster layers
>high_pass =brick(high_pass1, high_pass2, high_pass3)
>high_pass#Display attributes of high pass filtered output raster

## class      : RasterBrick
## dimensions : 265, 273, 72345, 3  (nrow, ncol, ncell, nlayers)
## resolution : 10, 10  (x, y)
## extent     : 301340, 304070, 2919700, 2922350  (xmin, xmax, ymin,
ymax)
## crs        : +proj=utm +zone=44 +datum=WGS84 +units=m +no_defs
+ellps=WGS84 +towgs84=0,0,0
## source     : memory
## names      : layer.1, layer.2, layer.3
## min values :   -2045,   -1791,    -288
## max values :    8173,    5274,    4036

>plotRGB(high_pass, stretch ='lin', axes =TRUE,
main ='High Pass Filter') #Plot FCC of high pass filtered image
```

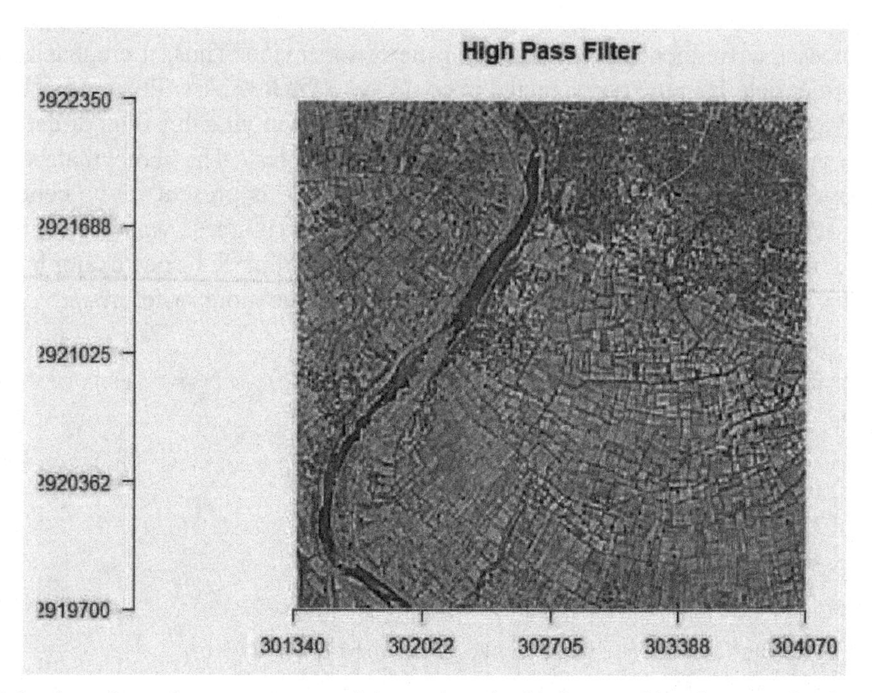

It is clear from the output raster image that the high pass filter has emphasized the high frequency areas and finer details, thereby, creating sharpening effect.

13.3. Edge enhancement filters

Edge enhancement filter delineates and emphasizes the edges in the raster image. Edges are characterized by sharp changes in the brightness values between two adjacent pixels. Edge enhancement filters can be directional *i.e.* enhance the edges in the particular specified direction e.g. horizontal, vertical, and diagonal filters, or non-directional *i.e.* enhance edges irrespective of the direction e.g. Laplacian filter. In R, matrix of the desired edge enhancement filter is created and applied to the raster image. Most commonly used edge enhancement filters with their respective matrix and R – codes are explained in the following section.

13.3.1. Linear edge enhancement filter

Linear edge enhancement filter enhances the edges by application of a directional first difference algorithm that approximates the first derivative between two adjacent pixels. These filters highlight gradient in particular direction as specified by the user. Linear edge enhancement filter matrix can be designed by placing -1 and +1 sign in the direction in which you want to enhance edges. In the following R – code example, we are going to design filter to enhance the edges in the east direction.

```
# Create Linear edge enhancement filter matrix for east direction
> eel =matrix(c(0,1,0,0,0,0,0,-1,0), nrow =3)
> eel #Display created filter matrix

##      [,1] [,2] [,3]
## [1,]    0    0    0
## [2,]    1    0   -1
## [3,]    0    0    0

# Apply high pass filter to each band of the input raster
> edge_linear1 =focal(img[[1]], w = eel) #1st band
> edge_linear2 =focal(img[[2]], w = eel) #2nd band
> edge_linear3 =focal(img[[3]], w = eel) #3rd band

# Create brick of the filtered raster layers
>edge_linear =brick(edge_linear1, edge_linear2, edge_linear3)
>edge_linear#Display attributes of filtered output raster

## class      : RasterBrick
## dimensions : 265, 273, 72345, 3  (nrow, ncol, ncell, nlayers)
## resolution : 10, 10  (x, y)
## extent     : 301340, 304070, 2919700, 2922350  (xmin, xmax, ymin, ymax)
## crs        : +proj=utm +zone=44 +datum=WGS84 +units=m +no_defs +ellps=WGS84 +towgs84=0,0,0
## source     : memory
## names      : layer.1, layer.2, layer.3
## min values :   -1567,    -915,    -462
## max values :    1569,     854,     506

>plotRGB(edge_linear, stretch ='lin', axes =TRUE,
main ='Linear Edge Enhancement') #Plot FCC of filtered image
```

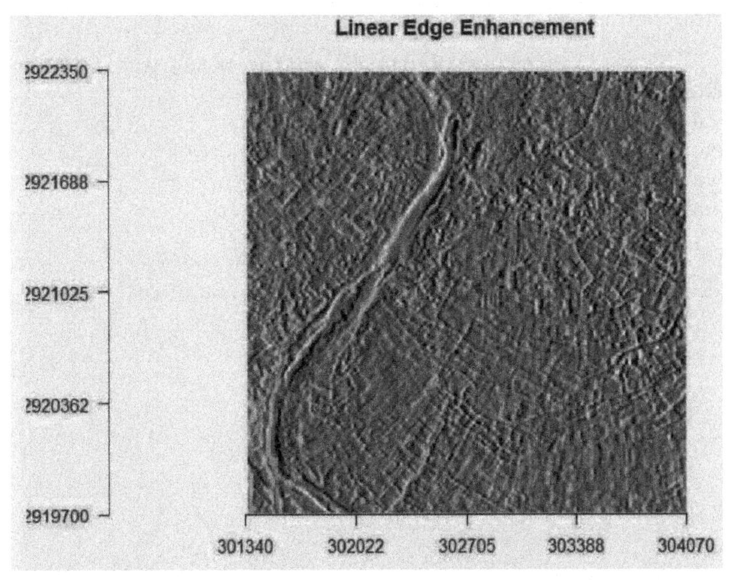

13.3.2. Prewitt filter

Prewitt filter uses two 3x3 kernels, each for horizontal and vertical direction, to calculate approximations of the derivaties. It is used for edge detection in vertical and horizontal directions separately. Edges are calculated by using difference between corresponding pixel'brightness values of a raster image. In the following R – code example, both horizontal and vertical mask of Prewitt filter are applied to the raster image.

```
# Create matrix for horizontal mask of Prewitt filter
>eeh =matrix(c(-1,0,1,-1,0,1,-1,0,1), nrow=3)
>eeh#Display horizontal mask of Prewitt filter matrix

##      [,1] [,2] [,3]
## [1,]   -1   -1   -1
## [2,]    0    0    0
## [3,]    1    1    1

# Apply horizontal Prewitt filter to each band of the input
raster
> edge_h1 =focal(img[[1]], w =eeh) #1st band
> edge_h2 =focal(img[[2]], w =eeh) #2nd band
> edge_h3 =focal(img[[3]], w =eeh) #3rd band

# Create brick of the filtered raster layers
>edge_horizontal =brick(edge_h1, edge_h2, edge_h3)
>edge_horizontal#Display attributes of filtered output raster

## class      : RasterBrick
## dimensions : 265, 273, 72345, 3  (nrow, ncol, ncell, nlayers)
## resolution : 10, 10  (x, y)
## extent     : 301340, 304070, 2919700, 2922350  (xmin, xmax, ymin,
ymax)
## crs         : +proj=utm +zone=44 +datum=WGS84 +units=m +no_defs
+ellps=WGS84 +towgs84=0,0,0
## source     : memory
## names      : layer.1, layer.2, layer.3
## min values :   -3624,   -2167,   -1060
## max values :    3369,    2109,    1072

# Plot FCC of horizontal Prewitt filtered image
>plotRGB(edge_horizontal, stretch ='lin', axes =TRUE,
main ='Horizontal mask of Prewitt Filter')
```

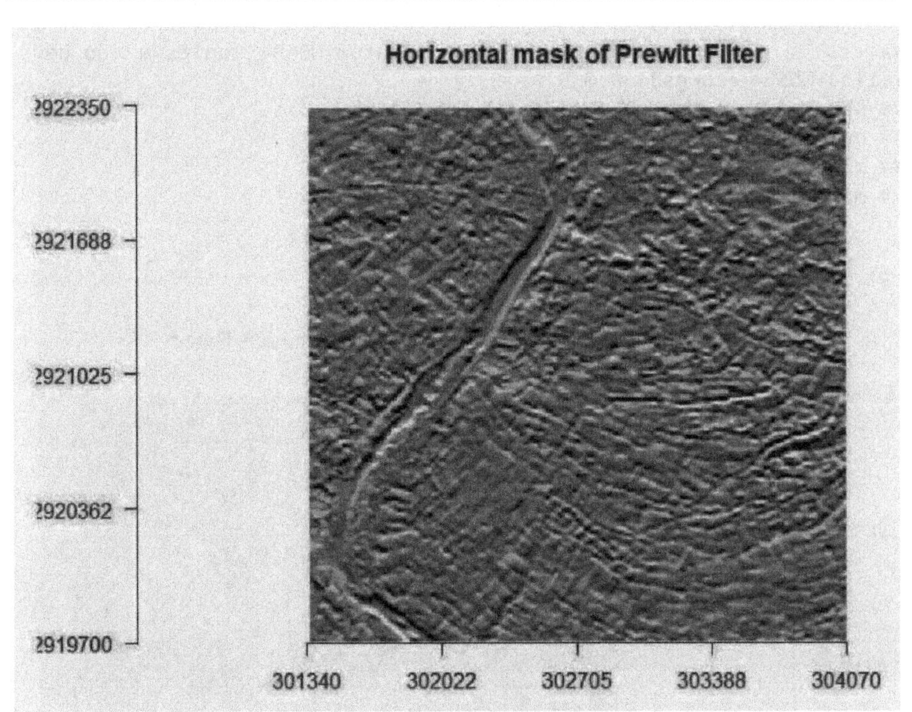

In the above image, we can see that the horizontal mask of Prewitt filter has detected the edges in the horizontal direction. Similarly, we can apply Prewitt filter in the vertical direction as shown in the following R – code example.

```
# Create matrix for vertical mask of Prewitt filter
>eev =matrix(c(-1,-1,-1,0,0,0,1,1,1), nrow=3)
>eev#Display vertical mask of Prewitt filter matrix

##      [,1] [,2] [,3]
## [1,]   -1    0    1
## [2,]   -1    0    1
## [3,]   -1    0    1

# Apply vertical Prewitt filter to each band of the input raster
> edge_v1 =focal(img[[1]], w =eev) #1st band
> edge_v2 =focal(img[[2]], w =eev) #2nd band
> edge_v3 =focal(img[[3]], w =eev) #3rd band

# Create brick of the filtered raster layers
>edge_vertical =brick(edge_v1, edge_v2, edge_v3)
>edge_vertical#Display attributes of filtered output raster

## class      : RasterBrick
## dimensions : 265, 273, 72345, 3  (nrow, ncol, ncell, nlayers)
## resolution : 10, 10  (x, y)
## extent     : 301340, 304070, 2919700, 2922350  (xmin, xmax, ymin,
ymax)
```

```
## crs         : +proj=utm +zone=44 +datum=WGS84 +units=m +no_defs
+ellps=WGS84 +towgs84=0,0,0
## source     : memory
## names      : layer.1, layer.2, layer.3
## min values :   -4244,   -2413,   -1209
## max values :    4558,    2393,    1147
```

```
>plotRGB(edge_vertical, stretch ='lin', axes =TRUE,
main ='Vertical mask of Prewitt Filter') # Plot FCC of output
```

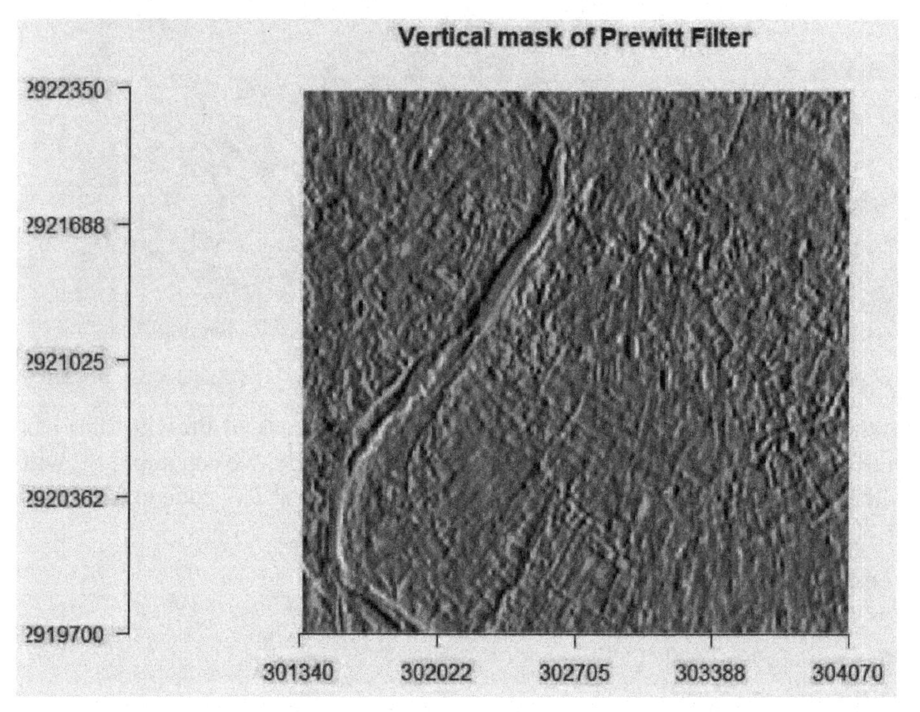

In the output raster image, we can see that the vertical mask of Prewitt filter has detected the edges in the vertical direction. Thus, it is clear that the Prewitt filter is very useful in detecting edges in horizontal and vertical direction.

13.3.3. Sobel filter

Sobel filter performs non-linear edge enhancement. It is similar to Prewitt filter in application as both compute approximations of the derivaties, but coefficients are not fixed in Sobel filter unlike Prewitt filter. Therefore, user can modify the coefficients provided that the filter mask or matrix retains derivative properties. The derivative mask has the following properties:

1. Opposite sign should be present in the mask.
2. Sum of mask should be zero.

3. More weights result in more edge detection in the image.

Sobel filters for both vertical and horizontal directions along with one filter having increased weight is applied to the raster image in the following R – code examples.

```
# Create matrix for vertical mask of Sobel filter
>sbv =matrix(c(-1,-2,-1,0,0,0,1,2,1), nrow=3)
>sbv#Display vertical mask of Sobel filter matrix

##      [,1] [,2] [,3]
## [1,]   -1    0    1
## [2,]   -2    0    2
## [3,]   -1    0    1

# Apply vertical mask of Sobel filter to each band of the input raster
> sobel_vert1 =focal(img[[1]], w=sbv) #1st band
> sobel_vert2 =focal(img[[2]], w=sbv) #2nd band
> sobel_vert3 =focal(img[[3]], w=sbv) #3rd band

# Create brick of the filtered raster layers
>sobel_vert =brick(sobel_vert1, sobel_vert2, sobel_vert3)
>sobel_vert#Display attributes of filtered output raster

## class      : RasterBrick
## dimensions : 265, 273, 72345, 3  (nrow, ncol, ncell, nlayers)
## resolution : 10, 10  (x, y)
## extent     : 301340, 304070, 2919700, 2922350  (xmin, xmax, ymin, ymax)
## crs         : +proj=utm +zone=44 +datum=WGS84 +units=m +no_defs +ellps=WGS84 +towgs84=0,0,0
## source     : memory
## names      : layer.1, layer.2, layer.3
## min values :   -5813,   -3248,   -1681
## max values :    6125,    3308,    1609

# Plot FCC of vertical mask of Sobel filtered image
>plotRGB(sobel_vert, stretch ='lin', axes =TRUE,
main ='Vertical mask of Sobel filter')
```

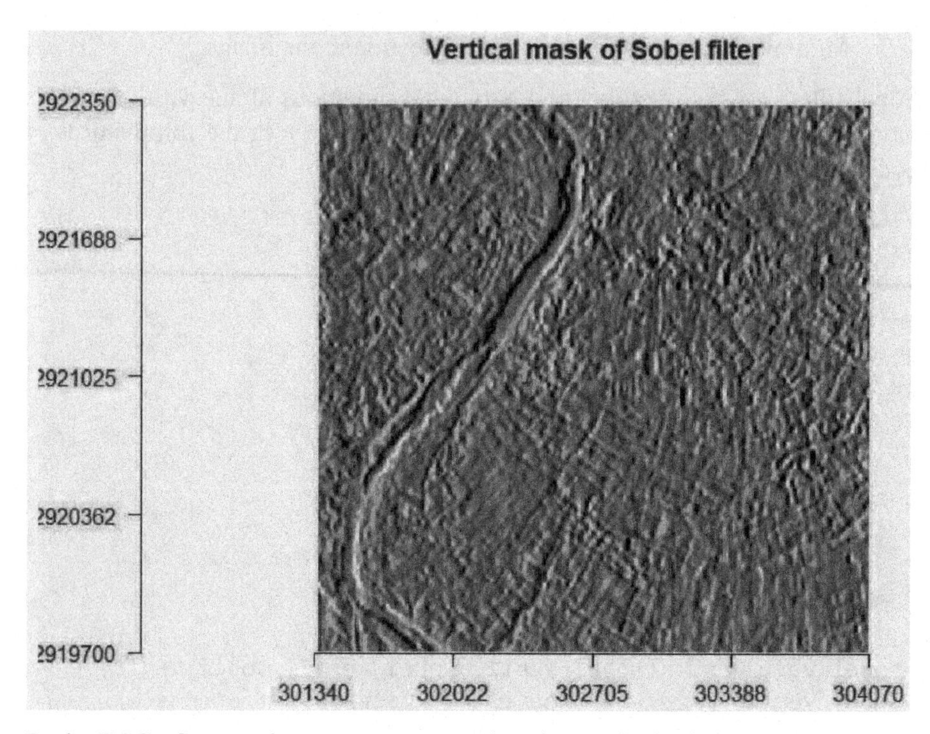

In the FCC of output image, we can see that the vertical edges are detected by the Sobel filter. Now we will apply horizontal mask of the Sobel filter.

```
# Create matrix for horizontal mask of Sobel filter
>sbh =matrix(c(-1,0,1,-2,0,2,-1,0,1), nrow=3)
>sbh#Display horizontal mask of Sobel filter matrix

##      [,1] [,2] [,3]
## [1,]   -1   -2   -1
## [2,]    0    0    0
## [3,]    1    2    1

# Apply horizontal mask of Sobel filter to each band of the
input raster
> sobel_hor1 =focal(img[[1]], w=sbh) #1st band
> sobel_hor2 =focal(img[[2]], w=sbh) #2nd band
> sobel_hor3 =focal(img[[3]], w=sbh) #3rd band

# Create brick of the filtered raster layers
>sobel_hor =brick(sobel_hor1, sobel_hor2, sobel_hor3)
>sobel_hor#Display attributes of filtered output raster

## class      : RasterBrick
## dimensions : 265, 273, 72345, 3  (nrow, ncol, ncell, nlayers)
## resolution : 10, 10  (x, y)
## extent     : 301340, 304070, 2919700, 2922350  (xmin, xmax, ymin,
ymax)
```

```
## crs          : +proj=utm +zone=44 +datum=WGS84 +units=m +no_defs
+ellps=WGS84 +towgs84=0,0,0
## source     : memory
## names      : layer.1, layer.2, layer.3
## min values :   -4911,   -2903,   -1512
## max values :    4559,    2888,    1443
```

```
# Plot FCC of horizontal mask of Sobel filtered image
>plotRGB(sobel_hor, stretch ='lin', axes =TRUE,
main ='Horizontal mask of Sobel filter')
```

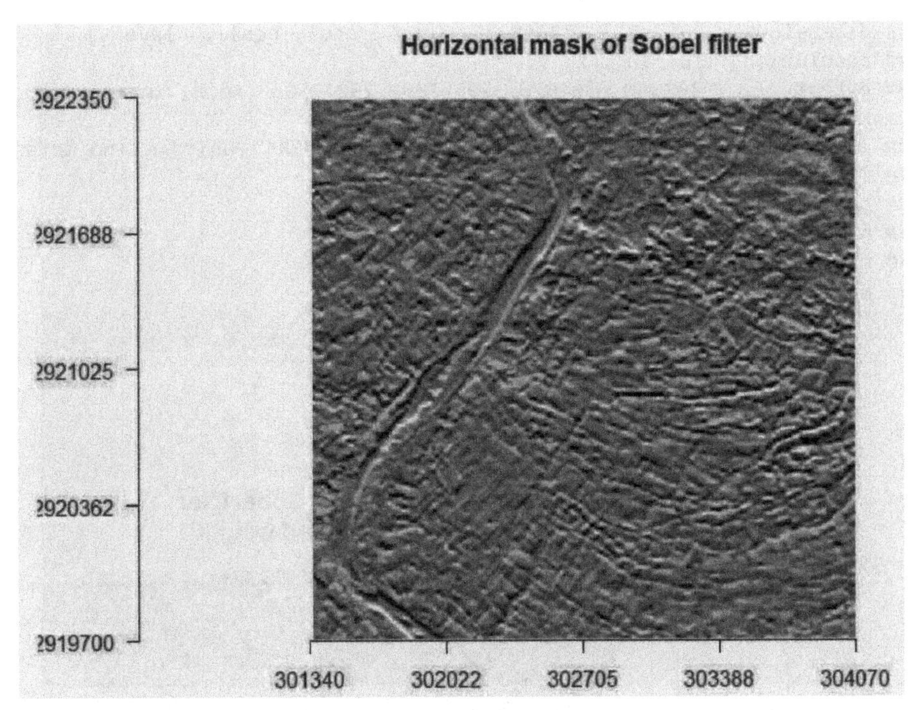

We can see in the FCC of output filtered image that horizontal edges are being detected by the Sobel filter. Now we will increase the weight of the horizontal mask of Sobel filter in the following R – code example.

```
# Create matrix for horizontal mask of Sobel filter with increased
weight
>sbw =matrix(c(-1,0,1,-5,0,5,-1,0,1), nrow=3)
>sbw#Display created Sobel filter matrix
```

```
##      [,1] [,2] [,3]
## [1,]   -1   -5   -1
## [2,]    0    0    0
## [3,]    1    5    1
```

```
# Apply Sobel filter with increased weight to each band of the
raster
> sobel_wt1 =focal(img[[1]], w =sbw) #1st band
> sobel_wt2 =focal(img[[2]], w =sbw) #2nd band
> sobel_wt3 =focal(img[[3]], w =sbw) #3rd band
# Create brick of the filtered raster layers
>sobel_wt =brick(sobel_wt1, sobel_wt2, sobel_wt3)
>sobel_wt#Display attributes of filtered output raster
## class      : RasterBrick
## dimensions : 265, 273, 72345, 3  (nrow, ncol, ncell, nlayers)
## resolution : 10, 10  (x, y)
## extent     : 301340, 304070, 2919700, 2922350  (xmin, xmax, ymin,
ymax)
## crs         : +proj=utm +zone=44 +datum=WGS84 +units=m +no_defs
+ellps=WGS84 +towgs84=0,0,0
## source     : memory
## names      : layer.1, layer.2, layer.3
## min values :   -8772,   -5321,   -2868
## max values :    8129,    5240,    2610
# Plot FCC of Sobel filtered image with increased weight
>plotRGB(sobel_wt, stretch ='lin', axes =TRUE,
main ='Horizontal mask of Sobel filter
        with increased weight')
```

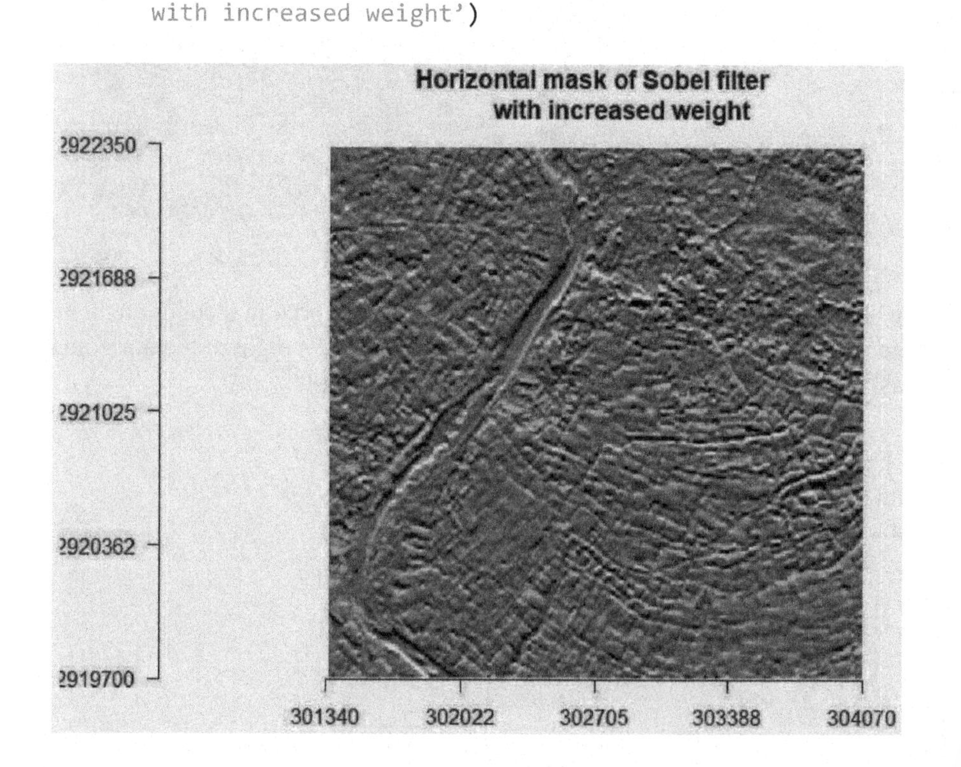

We can see in the FCC of the output filtered image that with increase in the weights of Sobel filter, more edges are being detected in the horizontal direction.

13.3.4. Laplacian filter

Laplacian filter is non-directional filter for edge detection. It computes second order derivatives of the raster image. Therefore, it helps us to determine whether the change in adjacent pixel values is an edge or a continuous progression. The kernel of Laplacian filter contains negative values in the cross pattern, and zero or positive values at corners. The matrix for applying Laplacian filter and its application on raster image is depicted in the given R – code example.

```
# Create matrix for Laplacian filter
> lap =matrix(c(0,-1,0,-1,4,-1,0,-1,0), nrow=3)
> lap #Display created Laplacian filter matrix

##      [,1] [,2] [,3]
## [1,]    0   -1    0
## [2,]   -1    4   -1
## [3,]    0   -1    0

# Apply Laplacian filter to each band of the input raster
> lap_filter1 =focal(img[[1]], w= lap) #1st band
> lap_filter2 =focal(img[[2]], w= lap) #2nd band
> lap_filter3 =focal(img[[3]], w= lap) #3rd band

# Create brick of the filtered raster layers
>lap_filter =brick(lap_filter1, lap_filter2, lap_filter3)
>lap_filter#Display attributes of filtered output raster

## class      : RasterBrick
## dimensions : 265, 273, 72345, 3  (nrow, ncol, ncell, nlayers)
## resolution : 10, 10  (x, y)
## extent     : 301340, 304070, 2919700, 2922350  (xmin, xmax, ymin, ymax)
## crs        : +proj=utm +zone=44 +datum=WGS84 +units=m +no_defs +ellps=WGS84 +towgs84=0,0,0
## source     : memory
## names      : layer.1, layer.2, layer.3
## min values :   -1553,   -1338,    -758
## max values :    1914,    1195,     892

# Plot FCC of Laplacian filtered image
>plotRGB(lap_filter, stretch ='lin', axes =TRUE,
main ='Laplacian filter')
```

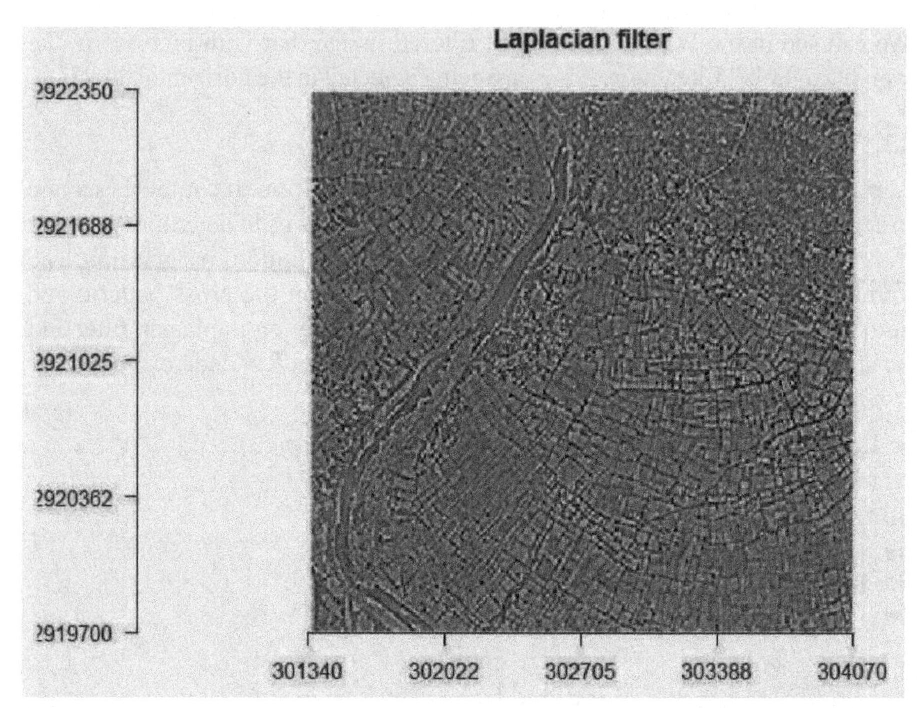

We can see in the FCC of output filtered image that Laplacian filter has highlighted the points, lines, and edges and has suppressed the uniform and smoothly varying regions.

13.3.5. Diagonal filter

Diagonal filter enhances edge in diagonal direction *i.e.* partial edge enhancement in both horizontal and vertical directions.

An example of a matrix of diagonal filter along with its application on raster image is depicted in the following R – code example.

```
# Create matrix for Diagonal filter
>eed =matrix(c(0,-1,-1,1,0,-1,1,1,0), nrow =3)
>eed#Display created Diagonal filter matrix

##      [,1] [,2] [,3]
## [1,]    0    1    1
## [2,]   -1    0    1
## [3,]   -1   -1    0

# Apply Diagonal filter to each band of the input raster
> edge_d1 =focal(img[[1]], w =eed) #1st band
> edge_d2 =focal(img[[2]], w =eed) #2nd band
> edge_d3 =focal(img[[3]], w =eed) #3rd band
```

```
# Create brick of the filtered raster layers
>edge_diagnol =brick(edge_d1, edge_d2, edge_d3)
>edge_diagnol#Display attributes of filtered output raster

## class      : RasterBrick
## dimensions : 265, 273, 72345, 3  (nrow, ncol, ncell, nlayers)
## resolution : 10, 10  (x, y)
## extent     : 301340, 304070, 2919700, 2922350  (xmin, xmax, ymin, 
ymax)
## crs        : +proj=utm +zone=44 +datum=WGS84 +units=m +no_defs 
+ellps=WGS84 +towgs84=0,0,0
## source     : memory
## names      : layer.1, layer.2, layer.3
## min values :   -3332,   -2171,   -1074
## max values :    3645,    2199,    1088

# Plot FCC of Diagonal filtered image
>plotRGB(edge_diagnol, stretch ='lin', axes =TRUE, 
main ='Diagonal Edge Enhancement')
```

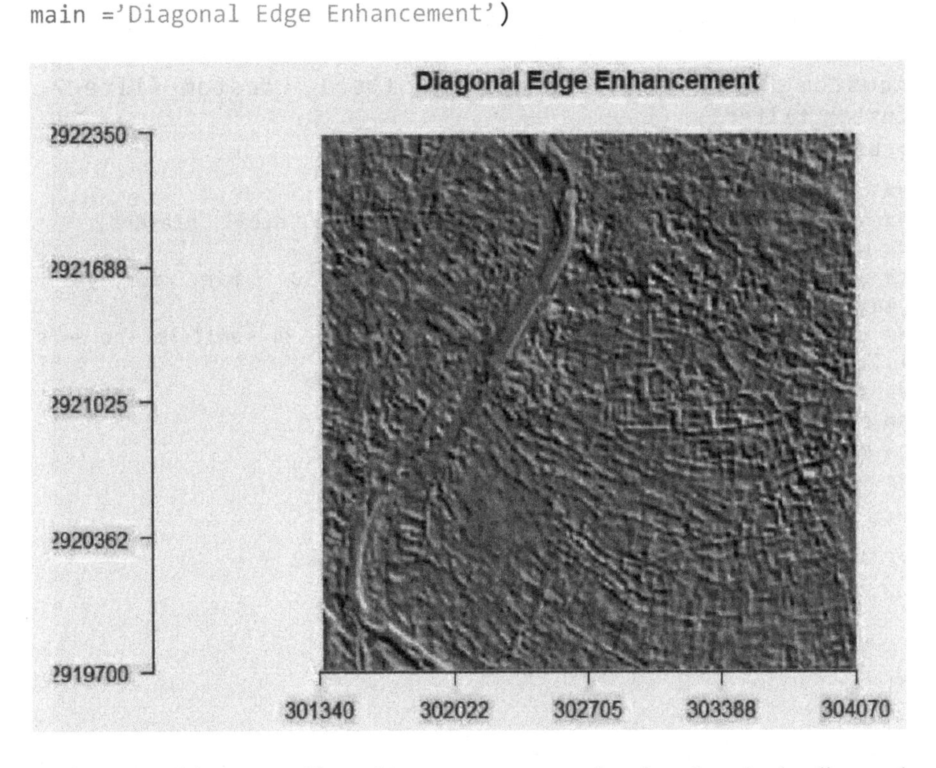

In the FCC of the output filtered image, we can see that the edges in the diagonal direction are being detected.

13.4. Customized filters

You can also create a filter customized according to your requirement by defining your own kernel matrix. In the following R – code example, we have created a customized high pass filter and applied it to the raster image.

```
# Create matrix for customized filter
> custom =matrix(c(1,-2,1,-2,5,-2,1,-2,1), nrow =3)
> custom #Display created customized filter matrix

##       [,1] [,2] [,3]
## [1,]    1   -2    1
## [2,]   -2    5   -2
## [3,]    1   -2    1

# Apply customized filter to each band of the input raster
> custom_filter1 =focal(img[[1]], w = custom) #1st band
> custom_filter2 =focal(img[[2]], w = custom) #2nd band
> custom_filter3 =focal(img[[3]], w = custom) #3rd band

# Create brick of the filtered raster layers
>custom_filter =brick(custom_filter1, custom_filter2, custom_filter3)
>custom_filter#Display attributes of filtered output raster

## class      : RasterBrick
## dimensions : 265, 273, 72345, 3  (nrow, ncol, ncell, nlayers)
## resolution : 10, 10  (x, y)
## extent     : 301340, 304070, 2919700, 2922350  (xmin, xmax, ymin, ymax)
## crs         : +proj=utm +zone=44 +datum=WGS84 +units=m +no_defs +ellps=WGS84 +towgs84=0,0,0
## source     : memory
## names      : layer.1, layer.2, layer.3
## min values :     272,     320,     865
## max values :    4571,    3212,    2614

# Plot FCC of customized filtered image
>plotRGB(custom_filter, stretch ='lin', axes =TRUE,
main ='Customized High pass filter')
```

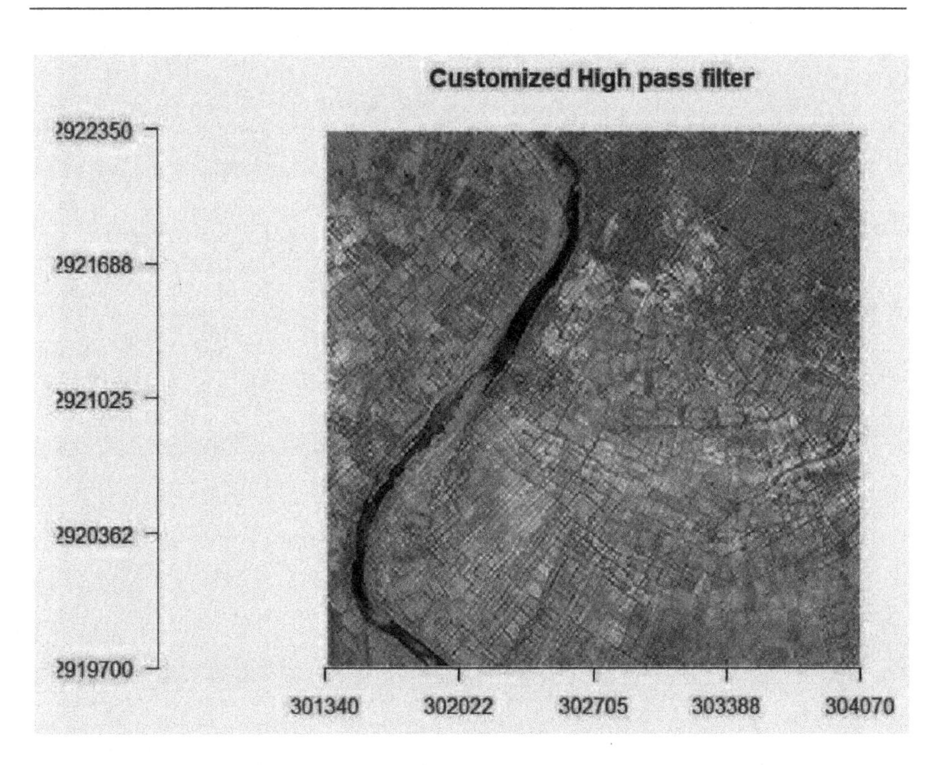
Customized High pass filter
2922350
2921688
2921025
2920362
2919700
301340
302022
302705
303388
304070

Chapter 14

Transformations

Raster transformations involve mathematical manipulation of raster layers to generate new raster images. These transformed raster images either contain the information of interest or highlight the features which are not discernible in the original image. Transformations are also performed to preserve the essential information in reduced number of dimensions or layers. In this chapter, we are going to discuss few most commonly used transformation tools listed below:

1. Computation of spectral indices
2. Performing algebra on raster images
3. Density slicing
4. Tasselled cap transformation
5. Principal component transformation
6. Pan sharpening

These transformation techniques with their respective R – code examples are explained in this chapter. First of all, import all the packages required to perform this exercise.

```
#Import packages
>library(raster)    #For importing and transformation of raster
data
>library(rasterVis) #For visualisation of raster data
>library(RStoolbox) #For transformation & visualization of raster
data
>library(ggplot2)   #For visualization of raster data with
'RStoolbox'
```

From the imported packages, `raster` and `RStoolbox` packages are required to perform transformations of raster data. For displaying the results in the form of appropriate graphics, `rasterVis` and `ggplot2` packages are required. Some plots are also created by combining function from `RStoolbox` package with ggplot2.

Import the input multiband raster image on which you want to perform transformations by using brick function.

```
#Import multiband raster
>setwd("E:\\...........\\Chapter14\\data")
>img =brick('reflectance_img.tif')
```

Now set names of individual bands in the imported multiband raster image with respect to their wavelengths. As this imported raster image belongs to Landsat 8 OLI & TIRS sensor, therefore, names are assigned to its bands as per sensor' specifications.

```
# Setting names of the individual bands
>names(img) #Display original names of the bands

##  [1] "reflectance_img.1"  "reflectance_img.2"  "reflectance_img.3"
##  [4] "reflectance_img.4"  "reflectance_img.5"  "reflectance_img.6"
##  [7] "reflectance_img.7"  "reflectance_img.8"  "reflectance_img.9"
## [10] "reflectance_img.10"

# Assign names to the bands as per their respective wavelengths
>names(img) =c('Coastal_aerosol', 'Blue', 'Green', 'Red',
'NIR', 'SWIR1', 'SWIR2','cirrus', 'TIRS1','TIRS2')
>names(img) #check the changed names

##  [1] "Coastal_aerosol" "Blue"            "Green"
##  [4] "Red"             "NIR"             "SWIR1"
##  [7] "SWIR2"           "cirrus"          "TIRS1"
## [10] "TIRS2"
```

Now display the attributes and standard False Colour Composite (FCC) of the imported raster image.

```
>img#Display attributes of imported raster

## class      : RasterBrick
## dimensions : 323, 357, 115311, 10  (nrow, ncol, ncell, nlayers)
## resolution : 30, 30  (x, y)
## extent     : 427785, 438495, 2836125, 2845815  (xmin, xmax, ymin,
ymax)
## crs         : +proj=utm +zone=44 +datum=WGS84 +units=m +no_defs
+ellps=WGS84 +towgs84=0,0,0
## source     : E:/......................................./reflectance_img.tif
## names      : Coastal_aerosol, Blue, Green, Red, NIR, SWIR1, SWIR2,
cirrus, TIRS1, TIRS2

#Plot FCC of imported raster image with 'RStoolbox' and 'ggplot2'
packages
>ggRGB(img, r=5, g=4, b=3, stretch ='lin', geom_raster =TRUE) +
ggtitle('Input Raster') +# give title to plot
```

```
labs(x='Longitude (m)', y='Latitude (m)') +#Axis Labels
theme(plot.title =element_text(hjust =0.5, #Align title in centre
size =30), #Adjust size of title
```

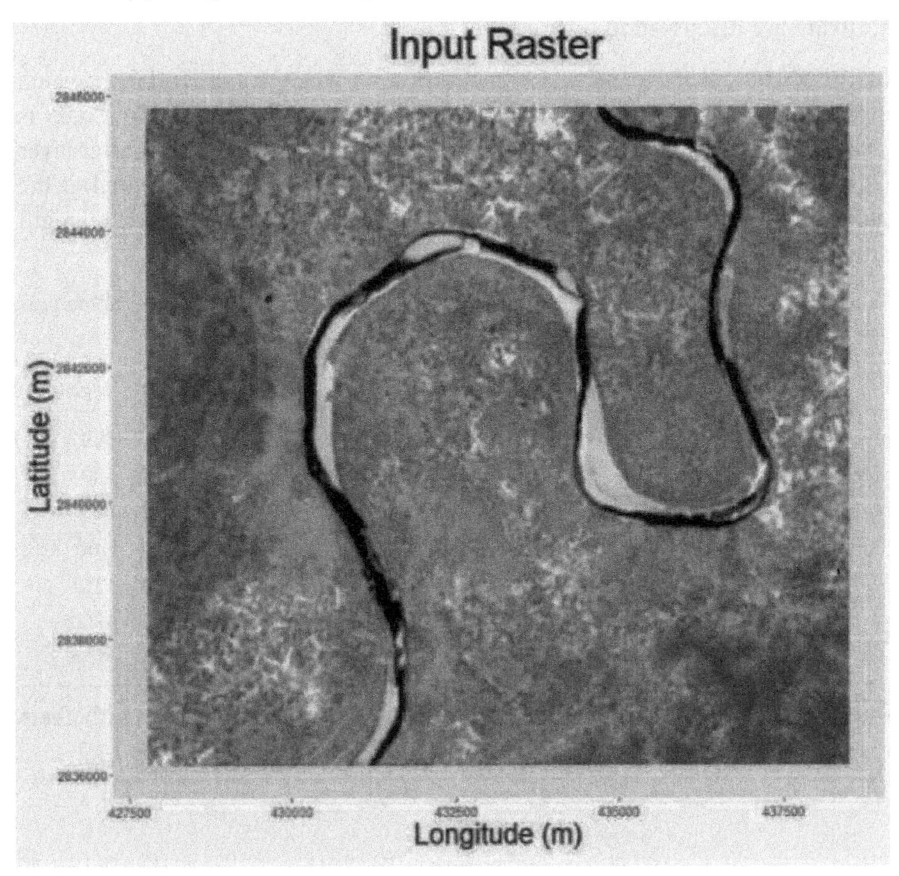


```
axis.title =element_text(size =20)) #Adjust size of axis labels
```

14.1. Computation of Spectral Indices

Spectral Index is a quantitative measure of features of interest, usually formed by combination of several spectral bands. The values of these spectral bands are subjected to mathematical operations like addition, subtraction, multiplication and division to yield a single value known as spectral index which is a better indicator of features' properties or provide unique and valuable information about the features as compared to the individual bands. Spectral indices enhance the radiometric response to a particular target feature and reduce the data dimensionality. These indices facilitate to perform comparison among different features or similar features present at different locations or time. These indices also reduce the noise caused by variations in illumination, influence of atmosphere and background of the target feature. There are several spectral indices available in the literature with their respective advantages and limitations. Some of these commonly used indices and their computation in R software are discussed in this section.

In `RStoolbox` package, there is a function known as `spectralIndices` which directly computes 23 indices from the multiband raster data if input data is provided for each index. The pixel-wise index value in the form of raster layer (if single index computed) or raster brick (if multiple indices computed) is the output of this function. The names of bands with their respective wavelength range used as input in this function are given in the table 14.1.

Table 14.1: Wavelength range of bands used in the *'spectralIndices'* function of *'RStoolbox'* package

Band	Description	Wavelength range (nm)
blue	Blue	450 – 520
green	Green	520 – 600
red	Red	630 – 690
nir	Near Infra-Red	700 – 1100
swir1	Shortwave Infrared	1100 – 1351
swir2	Shortwave Infrared	1400 – 1800
swir3	Shortwave Infrared	2000 – 2500

The list of indices computed by `spectralIndices` function with their respective formula, source and spectral bands are given in the table 14.2.

Table 14.2: List of spectral indices computed by the *'spectralIndices'* function of *'RStoolbox'* package

S. No.	Index	Description	Source	Bands	Formula
1.	RVI	Ratio Vegetation Index		red, nir	red/nir
2.	SR	Simple Ratio Vegetation Index	Birth (1968)	red, nir	nir/red
3.	NDVI	Normalised Difference Vegetation Index	Rouse (1974)	red, nir	(nir - red)/(nir + red)
4.	TVI	Transformed Vegetation Index	Deering (1975)	red, nir	sqrt((nir - red)/(nir + red) + 0.5)
5.	DVI	Difference Vegetation Index	Richardson (1977)	red, nir	s * nir - red
6.	WDVI	Weighted Difference Vegetation Index	Richardson (1977)	red, nir	nir - s * red
7.	EVI	Enhanced Vegetation Index	Huete (1999)	red, nir, blue	G * ((nir - red)/(nir + C1 * red - C2 * blue + L_evi))
8.	EVI2	Two-band Enhanced Vegetation Index	Jiang (2008)	red, nir	G * (nir - red)/(nir + 2.4 * red + 1)
9.	GEMI	Global Environmental Monitoring Index	Pinty (1992)	red, nir	(((nir^2 - red^2) * 2 + (nir * 1.5) + (red * 0.5))/(nir + red + 0.5)) * (1 - ((((nir^2 - red^2) * 2 + (nir * 1.5) + (red * 0.5))/(nir + red + 0.5)) * 0.25)) - ((red - 0.125)/(1 - red))
10.	GNDVI	Green Normalised Difference Vegetation Index	Gitelson (1998)	green, nir	(nir - green)/(nir + green)

Contd.

S. No.	Index	Description	Source	Bands	Formula
11.	NDWI	Normalised Difference Water Index	McFeeters (1996)	green, nir	(green - nir)/(green + nir)
12.	NDWI2	Normalised Difference Water Index	Gao (1996)	nir, swir2	(nir - swir2)/(nir + swir2)
13.	MNDWI	Modified Normalised Difference Water Index	Xu (2006)	green, swir2	(green - swir2)/(green + swir2)
14.	MSAVI	Modified Soil Adjusted Vegetation Index	Qi (1994)	red, nir	nir + 0.5 - (0.5 * sqrt((2 * nir + 1)^2 - 8 * (nir - (2 * red))))
15.	MSAVI2	Modified Soil Adjusted Vegetation Index 2	Qi (1994)	red, nir	(2 * (nir + 1) - sqrt((2 * nir + 1)^2 - 8 * (nir - red)))/2
16.	NBRI	Normalised Burn Ratio Index	Garcia (1991)	nir, swir3	(nir - swir3)/(nir + swir3)
17.	CTVI	Corrected Transformed Vegetation Index	Perry (1984)	red, nir	(NDVI + 0.5)/sqrt(abs(NDVI + 0.5))
18.	NDVIC	Corrected Normalised Difference Vegetation Index	Nemani (1993)	red, nir, swir2	(nir - red)/(nir + red) * (1 - ((swir2 - swir2ccc)/(swir2coc - swir2ccc)))
19.	NRVI	Normalised Ratio Vegetation Index	Baret (1991)	red, nir	(red/nir - 1)/(red/nir + 1)
20.	SATVI	Soil Adjusted Total Vegetation Index	Marsett (2006)	red, swir2, swir3	(swir2 - red)/(swir2 + red + L) * (1 + L) - (swir3/2)
21.	SAVI	Soil Adjusted Vegetation Index	Huete (1988)	red, nir	(nir - red) * (1 + L)/(nir + red + L)
22.	SLAVI	Specific Leaf Area Vegetation Index	Lymburger (2000)	red, nir, swir2	nir/(red + swir2)
23.	TTVI	Thiam's Transformed Vegetation Index	Thiam (1997)	red, nir	sqrt(abs((nir - red)/(nir + red) + 0.5))

Some indices listed above require additional coefficients whose description is given in the table 14.3.

Table 14.3: List of additional coefficients required for the computation of some indices in the *'spectralIndices'* function of *'RStoolbox'* package

Coefficient	Description	Affected Indices
s	slope of the soil line	DVI, WDVI
L_evi, C1, C2, G	various	EVI
L	soil brightness factor	SAVI, SATVI
swir2ccc	minimum swir2 value (completely closed forest canopy)	NDVIC
swir2coc	maximum swir2 value (completely open canopy)	NDVIC

If you want to compute any of the above listed indices, then mention name of that index in `index =` argument of **`spectralIndices`** function. The names of the bands in the input multiband raster image corresponding to the wavelengths listed above is also specified in arguments of the function. In this R – code example, we are going to compute Specific Leaf Area Vegetation Index (SLAVI). Hence, we put `index ='SLAVI'` in the **`spectralIndices`** function. As this index require red, nir and swir2 bands for its computation, therefore, put their names in the argument as `red ="Red"`, `nir ="NIR"` and `swir2 ="SWIR1"`. We are mentioning 'SWIR1' in swir2 argument because the wavelength range of SWIR1 in Landsat 8 OLI & TIRS sensor is from 1570 to 1650 nm which corresponds to the swir2 wavelength range of the given function. This function will give pixel-wise output of computed SLAVI in the form of a raster layer.

```
# Calculation of SLAVI index from the pre-defined available
code
>slavi =spectralIndices(img, red =”Red”, nir =”NIR”,
swir2 =”SWIR1", index =’SLAVI’)
>slavi#Display the properties of the computed SLAVI layer
```

```
## class      : RasterLayer
## dimensions : 323, 357, 115311  (nrow, ncol, ncell)
## resolution : 30, 30  (x, y)
## extent     : 427785, 438495, 2836125, 2845815  (xmin, xmax, ymin,
ymax)
## crs        : +proj=utm +zone=44 +datum=WGS84 +units=m +no_defs
+ellps=WGS84 +towgs84=0,0,0
## source     : memory
## names      : SLAVI
## values     : 0.4435838, 1.352832  (min, max)
```

```
#Display distribution of SLAVI values
>histogram(slavi, col =’wheat’,
```

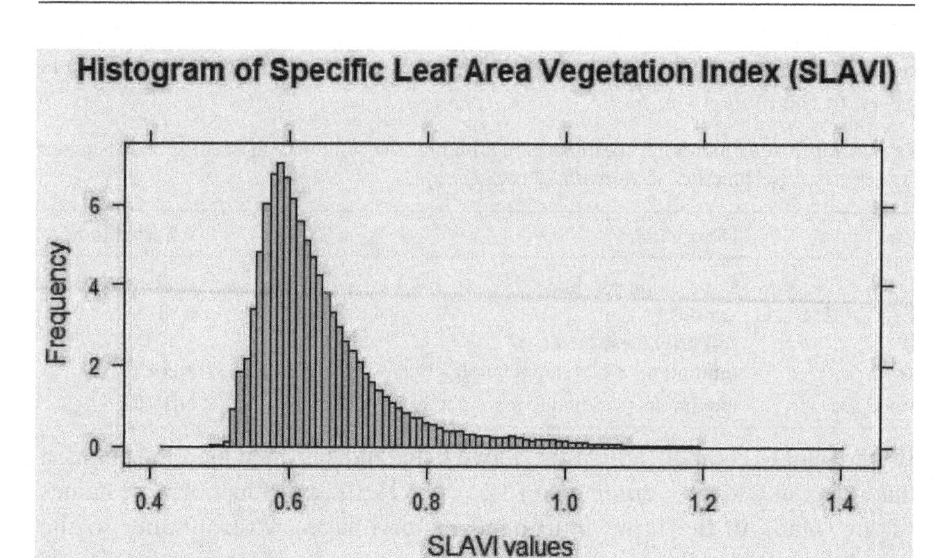

In the histogram, we can see that most of the pixels in SLAVI are in the range of 0.5 – 0.8. Therefore, histogram gives idea of the distribution of pixel values in the computed index. We can see the spatial distribution of SLAVI by plotting its raster.

```
main ='Histogram of Specific Leaf Area Vegetation Index (SLAVI)',
xlab ='SLAVI values', ylab ='Frequency')

#Display raster plot of SLAVI
>ggR(slavi, geom_raster =TRUE) +#plot with ggplot2
      ggtitle('Specific Leaf Area Vegetation Index') +#title
of the plot
      labs(x ='Longitude (m)', y ='Latitude (m)') +#axis labels
      theme(plot.title =element_text(hjust =0.5, #Align title
in centre
size =30), #size of title
      axis.title =element_text(size =20), #size of axis labels
      legend.key.size =unit(1, "cm"), #size of legend
      legend.title =element_text(size =20), #size of legend
title
legend.text =element_text(size =15))+#size of legend text
#Provide customized gradient colours
      scale_fill_gradientn(name ='SLAVI values',
colours =c("white", "darkgreen"))
```

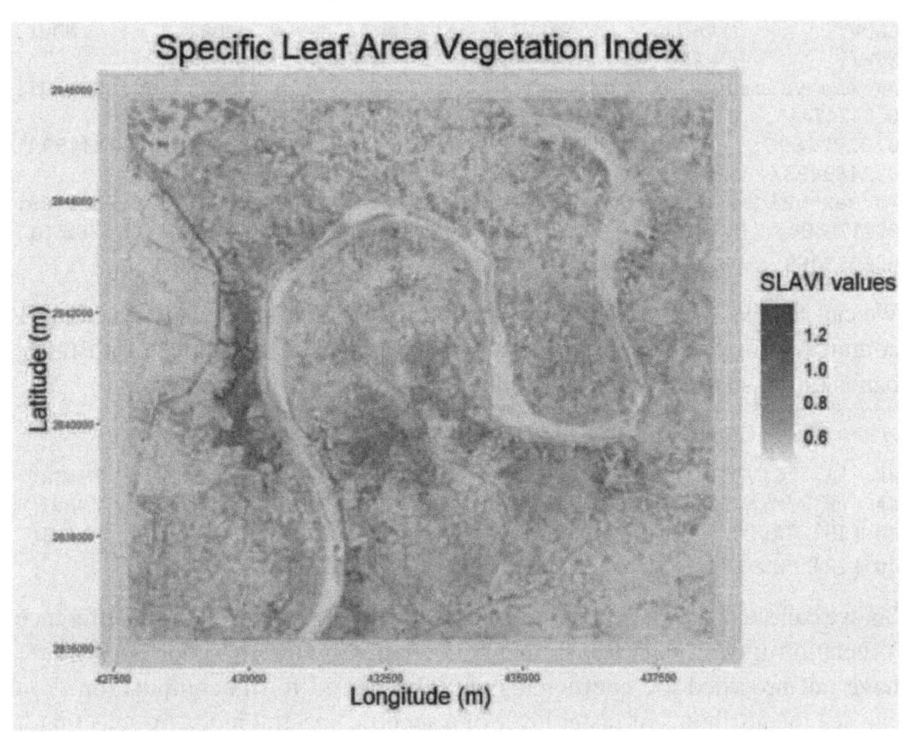

If you want to calculate all the spectral indices available in the `spectralIndices` function at one time, then you have to provide all the arguments and coefficients in the function. We don't have coefficient values for 'swir2ccc' and 'swir2coc', therefore, all the spectral indices except Corrected Normalised Difference Vegetation Index (NDVIC) will be computed by using the given R – code.

```
#Calculate all possible indices for which data is provided
> SI =spectralIndices(img, blue =”Blue”, green =”Green”, red
=”Red”,
nir =”NIR”, swir2 =”SWIR1", swir3 =”SWIR2",
coefs =list(L =0.5, G =2.5, L_evi =1,
C1 =6, C2 =7.5, s =1))

> SI #Display attributes of computed Spectral Indices

## class      : RasterBrick
## dimensions : 323, 357, 115311, 22  (nrow, ncol, ncell, nlayers)
## resolution : 30, 30  (x, y)
## extent     : 427785, 438495, 2836125, 2845815  (xmin, xmax, ymin,
ymax)
## crs         : +proj=utm +zone=44 +datum=WGS84 +units=m +no_defs
+ellps=WGS84 +towgs84=0,0,0
## source     : memory
## names      :  CTVI,        DVI,        EVI,       EVI2,       GEMI,
```

```
GNDVI,      MNDWI,      MSAVI,      MSAVI2,      NBRI,      NDVI,
NDWI,      NDWI2,      NRVI,      RVI, ...
## min values :  0.59022495, -0.03497455, -0.10214537, -0.20988705,
0.24247443, -0.19525144, -0.38686679, -0.24799654, -0.05588930,      -
0.02929896, -0.15163451, -0.52963722, -0.17516224, -0.60134594,
0.24894937, ...
## max values :   1.0494503,   0.2687274,   0.6779691,   0.9999509,
0.7179404,  0.5296372,  0.3527123,  0.2376608,  0.4094332,  0.6362561,
0.6013459,   0.1952514,   0.3420649,   0.1516345,   1.3574745, ...
```

We can see that the output of this function is a raster brick in which each layer contains values of a specific index. You can display the names of all raster bands or the spectral indices computed by this function as follows.

```
>names(SI) #Display list of indices computed
```

```
##  [1] "CTVI"   "DVI"     "EVI"    "EVI2"   "GEMI"   "GNDVI"  "MNDWI"
##  [8] "MSAVI"  "MSAVI2"  "NBRI"   "NDVI"   "NDWI"   "NDWI2"  "NRVI"
## [15] "RVI"    "SATVI"   "SAVI"   "SLAVI"  "SR"     "TVI"    "TTVI"
## [22] "WDVI"
```

So, we can see that all spectral indices except Corrected Normalised Difference Vegetation Index (NDVIC) are being computed by the R – code because we have not provided the coefficients values required for its computation. You can see the attributes of raster layer of a specific spectral index by selecting it using $ sign.

```
> SI$NDWI #Display properties of raster layer of a specific
index
```

```
## class      : RasterLayer
## dimensions : 323, 357, 115311  (nrow, ncol, ncell)
## resolution : 30, 30  (x, y)
## extent     : 427785, 438495, 2836125, 2845815  (xmin, xmax, ymin,
ymax)
## crs         : +proj=utm +zone=44 +datum=WGS84 +units=m +no_defs
+ellps=WGS84 +towgs84=0,0,0
## source     : memory
## names      : NDWI
## values     : -0.5296372, 0.1952514  (min, max)
```

We can see the raster plots of all the computed indices at once by using `levelplot` function of `rasterVis` package.

```
#Plot all the computed indices together
>levelplot(SI, main ='Maps of Spectral Indices',
xlab ='Longitude (m)', ylab ='Latitude (m)')
```

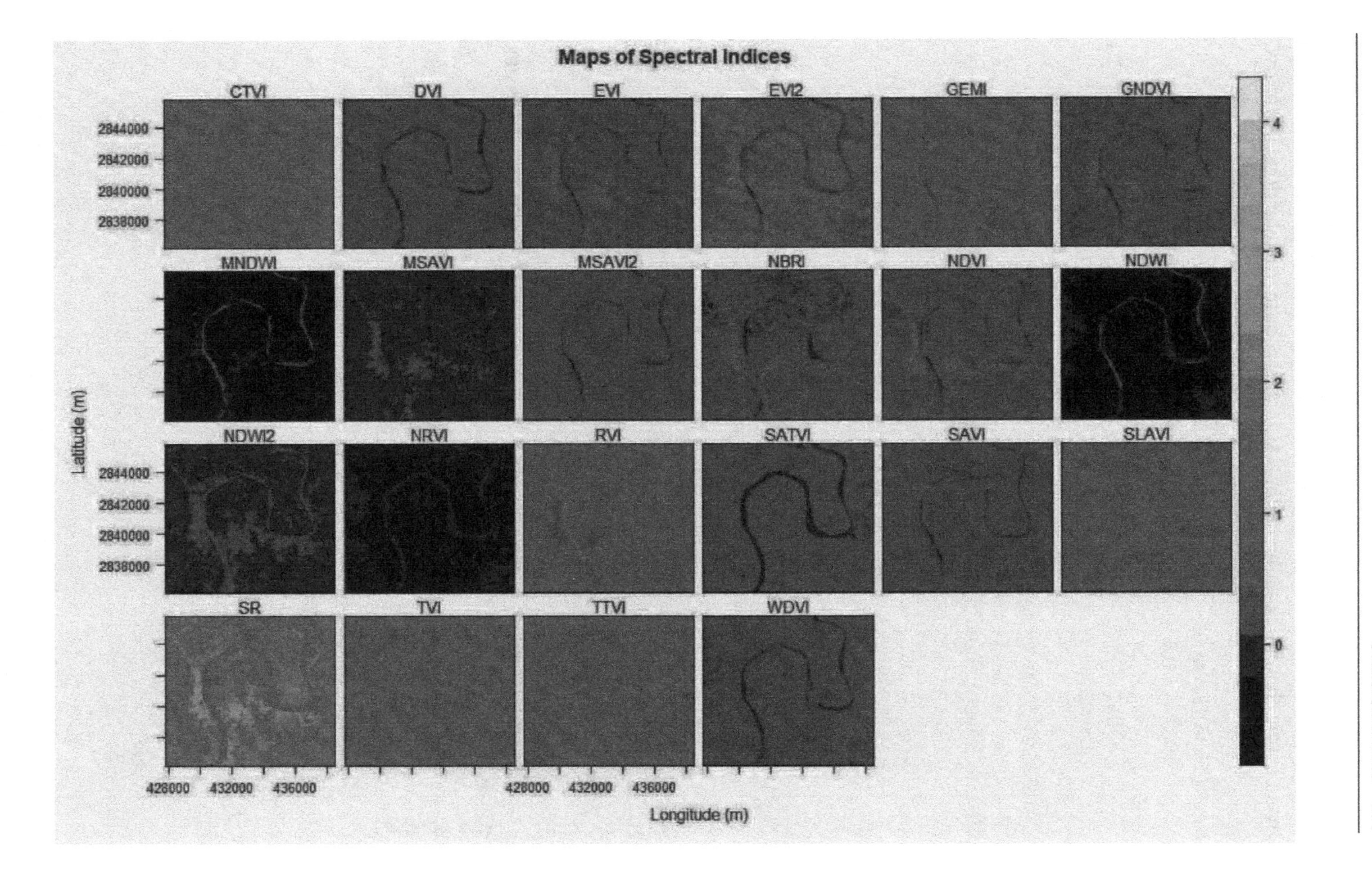
Maps of Spectral Indices
CTVI
DVI
EVI
EVI2
GEMI
GNDVI
MNDWI
MSAVI
MSAVI2
NBRI
NDVI
NDWI
NDWI2
NRVI
RVI
SATVI
SAVI
SLAVI
SR
TVI
TTVI
WDVI
2844000
2842000
2840000
2838000
428000
432000
436000
Latitude (m)
Longitude (m)
4
3
2
1
0

A single spectral index from the raster brick of computed indices can be plotted by selecting that index name using $ sign and plotting can be done as follow.

```
#Plot single index from SI
>ggR(SI$TVI, geom_raster =TRUE) +#plot with ggplot2
      ggtitle('Transformed Vegetation Index') +#title of the plot
      labs(x ='Longitude (m)', y ='Latitude (m)') +#axis labels
      theme(plot.title =element_text(hjust =0.5, #Align title in centre
size =30), #size of title
      axis.title =element_text(size =20), #size of axis labels
      legend.key.size =unit(1, "cm"), #size of legend
      legend.title =element_text(size =20), #size of legend title
      legend.text =element_text(size =15))+#size of legend text
#Provide customized gradient colours
      scale_fill_gradientn(name ='TVI values',
colours =c("skyblue", "wheat", "darkgreen"))
```

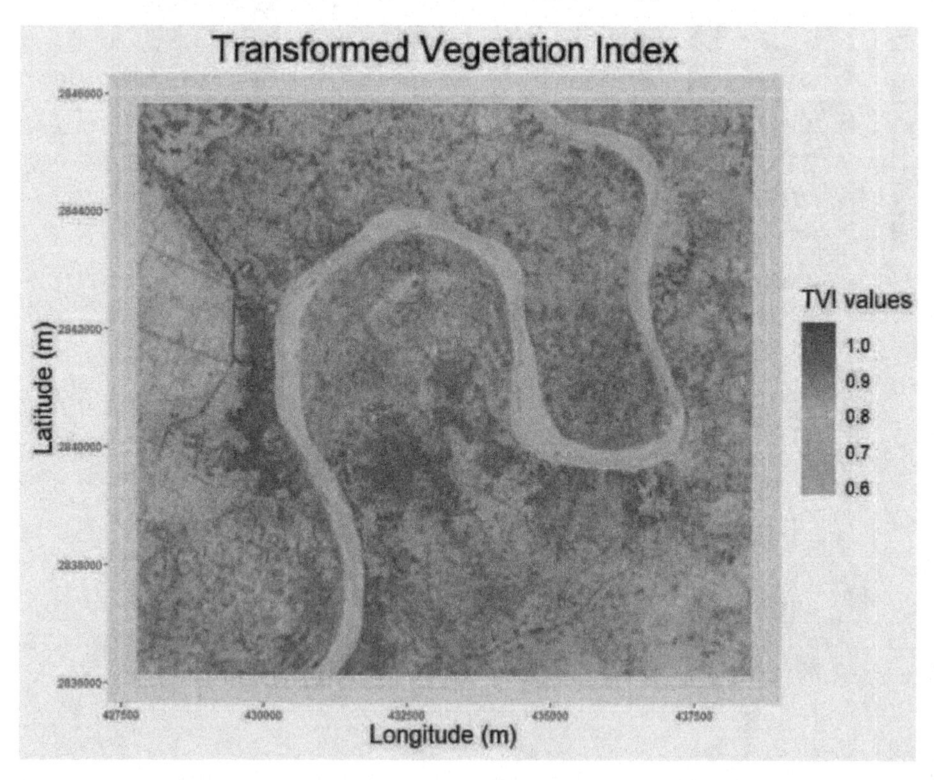

If the spectral index you want to compute is not available in the `spectralIndices` command or you want to formulate your own spectral index, then you have to define a function to compute it. This defined function can be applied to the raster images by using `overlay` command. In the following R – code example, we are defining a function for computation of Normalized Difference Vegetation Index (NDVI) whose mathematical formula is given below.

$$\text{NDVI} = \frac{NIR - Red}{NIR + Red}$$

In R, function for its computation is defined by using `function` command and specifying its formula in the function. This function is applied to the bands defining NIR (5th band) and Red (4th band) bands of the input raster image by using `overlay` command.

```
#Define function to compute NDVI
> vi =function(x, y) {
       (x-y)/(x+y)
       }


#Apply function on raster layers to compute NDVI
>ndvi =overlay(img[[5]], img[[4]], fun = vi)
>ndvi#Display properties of computed NDVI

## class      : RasterLayer
## dimensions : 323, 357, 115311  (nrow, ncol, ncell)
## resolution : 30, 30  (x, y)
## extent     : 427785, 438495, 2836125, 2845815  (xmin, xmax, ymin,
ymax)
## crs          : +proj=utm +zone=44 +datum=WGS84 +units=m +no_defs
+ellps=WGS84 +towgs84=0,0,0
## source     : memory
## names      : layer
## values     : -0.1516345, 0.6013459  (min, max)
```

The raster layer containing pixel-wise computed NDVI can be plotted by using following R – code.

```
#Plot raster layer of computed NDVI
>ggR(ndvi, geom_raster =TRUE) +#Plot with ggplot2
       ggtitle('NDVI plot') +#Title of plot
       labs(x='Longitude (m)', y='Latitude (m)')+#Axis labels
       theme(plot.title =element_text(hjust =0.5, #Align title
in centre
size =30), #Size of title
```

```
      axis.title =element_text(size =20), #Size of axis labels
      legend.key.size =unit(1, "cm"), #Size of legend
      legend.title =element_text(size =20), #Size of legend
title
      legend.text =element_text(size =15))+#Size of legend text
#Provide customized gradient colours to the NDVI values
      scale_fill_gradientn(name ='NDVI values',
colours =c("white", "darkgreen"))
```

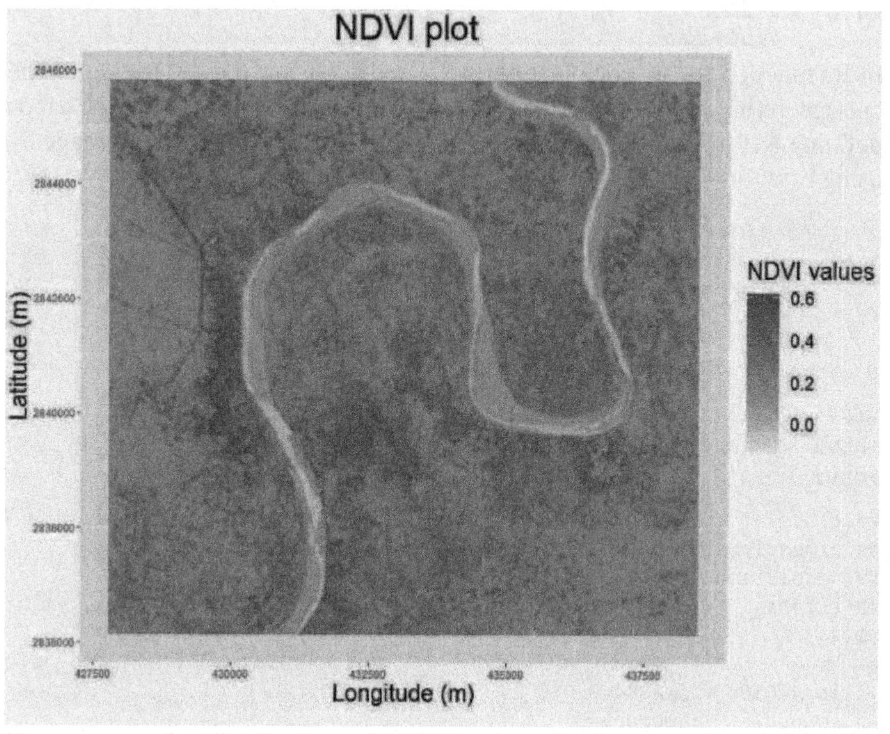

You can see the distribution of NDVI values in the raster by plotting its histogram by using following R – code.

```
#view histogram of NDVI data
>hist(ndvi, main ="Distribution of NDVI", #title of histogram
      xlab ="NDVI", ylab ="Frequency", #axis labels
      col ="sky blue", #colour of bins
      xlim =c(-1, 1), breaks =20, #limits and intervals of x-
axis
      xaxt ='n') #disable default axis
#provide customized axis to the histogram
>axis(side =1, at =seq(-1, 1, 0.1), labels =seq(-1, 1, 0.1))
```

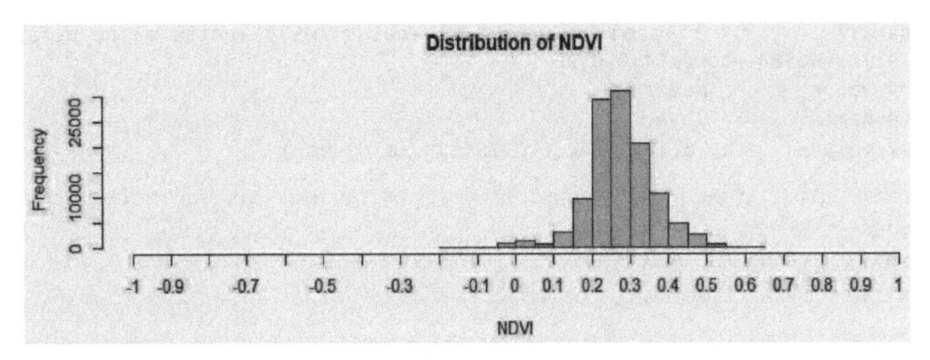

The histogram shows that most of the pixels in the raster image has NDVI values ranging from 0.2 to 0.4.

Function can also be defined directly for raster image in R without using `overlay` command. In the following R – code example, we are defining a function for the computation of Enhanced Vegetation Index (EVI) having following formula for the raster image.

$$\text{EVI} = 2.5\left(\frac{NIR - Red}{NIR + 6 * Red - 7.5 * Blue + 1}\right)$$

In the function, you have to give '`img`' variable for raster image so that function is directly applied over it as shown below.

```
#Define function for computation of Enhanced Vegetation Index
(EVI)
>evi =function(img, i, j, k) {

            bi =img[[i]]
            bj =img[[j]]
            bk =img[[k]]
            evi =2.5*((bi-bj)/(bi+(6*bj)-(7.5*bk)+1))
            return(evi)
        }


#Apply defined function by giving image name & band numbers
corresponding to respective wavelengths

> EVI =evi(img =img, 5,4,2)

> EVI #Display the properties of computed EVI raster layer

## class      : RasterLayer
## dimensions : 323, 357, 115311  (nrow, ncol, ncell)
## resolution : 30, 30  (x, y)
## extent     : 427785, 438495, 2836125, 2845815  (xmin, xmax, ymin,
ymax)
```

```
## crs         : +proj=utm +zone=44 +datum=WGS84 +units=m +no_defs
+ellps=WGS84 +towgs84=0,0,0
## source      : memory
## names       : layer
## values      : -0.1021454, 0.6779691  (min, max)
```

```
#Plot histogram to display the distribution of EVI values in
raster image
>histogram(EVI, col ='skyblue', main ='Histogram of EVI',
xlab ='EVI values', ylab ='Frequency')
```

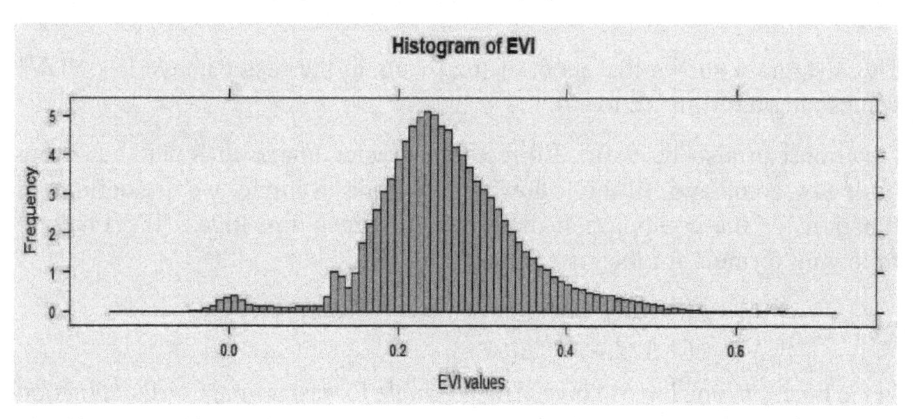

The histogram shows that most of the pixels have EVI values ranging from 0.1 to 0.4. The spatial distribution of EVI can be displayed by plotting its raster and applying gradient colours to its values by using following R – code.

```
#Plot EVI raster layer
>ggR(EVI, geom_raster =TRUE) +#plot with ggplot2
      ggtitle('EVI plot') +#title of the plot
      labs(x='Longitude (m)', y='Latitude (m)')+#axis labels
      theme(plot.title =element_text(hjust =0.5, #Align title
in centre
      size =30), #size of title
      axis.title =element_text(size =20), #size of axis labels
      legend.key.size =unit(1, "cm"), #size of legend
      legend.title` =element_text(size =20), #size of legend
title
      legend.text =element_text(size =15))+#size of legend text
#Provide customized gradient colours
      scale_fill_gradientn(name ='EVI values', #name of scalebar
na.value ='white', #colour of no data values
colours =c("white", "darkgreen")) #gradient colour
```

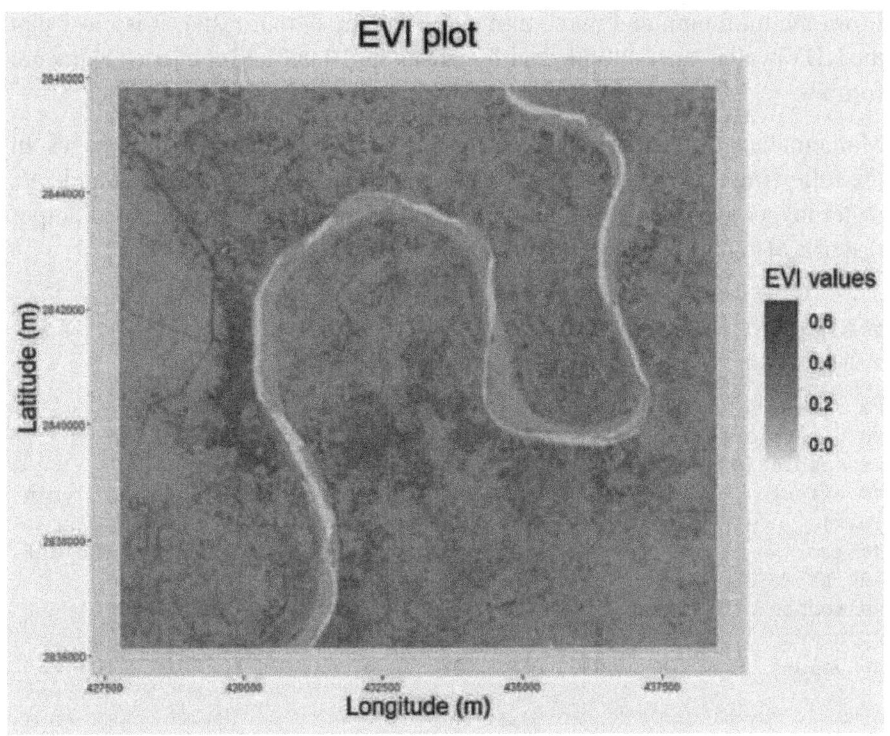

In this way, spectral indices can be computed through R software.

14.2. Performing algebra on raster images

Mathematical operations can be performed on raster layers by using `raster` package.

In the following R – code example, we are multiplying the NDVI values in the raster layer with scale factor 10,000 and converting it to integer format rather than float by using `as.integer` command.

```
# Multiplying NDVI values with 10000 scale factor
>ndvi_int =as.integer(ndvi*10000)
>ndvi_int#Display attributes of output

## class       : RasterLayer
## dimensions : 323, 357, 115311  (nrow, ncol, ncell)
## resolution : 30, 30  (x, y)
## extent     : 427785, 438495, 2836125, 2845815  (xmin, xmax, ymin,
ymax)
## crs         : +proj=utm +zone=44 +datum=WGS84 +units=m +no_defs
+ellps=WGS84 +towgs84=0,0,0
## source     : memory
## names      : layer
## values     : -1516, 6013  (min, max)
```

From the minimum and maximum values of the output raster, it is clear that the NDVI values are multiplied with scale factor 10,000 and converted to integer format.

Mathematical operations can be performed on many different raster images. In the following R – code example, we are performing algebra on NDVI and EVI raster layers to compute a combined hypothetical index. The computed output raster is also plotted to see the effect of applied algebra.

```
# Applying an algebra involving multiple raster images
> algebra1 =((ndvi+EVI)*1000)/2
> algebra1 #Display attributes of output
```

```
## class       : RasterLayer
## dimensions : 323, 357, 115311  (nrow, ncol, ncell)
## resolution : 30, 30  (x, y)
## extent      : 427785, 438495, 2836125, 2845815  (xmin, xmax, ymin,
ymax)
## crs          : +proj=utm +zone=44 +datum=WGS84 +units=m +no_defs
+ellps=WGS84 +towgs84=0,0,0
## source      : memory
## names       : layer
## values      : -126.8899, 635.3407  (min, max)
```

```
#Plot the output of applied algebra
>ggR(algebra1, geom_raster =TRUE) +#plot with ggplot2
       ggtitle('Plot of Algebra 1') +#title of plot
       labs(x ='Longitude (m)', y ='Latitude (m)') +#axis labels
       theme(plot.title =element_text(hjust =0.5, #Align title
in centre
size =30), #size of title
       axis.title =element_text(size =20), #size of axis labels
       legend.key.size =unit(1, "cm"), #size of legend
       legend.title =element_text(size =20), #size of legend
title
       legend.text =element_text(size =15))+#size of legend text
#Provide customized gradient colours
       scale_fill_gradientn(colours =c("white", "brown",
"green"))
```

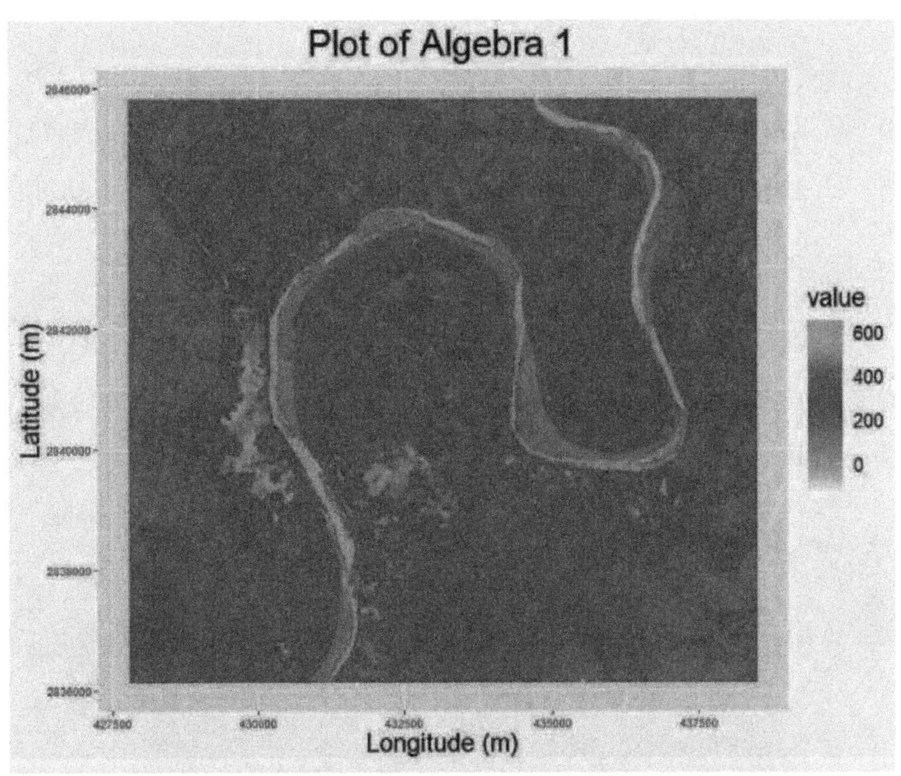

Similarly, a mathematical operation involving different bands of a multiband raster image can be performed as shown in the following R – code example.

```
#Applying algebra to different bands of a multiband raster
> algebra2 =(((img[[7]]-img[[5]])/(img[[7]]+img[[5]]))*100)
> algebra2 #Display attributes of output
```

```
## class      : RasterLayer
## dimensions : 323, 357, 115311  (nrow, ncol, ncell)
## resolution : 30, 30  (x, y)
## extent     : 427785, 438495, 2836125, 2845815  (xmin, xmax, ymin, ymax)
## crs         : +proj=utm +zone=44 +datum=WGS84 +units=m +no_defs +ellps=WGS84 +towgs84=0,0,0
## source     : memory
## names      : layer
## values     : -63.62561, 2.929896  (min, max)
```

```
#Plot the output of applied algebra
>ggR(algebra2, geom_raster =TRUE) +#plot with ggplot2
      ggtitle('Plot of Algebra 2') +#title of plot
      labs(x ='Longitude (m)', y ='Latitude (m)') +#axis labels
      theme(plot.title =element_text(hjust =0.5, #Align title in centre
```

```
      size =30), #size of title
      axis.title =element_text(size =20), #size of axis labels
      legend.key.size =unit(1, "cm"), #size of legend
      legend.title =element_text(size =20), #size of legend
title
      legend.text =element_text(size =15)) #size of legend text
```

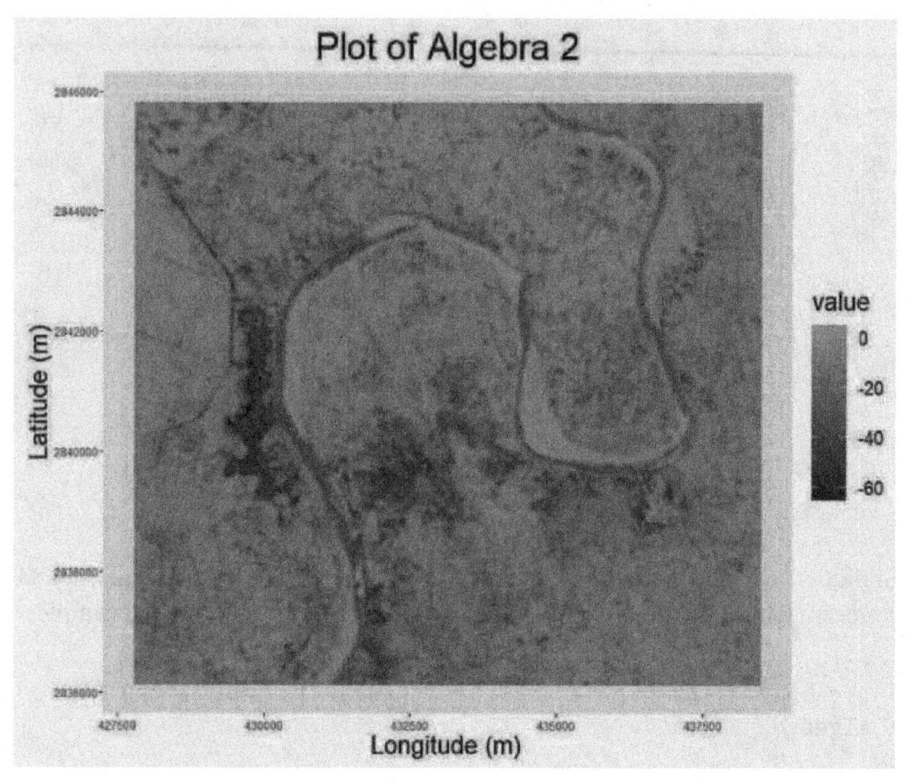

If you want to perform some mathematical operation involving all pixels of a raster layer, then you have to define a function. In the following R – code example, we are applying an algebra for rescaling the NDVI values in a raster layer in the scale of 0 to 100.

```
#Applying algebra involving all pixels of a raster layer
#Define function for rescaling pixel values between 0 to 100
> rescale =function(x) {
            y =((x -min(x))/(max(x)-min(x))*100)
            return(y)
            }


#Applying defined function to raster
> algebra3 =calc(ndvi, fun = rescale)
> algebra3 #Display attributes of output
```

```
## class      : RasterLayer
## dimensions : 323, 357, 115311  (nrow, ncol, ncell)
## resolution : 30, 30  (x, y)
## extent     : 427785, 438495, 2836125, 2845815  (xmin, xmax, ymin,
ymax)
## crs        : +proj=utm +zone=44 +datum=WGS84 +units=m +no_defs
+ellps=WGS84 +towgs84=0,0,0
## source     : memory
## names      : layer
## values     : 0, 100  (min, max)
```

```
#Plot the output of applied algebra
>ggR(algorithm3, geom_raster =TRUE) +#plot with ggplot2
      ggtitle('Plot of Algebra 3') +#title of plot
      labs(x ='Longitude (m)', y ='Latitude (m)') +#axis labels
      theme(plot.title =element_text(hjust =0.5, #Align title in centre
      size =30), #size of title
      axis.title =element_text(size =20), #size of axis labels
      legend.key.size =unit(1, "cm"), #size of legend
      legend.title =element_text(size =20), #size of legend title
      legend.text =element_text(size =15)) +#size of legend text
      scale_fill_gradientn(colours =c("black", "wheat", "darkgreen"))
```

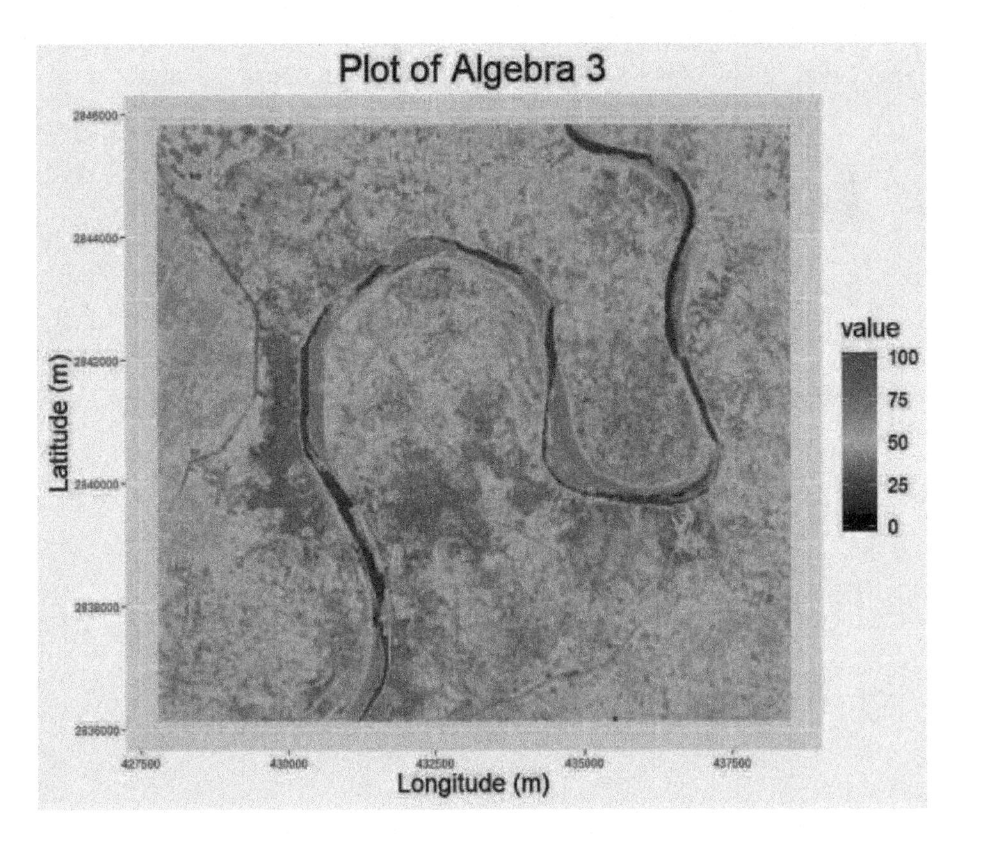

So, you can see in the raster plot that NDVI values have been rescaled to the range of 0 to 100. The features having least NDVI like water bodies have been assigned 0 value and the features with highest NDVI like healthy and dense vegetation have been assigned 100 value.

In this way, you can apply any algebraic function on the raster layers to get the desired output raster.

14.3. Thresholding

Thresholding is the process of retaining the pixels having values within the threshold limits specified by the user in the input raster image. For example, the NDVI values range from -1 to +1. The pixels with NDVI values more than 0.3 are most probably vegetation. If we want only vegetation pixels in the output raster, then thresholding can be done in R by using `reclassify` command of `raster` package. In this function, we are assigning no data *i.e.* NA to the pixel values ranging from -" to 0.3 as shown below.

```
# Thresholding to get pixels of NDVI values > 0.3 i.e. vegetation
pixels
> veg =reclassify(ndvi, cbind(-Inf, 0.3, NA))
# Plot the thresholding output raster containing vegetation
>ggR(veg, geom_raster =TRUE) +#plot with ggplot2
      ggtitle('Vegetation') +#title of plot
      labs(x ='Longitude (m)', y ='Latitude (m)') +#axis labels
      theme(plot.title =element_text(hjust =0.5, #Align title
in centre
      size =30), #size of title
      axis.title =element_text(size =20), #size of axis labels
      legend.key.size =unit(1, "cm"), #size of legend
      legend.title =element_text(size =20), #size of legend
title
      legend.text =element_text(size =15))+#size of legend text
#Provide customized gradient colours
      scale_fill_gradientn(name ='NDVI values',
      colours =c("lightgreen", "darkgreen"))
```

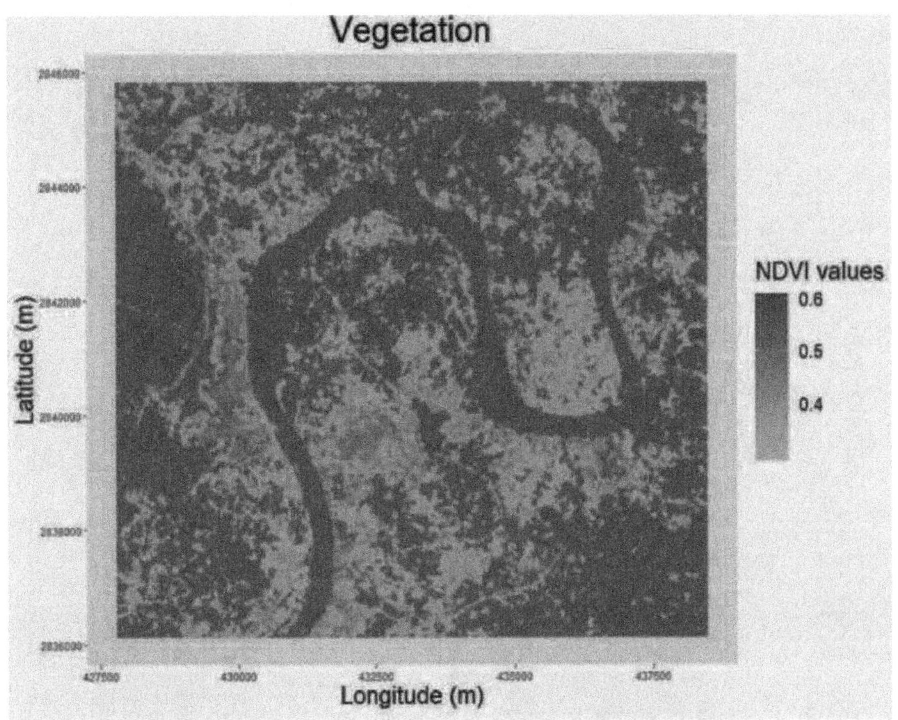

In the output raster plot, you can view the spatial distribution of vegetation. The vegetation pixels having more NDVI values or more intensity of green colour is depicting healthy or dense vegetation.

If you want to retain pixels with NDVI values ranging from 0.1 to 0.3 which most probably belongs to soil feature, then thresholding operation can be performed. In the following R – code, we have assigned value '1' to all the pixels in the range of 0.1 to 0.3, and NA to the pixels above and below that NDVI range.

```
#Thresholding to get NDVI values between 0.1 - 0.3 i.e. soil
pixels
> soil =reclassify(ndvi, c(-Inf, 0.1, NA, 0.1, 0.3, 1, 0.2, Inf,
NA))
# Plot the thresholding output raster containing soil
>ggR(soil, geom_raster =TRUE) +#plot with ggplot2
      ggtitle('Soil') +#title of plot
      theme(plot.title =element_text(hjust =0.5, #Align title
in centre
      size =30), #size of title
      legend.position ="none", #remove legend bar
      axis.title =element_text(size =20)) +#size of axis labels
      scale_fill_gradientn(colours ="wheat3", na.value ="white") +
      labs(x ='Longitude (m)', y ='Latitude (m)') #axis labels
```

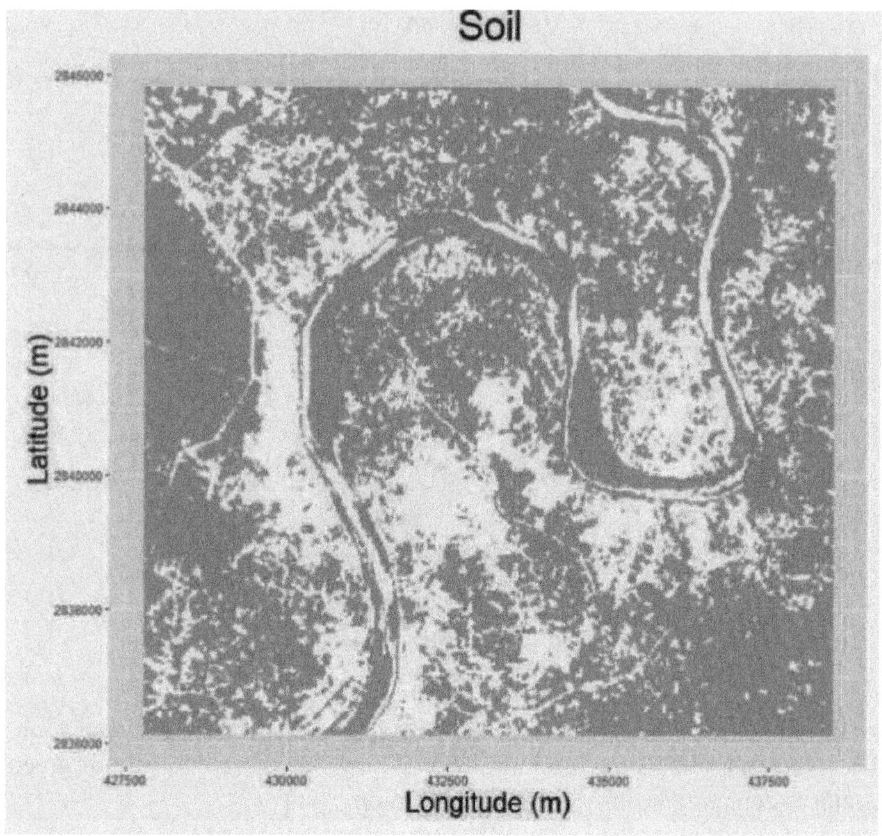

14.4. Density Slicing

Density slicing is the process of slicing or dividing the pixel values of input raster data into different ranges and a specified value or colour is assigned to each range in the output raster. Density slicing can be performed in R by using `reclassify` function of `raster` package. In `reclassify` function, input raster image and matrix for reclassification are supplied as arguments. The matrix for reclassification has three columns. First column and second column has lowest and highest limit of a particular range whereas third column contains value to be assigned to that range. In the following R – code example, we are density slicing the NDVI values into four ranges or classes.

```
#Prepare matrix for reclassification of NDVI values
>ds_matrix =matrix(c(-0.25,0.1,1,0.1,0.3,2,0.3,0.5,3,0.5,0.75,4),
      ncol =3 , byrow =TRUE)
>ds_matrix#Display matrix for reclassification

##       [,1] [,2] [,3]
## [1,] -0.25 0.10    1
## [2,]  0.10 0.30    2
```

```
## [3,]  0.30 0.50    3
## [4,]  0.50 0.75    4
```

```
> ndvi_2 =round(ndvi, digits =2) #round off NDVI values to two digits
#Density slicing of NDVI raster
>ds_ndvi =reclassify(ndvi_2, ds_matrix)
>ds_ndvi#Display attributes of density sliced output layer
```

```
## class      : RasterLayer
## dimensions : 323, 357, 115311  (nrow, ncol, ncell)
## resolution : 30, 30  (x, y)
## extent     : 427785, 438495, 2836125, 2845815  (xmin, xmax, ymin, ymax)
## crs        : +proj=utm +zone=44 +datum=WGS84 +units=m +no_defs +ellps=WGS84 +towgs84=0,0,0
## source     : memory
## names      : layer
## values     : 1, 4  (min, max)
```

```
#Plot density sliced output raster layer
>ggR(ds_ndvi, geom_raster =TRUE) +#plot with ggplot2
      ggtitle('Density slicing of NDVI') +#title of plot
      labs(x ='Longitude (m)', y ='Latitude (m)') +#axis labels
      theme(plot.title =element_text(hjust =0.5, #Align title in centre
      size =30), #size of title
      axis.title =element_text(size =20), #size of axis labels
      legend.key.size =unit(1, "cm"), #size of legend
      legend.title =element_text(size =20), #size of legend title
      legend.text =element_text(size =15))+#size of legend text
#Provide customized gradient colours
      scale_fill_gradientn(colours =c("turquoise2","burlywood","green3", "darkgreen"))
```

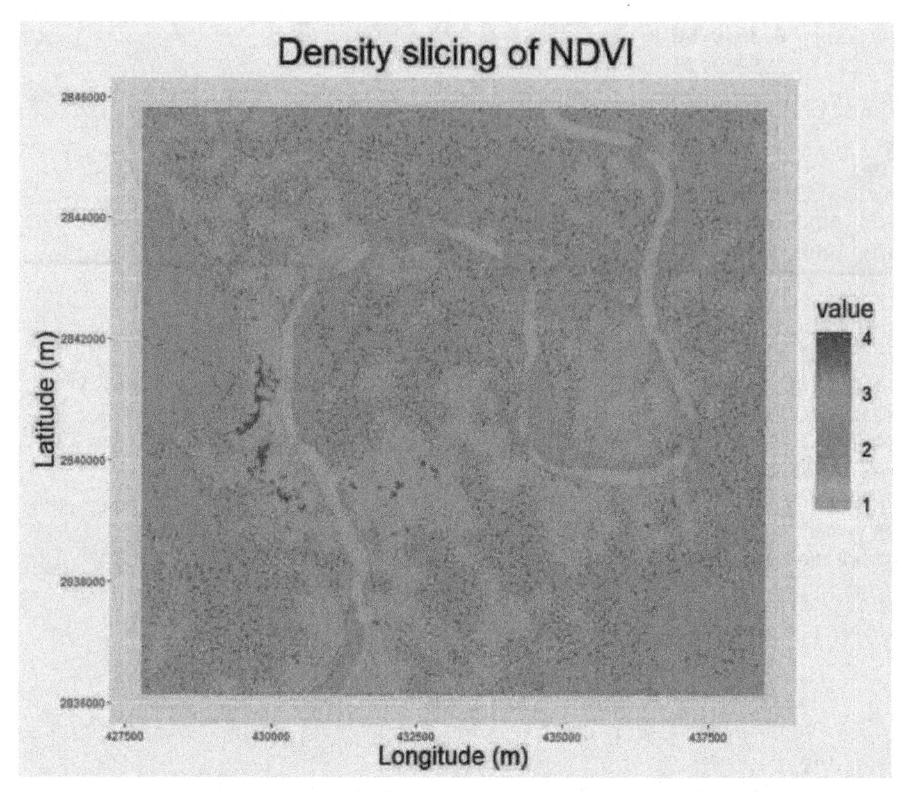

So, the values in the third column of the matrix are assigned to the range specified by the first and second column in the output raster layer. The raster plot shows that the given R – code has successfully performed the density slicing operation.

14.5. Tasseled cap transformation

Tasseled cap transformation is the orthogonal transformation of raster data to a new coordinate system to represent soil line and vegetation. It is performed by taking linear combinations of the original bands. This transformation got its name of "tasseled cap" due to cap like shape of graph between Red and Near Infra-red bands.

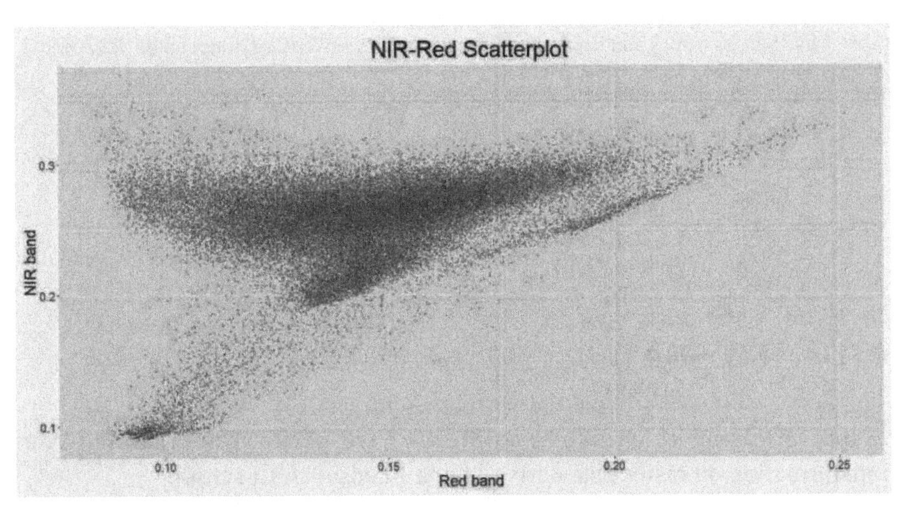

The triangular region in the graph shows the vegetation in various stages of growth. Therefore, this transformation is very useful for vegetation monitoring and mapping. The tasseled cap transformation reduces the amount of data from several bands of a raster image to three primary components known as brightness, greenness and wetness. The primary axis of the newly transformed coordinates is called as *brightness* which is the weighted sum of reflectance of all spectral bands and accounts for maximum variability in the raster data. Brightness represents variations in soil background reflectance. The second component is orthogonal to the first component which is known as *greenness*. Greenness indicates the variations in the vigour of vegetation. The last component which is orthogonal to the first two components is *wetness*. Wetness is associated with water, soil moisture and moist features. Thus, tasselled cap transformation is very beneficial in detecting and comparing variations in vegetation, soil and man-made features at different time periods.

In R software, tasselled cap transformation on raster band can be performed by using `tasseledCap` function of `RStoolbox` package. You have to assign the name of sensor of the input data in 'sat' argument. This function can be applied to few sensors only whose names and band numbers required for tasselled cap transformation are given in the table 14.4. The input raster data should be in reflectance units.

Table 14.4: Sensor names and band numbers on which '*tasseledCap*' function of '*RStoolbox*' package is applicable

S.No.	Sensor name	Band numbers
1.	Landsat4 TM	1,2,3,4,5,7
2.	Landsat5 TM	1,2,3,4,5,7
3.	Landsat7 ETM	1,2,3,4,5,7
4.	Landsat8 OLI	2,3,4,5,6,7
5.	MODIS	1,2,3,4,5,6,7
6.	QuickBird	2,3,4,5
7.	Spot 5	2,3,4,5
8.	RapidEye	1,2,3,4,5

In the following R – code example, we are performing Tasseled cap transformation on raster data derived from Landsat8 OLI sensor.

```
#Tasseled Cap transformation
>img_tc =tasseledCap(img[[2:7]], sat ="Landsat8OLI")
>img_tc#display attributes of output raster

## class      : RasterBrick
## dimensions : 323, 357, 115311, 3  (nrow, ncol, ncell, nlayers)
## resolution : 30, 30  (x, y)
## extent     : 427785, 438495, 2836125, 2845815  (xmin, xmax, ymin, ymax)
## crs         : +proj=utm +zone=44 +datum=WGS84 +units=m +no_defs +ellps=WGS84 +towgs84=0,0,0
## source     : memory
## names      :  brightness,   greenness,     wetness
## min values :  0.19932130, -0.07955372, -0.18452387
## max values :  0.68631011,  0.14740873,  0.08797207
```

So, we can see from the attributes of the output that this function has computed all the three components *i.e.* Brightness, Greenness and Wetness in the form of separate raster layers in a raster brick. Now we can plot these raster images to see the variations of brightness, greenness and wetness values corresponding to soil, vegetation, and moisture, respectively through the following R – codes.

```
# Plot Brightness raster layer
>ggR(img_tc$brightness, geom_raster =TRUE) +#plot with ggplot2
      ggtitle('Tasseled Cap Transformation - Brightness') +#title
      labs(x ='Longitude (m)', y ='Latitude (m)') +#axis labels
      theme(plot.title =element_text(hjust =0.5, #Align title in centre
      size =30), #size of title
      axis.title =element_text(size =20), #size of axis labels
      legend.key.size =unit(1, "cm"), #size of legend
      legend.title =element_text(size =20), #size of legend title
      legend.text =element_text(size =15))+#size of legend text
#Provide customized gradient colours
      scale_fill_gradientn(colours =c("white", "wheat", "brown"))
```

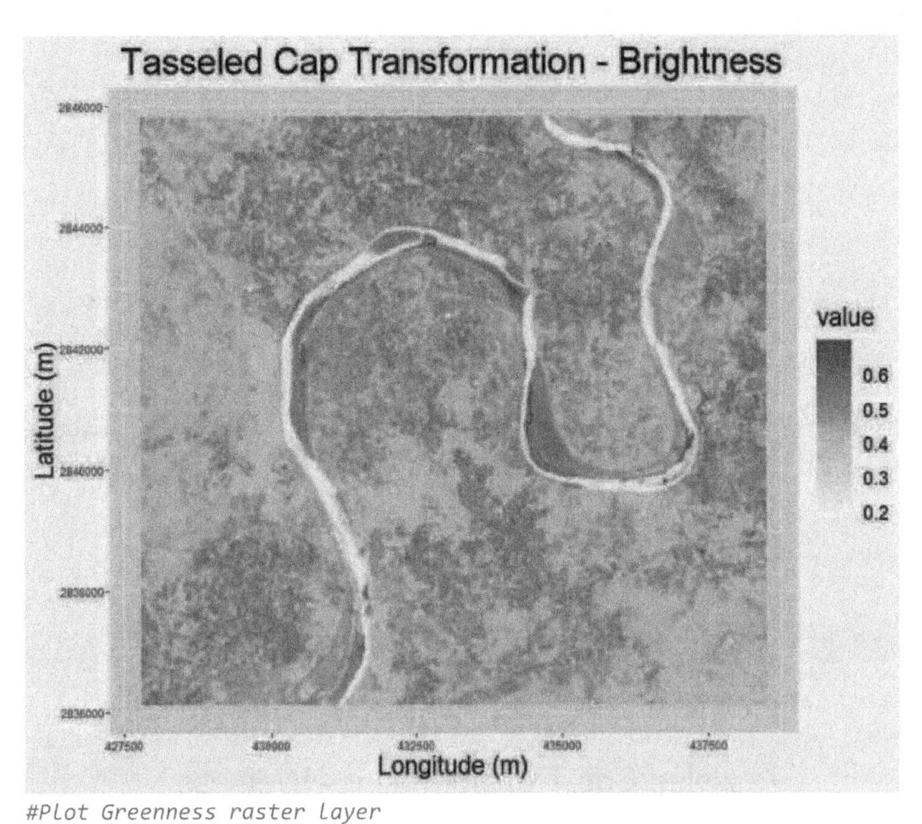


```
#Plot Greenness raster layer
>ggR(img_tc$greenness, geom_raster =TRUE) +#plot with ggplot2
       ggtitle('Tasseled Cap Transformation - Greenness') +#title
       labs(x ='Longitude (m)', y ='Latitude (m)') +#axis labels
       theme(plot.title =element_text(hjust =0.5, #Align title in centre
       size =30), #size of title
       axis.title =element_text(size =20), #size of axis labels
       legend.key.size =unit(1, "cm"), #size of legend
       legend.title =element_text(size =20), #size of legend title
       legend.text =element_text(size =15))+#size of legend text
#Provide customized gradient colours
       scale_fill_gradientn(colours =c("white", "darkgreen"))
#Plot Wetness raster layer
>ggR(img_tc$wetness, geom_raster =TRUE) +#plot with ggplot2
       ggtitle('Tasseled Cap Transformation - Wetness') +#title
       labs(x ='Longitude (m)', y ='Latitude (m)') +#axis labels
       theme(plot.title =element_text(hjust =0.5, #Align title in centre
       size =30), #size of title
       axis.title =element_text(size =20), #size of axis labels
       legend.key.size =unit(1, "cm"), #size of legend
       legend.title =element_text(size =20), #size of legend title
       legend.text =element_text(size =15))+#size of legend text
#Provide customized gradient colours
       scale_fill_gradientn(colours =c("skyblue", "darkblue"))
```

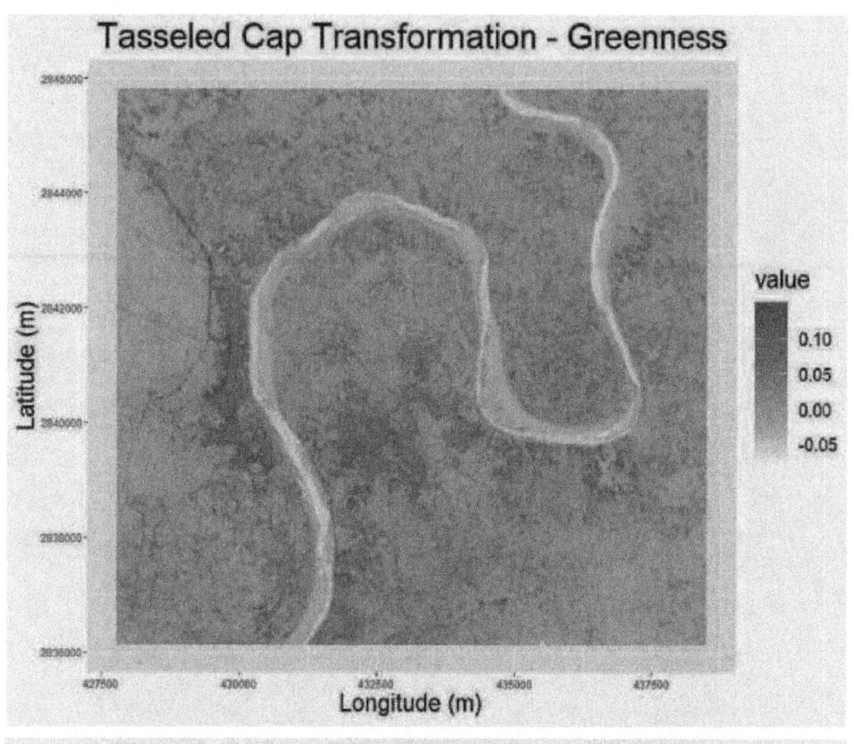
Tasseled Cap Transformation - Greenness
2846000
2844000
2842000
2840000
2838000
2836000
Latitude (m)
427500
430000
432500
435000
437500
Longitude (m)
value
0.10
0.05
0.00
-0.05

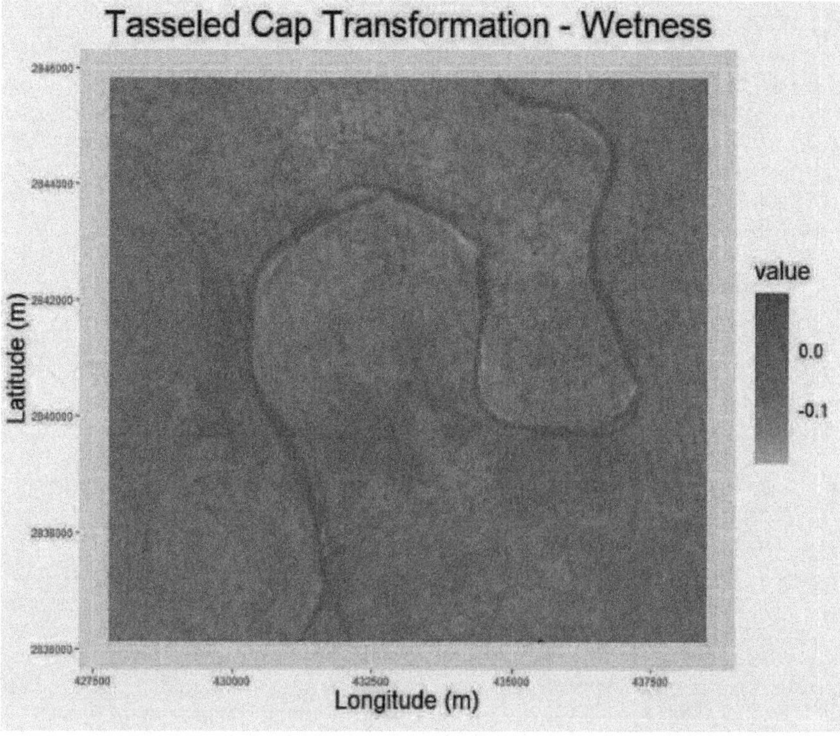
Tasseled Cap Transformation - Wetness
2846000
2844000
2842000
2840000
2838000
2836000
Latitude (m)
427500
430000
432500
435000
437500
Longitude (m)
value
0.0
-0.1

Similarly, you can perform Tasseled cap transformation in the other sensors listed in the table above and get the brightness, greenness and wetness values.

14.6. Principal Component Analysis

Principal Component Analysis (PCA) is the mathematical transformation of multiple variables into linear combinations of orthogonal components known as Principal Components with decreasing order of variance. This method transforms the brightness values in the input raster bands from its feature space to a new multivariate attribute space whose axes are rotated to represent the variance in each variable of the data. The first axis explains maximum variance. The second axis is orthogonal to the first axis and it explains remaining variance which was not explained by the first axis, and so on. These axes are known as Principal components. It is also clear from the following figure depicting feature space of two bands and axes of principal components with respect to the original axes.

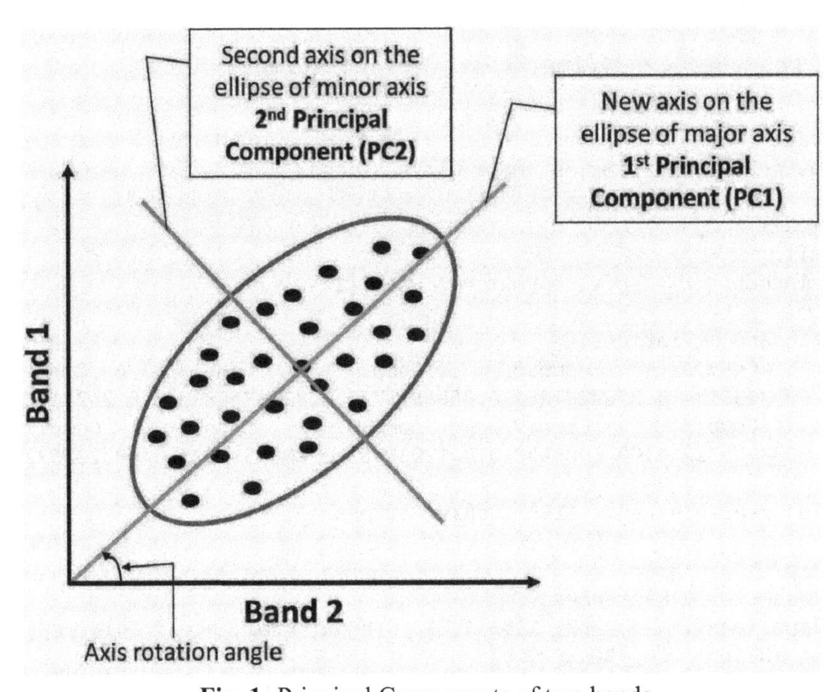

Fig. 1: Principal Components of two bands

So, from the figure 1, we can see that principal components are orthogonal and uncorrelated to each other. The direction of principal component is the eigen vector and its magnitude is the eigen value. The axis rotation angle is the angle used in transformation. These eigenvectors, eigenvalues, and computed covariance matrix (correlation matrix in case of normalized bands) of the multiband input raster are used to develop a linear formula defining the shift and rotation. This formula is applied to transform values of all pixels in the input raster relative to the new axis in order to create transformed output raster. Principal component analysis is mainly done to eliminate the redundancy in data as first few principal components explains most of the variance in the multiband input raster data.

In R, Principal Component Analysis (PCA) on multiband raster images can be performed by using `rasterPCA` function in `RStoolbox` package. If you assign the number of samples in `nSamples` argument, then PCA will be performed on random sample of pixels and then predicted for the full raster. If you don't assign that argument, then all the pixels will be considered which is more accurate. You can assign the number of principal components which you want to be computed in the `nComp` argument. The maximum number of principal components which we can compute is equal to the total number of bands in the input raster. This is clear in the given R – code example.

```
# Principal Component Analysis for multiband raster image
>img_pca =rasterPCA(img, nComp =nlayers(img))
>img_pca#Display attributes of computed PCA
```

```
## $call
## rasterPCA(img = img, nComp = nlayers(img))
##
## $model
## Call:
## princomp(cor = spca, covmat = covMat[[1]])
##
## Standard deviations:
##       Comp.1       Comp.2       Comp.3       Comp.4       Comp.5
## 1.6093138744 0.0760287001 0.0608855732 0.0246748847 0.0125221770
##       Comp.6       Comp.7       Comp.8       Comp.9      Comp.10
## 0.0038678312 0.0025990986 0.0013035783 0.0005530025 0.0002226722
##
##  10  variables and  115311 observations.
##
## $map
## class      : RasterBrick
## dimensions : 323, 357, 115311, 10  (nrow, ncol, ncell, nlayers)
## resolution : 30, 30  (x, y)
## extent     : 427785, 438495, 2836125, 2845815  (xmin, xmax, ymin,
ymax)
## crs        : +proj=utm +zone=44 +datum=WGS84 +units=m +no_defs
+ellps=WGS84 +towgs84=0,0,0
## source     : memory
## names      :           PC1,           PC2,           PC3,           PC4,
PC5,          PC6,          PC7,          PC8,          PC9,          PC10
## min values : -6.354192048, -0.596244024, -0.321730753, -0.133305550,
```

```
-0.047752881, -0.017172607, -0.017649629, -0.014416034, -0.002975797,
-0.001035999
## max values : 4.773473962, 0.683096881, 0.265656571, 0.102603938,
0.121832144, 0.017055868, 0.016002283, 0.008836472, 0.003603360,
0.001099577
##
##
## attr(,"class")
## [1] "rasterPCA" "RStoolbox"
```

So, we can see that this R – code has created a raster brick of 10 principal components. The standard deviation and proportion of variance explained by each principal component can be displayed using `summary` function on the PCA model.

```
#Display importance of each principal component
>summary(img_pca$model)
```

```
## Importance of components:
##                           Comp.1      Comp.2      Comp.3       Comp.4
## Standard deviation     1.6093139 0.076028700 0.060885573 0.0246748847
## Proportion of Variance 0.9960476 0.002223073 0.001425698 0.0002341579
## Cumulative Proportion  0.9960476 0.998270697 0.999696395 0.9999305525
##                               Comp.5        Comp.6        Comp.7
## Standard deviation     0.01252217705 0.00386783116 0.00259909855
## Proportion of Variance 0.00006030569 0.00000575352 0.00000259803
## Cumulative Proportion  0.99999085823 0.99999661175 0.99999920978
##                              Comp.8          Comp.9          Comp.10
## Standard deviation     0.001303578311 0.0005530025235 0.00022267220529
## Proportion of Variance 0.000000653541 0.0000001176123 0.00000001906912
## Cumulative Proportion  0.999999863319 0.9999999809309 1.00000000000000
```

So, it is clear from the output that first principal component explains 99.6% of variance in the raster data. The proportion of variance decreases in the succeeding principal components. You can display the proportion of variance explained by each principal component in graphical form by plotting screeplot using following code.

```
#Create screeplot of principal components
>screeplot(img_pca$model, main ="Screeplot of Principal
Components")
```

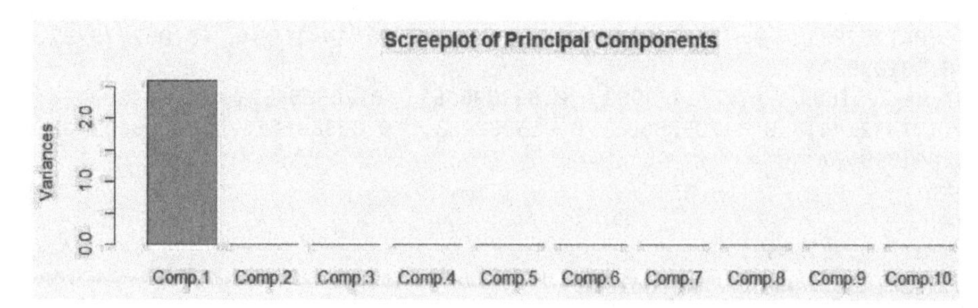

The screeplot clearly depicts that first principal component (PC1) is representing maximum variance. We can plot the raster of first principal component by using the following R – code.

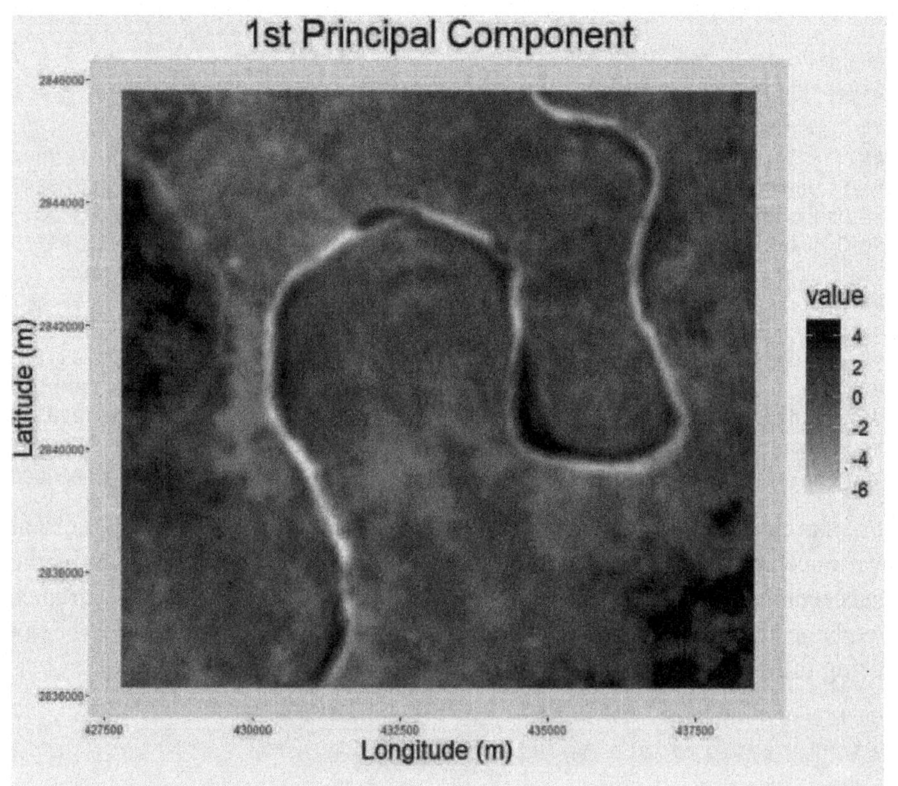

All the principal components (PCs) can be plotted at once by using following R – code.

```
#Plot all the principal components
>levelplot(img_pca$map, main ='Principal Components',
      xlab ='Longitude (m)', ylab ='Latitude (m)')
```

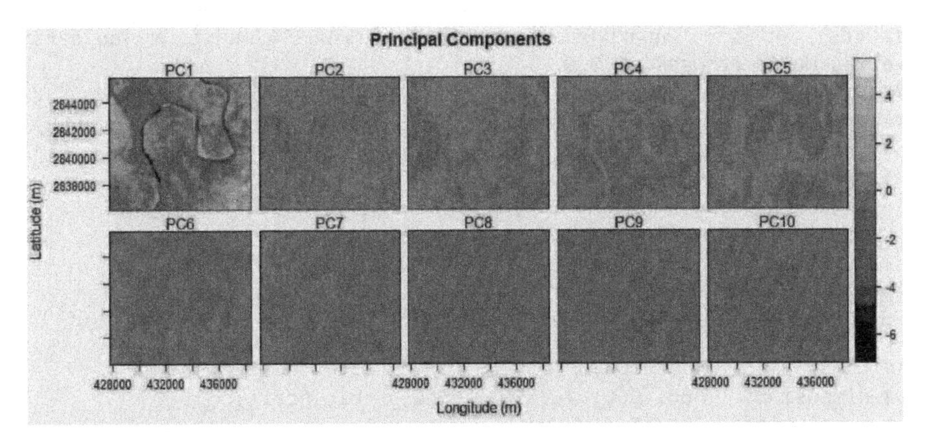

From the above plot, it is clear that PC1 is having maximum variance whereas 9th and 10th principal components (PC9 and PC10) are having negligible variance. So, the information of ten bands in the raster image can be effectively explained by first principal component.

14.7. Pan sharpening

Pan sharpening is the process of merging coarser resolution multispectral raster image with finer resolution panchromatic raster band to create a new multiband raster dataset having spectral properties of multispectral bands with spatial resolution of panchromatic band. Pan sharpening is done to increase spatial resolution of multispectral raster image in order to increase its visualisation. There are several methods available for pan sharpening like PCA, Brovey, Gram-Schmidt, Intensity – Hue – Saturation, etc.

In R software, pan sharpening can be performed by `panSharpen` function in `RStoolbox` package. The name of input image is assigned in `img =` argument and panchromatic image in `pan =` argument. There are only three methods available in R for pan sharpening *i.e.* Principal Component Analysis, Brovey and Intensity – Hue – Saturation method. We will perform pan sharpening with each available method in R in this section. First of all, lets import both multispectral raster of 30 m spatial resolution and panchromatic raster of 15 m resolution. Both the raster datasets should have pixel values in the same scale.

```
#Import 30 m multispectral raster image for pan sharpening
>ps_input =brick('pan_sharpening_input.tif')
>ps_input# Display attributes of imported raster
```

```
## class      : RasterBrick
## dimensions : 333, 333, 110889, 4  (nrow, ncol, ncell, nlayers)
## resolution : 30, 30  (x, y)
## extent     : 400005, 409995, 2800005, 2809995  (xmin, xmax, ymin,
ymax)
```

```
## crs        : +proj=utm +zone=44 +datum=WGS84 +units=m +no_defs
+ellps=WGS84 +towgs84=0,0,0
## source     : E:/............................................./pan_sharpening_input.tif
## names      : pan_sharpening_input.1, pan_sharpening_input.2,
pan_sharpening_input.3, pan_sharpening_input.4
## min values : 10463,       9852,        9074,        9300
## max values : 17455,       17995,       19205,       23548
```

```
#Import panchromatic band of 15 m spatial resolution
>pan_band =raster('Pan_band_15m.tif')
>pan_band#Display attributes of imported panchromatic band
```

```
## class      : RasterLayer
## dimensions : 666, 666, 443556  (nrow, ncol, ncell)
## resolution : 15, 15  (x, y)
## extent     : 399997.5, 409987.5, 2799998, 2809988  (xmin, xmax, ymin,
ymax)
## crs        : +proj=utm +zone=44 +datum=WGS84 +units=m +no_defs
+ellps=WGS84 +towgs84=0,0,0
## source     : E:/............................................./Pan_band_15m.tif
## names      : Pan_band_15m
## values     : 9376, 19793  (min, max)
```

The panchromatic raster layer can be plotted by using following R – code.

```
#Plot imported panchromatic band
>ggR(pan_band, stretch ='lin') +#plot with ggplot2
      ggtitle('Panchromatic band (15 m)') +#title of plot
      labs(x ='Longitude (m)', y ='Latitude (m)') +#axis labels
      theme(plot.title =element_text(hjust =0.5, #Align title
in centre
      size =30), #Adjust size of title
      axis.title =element_text(size =20)) #size of axis labels
```

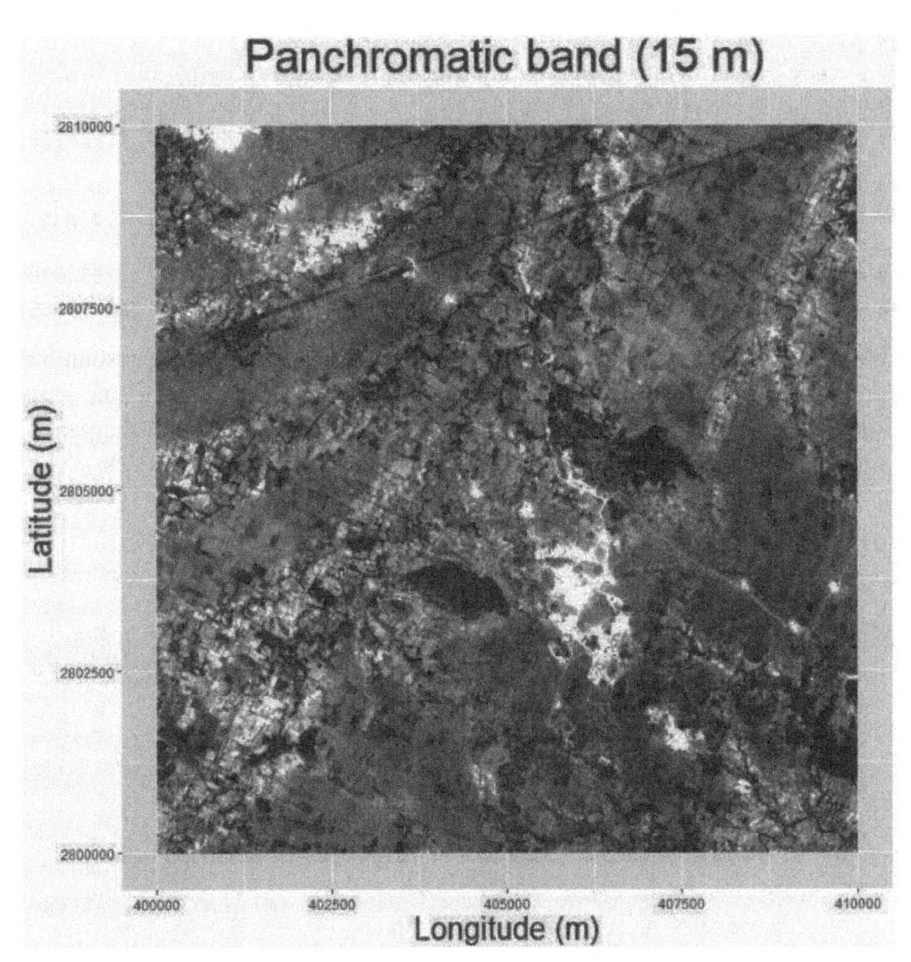

Now we will perform pan sharpening by all the three methods available in R.

14.7.1. Principal Component Analysis (PCA) method

The theory of Principal Component Analysis (PCA) is explained earlier in the section 14.6. In this method PCA is performed on multispectral raster image. The first principal component is then swapped for the panchromatic band and the PCA is rotated backwards. It is performed in R by using `method ='pca'` argument in `panSharpen` function. The advantage of this method is that it pan-sharpens all the bands in the multispectral raster image.

```
#Pan sharpening by PCA method
>ps_pca =panSharpen(img =ps_input, pan =pan_band, method ='pca')
>ps_pca#Display attributes of pan-sharpened raster

## class      : RasterBrick
## dimensions : 666, 666, 443556, 4  (nrow, ncol, ncell, nlayers)
```

```
## resolution : 15, 15  (x, y)
## extent     : 399997.5, 409987.5, 2799998, 2809988 (xmin, xmax, ymin,
ymax)
## crs          : +proj=utm +zone=44 +datum=WGS84 +units=m +no_defs
+ellps=WGS84 +towgs84=0,0,0
## source     : memory
## names      : pan_sharpening_input.1_pan, pan_sharpening_input.2_pan,
pan_sharpening_input.3_pan, pan_sharpening_input.4_pan
## min values :  8721.067,      7228.862,        4734.816,      9755.690
## max values : 17155.96,       17871.61,        18922.92,      22283.65
```

From the attributes of the pan-sharpened raster, it is clear that spatial resolution of multispectral raster is increased from 30 m to 15 m. Now we can plot the True Colour Composites (TCC) of both input multispectral raster and pan-sharpened raster to see the difference in visualisation by using following R – codes.

```
#Plot TCC of input multispectral raster of 30 m spatial resolution
>ggRGB(ps_input, r =3, g=2, b=1, stretch ='lin') +
      ggtitle('Input image for Pan sharpening (30 m resolution)')
+
      labs(x ='Longitude (m)', y ='Latitude (m)') +#axis labels
      theme(plot.title =element_text(hjust =0.5, #Align title
in centre
      size =30), #Adjust size of title
      axis.title =element_text(size =20)) #size of axis labels
#Plot TCC of pan sharpened raster
>ggRGB(ps_pca, r =3, g=2, b=1, stretch ='lin') +
      ggtitle('Pan sharpening by PCA') +#plot title
      labs(x ='Longitude (m)', y ='Latitude (m)') +#axis labels
      theme(plot.title =element_text(hjust =0.5, #Align title
in centre
      size =30), #Adjust size of title
      axis.title =element_text(size =20)) #size of axis labels
```

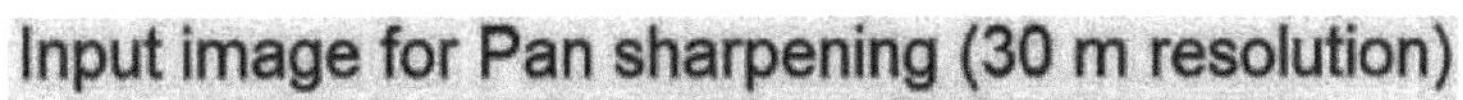
Input image for Pan sharpening (30 m resolution)

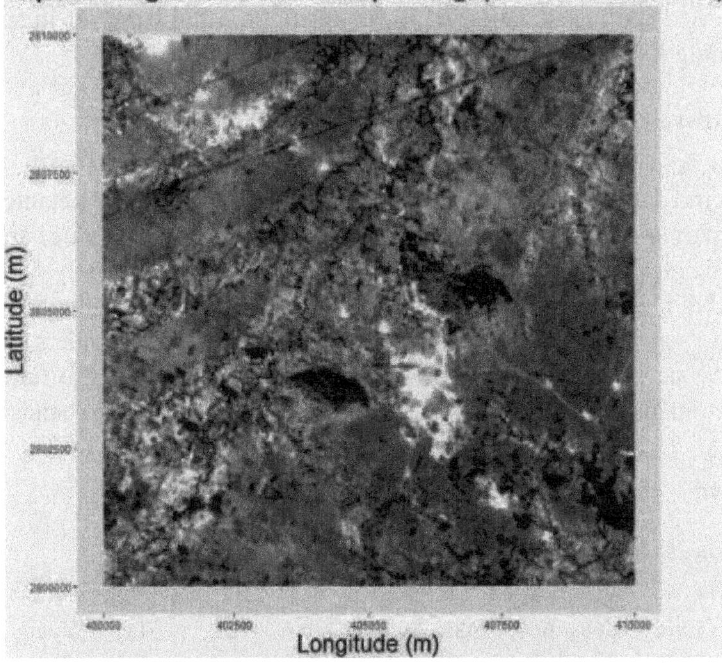
Latitude (m)
Longitude (m)

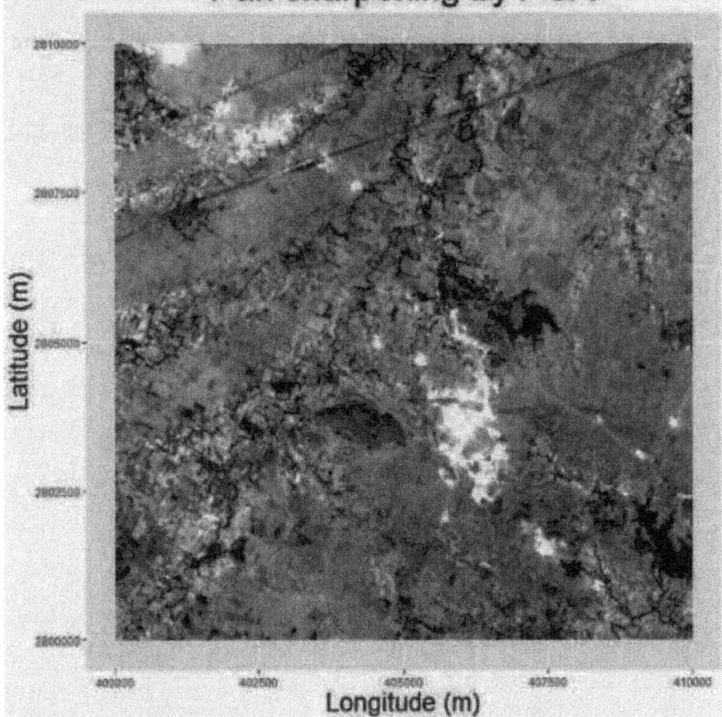
Pan sharpening by PCA
Latitude (m)
Longitude (m)

From the TCCs of both the raster images, we can conclude that pan sharpening has increased the visualization of lower resolution multispectral image by transforming it to the higher spatial resolution.

14.7.2. Brovey method

Brovey method performs pan-sharpening by multiplying each pixel of the input multispectral data with the ratio of corresponding panchromatic pixel intensity to the sum of multispectral intensities of pixels in red, green and blue bands. It can also be modified for near-infrared band, but it is available only for blue, green and red bands in R software. This can be applied by assigning `method ='brovey'` argument in **panSharpen** function. The band numbers of red, green and blue wavelengths are also assigned in the argument of this function. This function gives the pan-sharpened raster output for these three bands only.

```
#Pan sharpening by Brovey method
>ps_brovey =panSharpen(img =ps_input, pan =pan_band,
                       r =3, g =2, b =1, method ='brovey')
>ps_brovey#Display attributes of pan-sharpened raster

## class      : RasterBrick
## dimensions : 666, 666, 443556, 3  (nrow, ncol, ncell, nlayers)
## resolution : 15, 15  (x, y)
## extent : 399997.5, 409987.5, 2799998, 2809988  (xmin, xmax, ymin,
ymax)
## crs          : +proj=utm +zone=44 +datum=WGS84 +units=m +no_defs
+ellps=WGS84 +towgs84=0,0,0
## source     : memory
## names     : pan_sharpening_input.1_pan, pan_sharpening_input.2_pan,
pan_sharpening_input.3_pan
## min values : 3270.699,        3122.848,         2884.092
## max values : 6348.279,        6556.499,         6888.222

#Plot TCC of pan sharpened raster output
>ggRGB(ps_ihs, r =3, g=2, b=1, stretch ='lin') +
       ggtitle('Pan sharpening by Brovey method') +#plot title
       labs(x ='Longitude (m)', y ='Latitude (m)') +#axis labels
       theme(plot.title =element_text(hjust =0.5, #Align title
in centre
       size =30), #Adjust size of title
       axis.title =element_text(size =20)) #size of axis labels
```

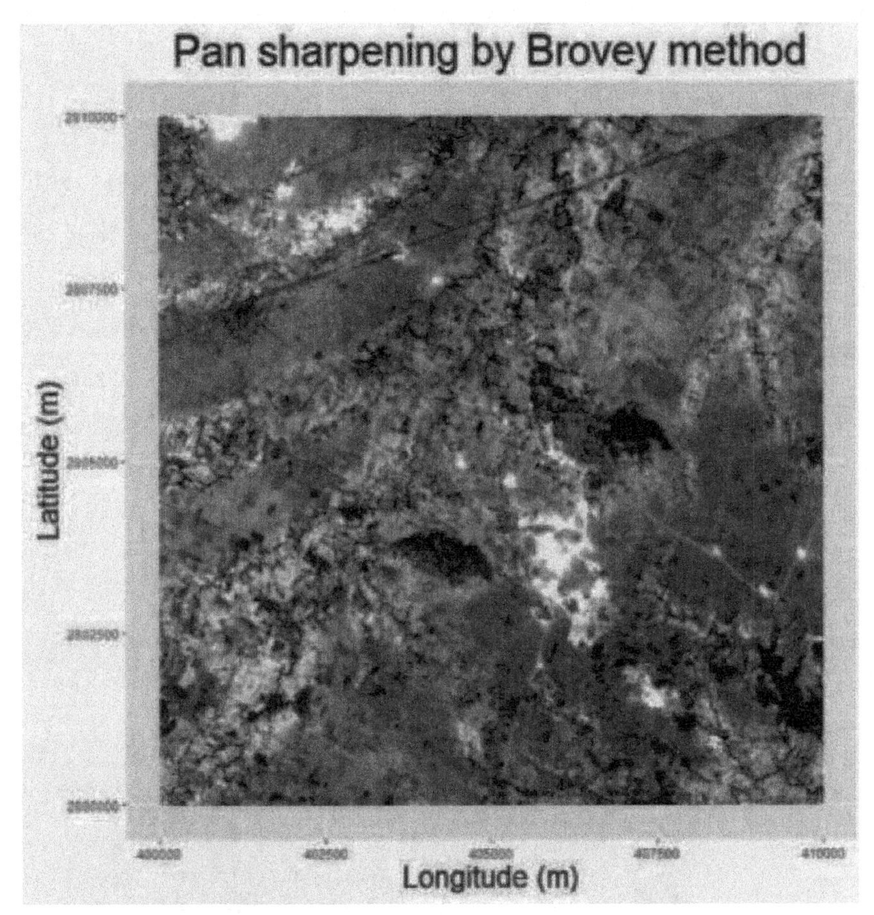

14.7.3. Intensity Hue Saturation (IHS) method

The Intensity Hue Saturation (HIS) method of pan-sharpening converts the multispectral image from RGB to intensity, hue, and saturation. Most of the spatial information of RGB composite is present in the Intensity component whereas Hue and Saturation components have spectral information. The histogram of panchromatic band is matched with the histogram of intensity component. Intensity component is then replaced with panchromatic image and inverse transform is applied to obtain RGB image with spatial details of panchromatic band of higher spatial resolution. This method is only applicable to red, green and blue bands of multispectral raster image in R software. This can be performed by assigning `method ='ihs'` argument and band numbers corresponding to red, green and blue wavelengths in **panSharpen** function.

```
#Pan sharpening by Intensity-Hue-Saturation method
>ps_ihs =panSharpen(img =ps_input, pan =pan_band, r =3, g =2, b
=1, method = 'ihs')
```

```
>ps_ihs#Display attributes of pan-sharpened raster

## class      : RasterBrick
## dimensions : 666, 666, 443556, 3  (nrow, ncol, ncell, nlayers)
## resolution : 15, 15  (x, y)
## extent    : 399997.5, 409987.5, 2799998, 2809988  (xmin, xmax, ymin,
ymax)
## crs         : +proj=utm +zone=44 +datum=WGS84 +units=m +no_defs
+ellps=WGS84 +towgs84=0,0,0
## source     : memory
## names     : pan_sharpening_input.1_pan, pan_sharpening_input.2_pan,
pan_sharpening_input.3_pan
## min values : 9020.333,        9669.667,         10238.333
## max values : 18517.00,        17467.33,         17262.33

#Plot TCC of pan sharpened raster
>ggRGB(ps_ihs, r =3, g=2, b=1, stretch ='lin') +
      ggtitle('Pan sharpening by IHS method') +#plot title
      labs(x ='Longitude (m)', y ='Latitude (m)') +#axis labels
      theme(plot.title =element_text(hjust =0.5, #Align title
in centre
      size =30), #Adjust size of title
      axis.title =element_text(size =20)) #size of axis labels
```

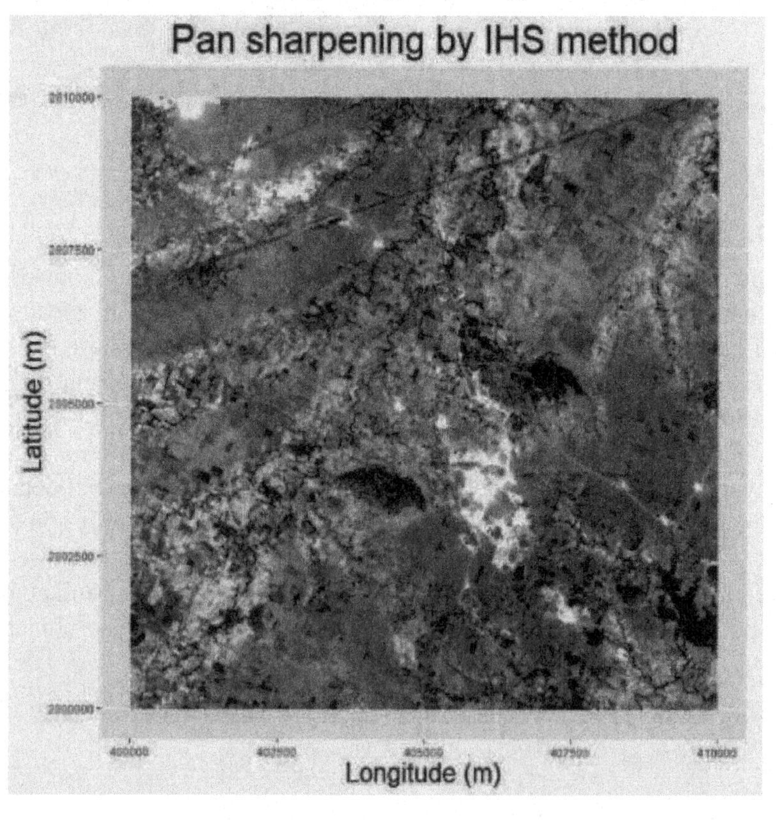

Chapter 15

Unsupervised Classification

Unsupervised classification is the technique of classification of raster data in which no prior information is provided by the user. Raster data is classified based on their spectral characteristics such that the pixels having similar spectral properties in the multispectral feature space are grouped together in one class. Unsupervised classification is mainly performed by clustering algorithms. The purpose of these algorithms is to minimize within cluster variability and maximize between clusters variability. The user must interpret and label the classes after performing unsupervised classification. This method of classification requires minimum data which makes it easy and quick, but user has toput efforts in identification of the classes. This method could be useful when user doesn't have any prior knowledge about the study area. This method removes subjectivity and enables us to get the unique classes which might be unrecognized by the user. K-means is the most commonly used method for performing unsupervised classification. In this chapter, we have discussed K-means, CLARA and unsupervised random forest method for classification of raster data in R.

For doing unsupervised classification of raster data in R, following packages are required.

```
#Import required packages
>library(raster)          #For importing raster data
>library(ggplot2)         #For visualization of raster data
>library(RStoolbox) #For K-means classification and visualization
>library(cluster)         #For CLARA classification
>library(randomForest)    #For random forest classification
```

Now let's import the input raster data on which we want to perform unsupervised classification and plot its False Colour Composite (FCC) image.

```
>setwd("E:/........./Chapter15/data")#Set working directory
#Import Input raster
```

```
>rast_img =brick('class_input.tif')
#Set the names of the bands
>names(rast_img) =c('Blue', 'Green', 'Red', 'NIR', 'SWIR', 'TIRS')
>rast_img                    #Display attributes of imported raster
```

```
## class      : RasterBrick
## dimensions : 678, 772, 523416, 6  (nrow, ncol, ncell, nlayers)
## resolution : 30, 30  (x, y)
## extent     : 432345, 455505, 2905185, 2925525  (xmin, xmax, ymin,
ymax)
## crs          : +proj=utm +zone=44 +datum=WGS84 +units=m +no_defs
+ellps=WGS84 +towgs84=0,0,0
## source     : E:/....................................../class_input.tif
## names      : Blue, Green, Red, NIR, SWIR, TIRS
```

```
#Plot FCC of input raster
>ggRGB(rast_img, r =4, g=3, b=2, stretch ='lin') +#plot with
ggplot2
      ggtitle('Input Image') +#title of plot
      labs(x ='Longitude (m)', y ='Latitude (m)') +#axis labels
      theme(plot.title =element_text(hjust =0.5, #Align title
in centre
      size =30), #size of title
      axis.title =element_text(size =20)) #size of axis labels
```

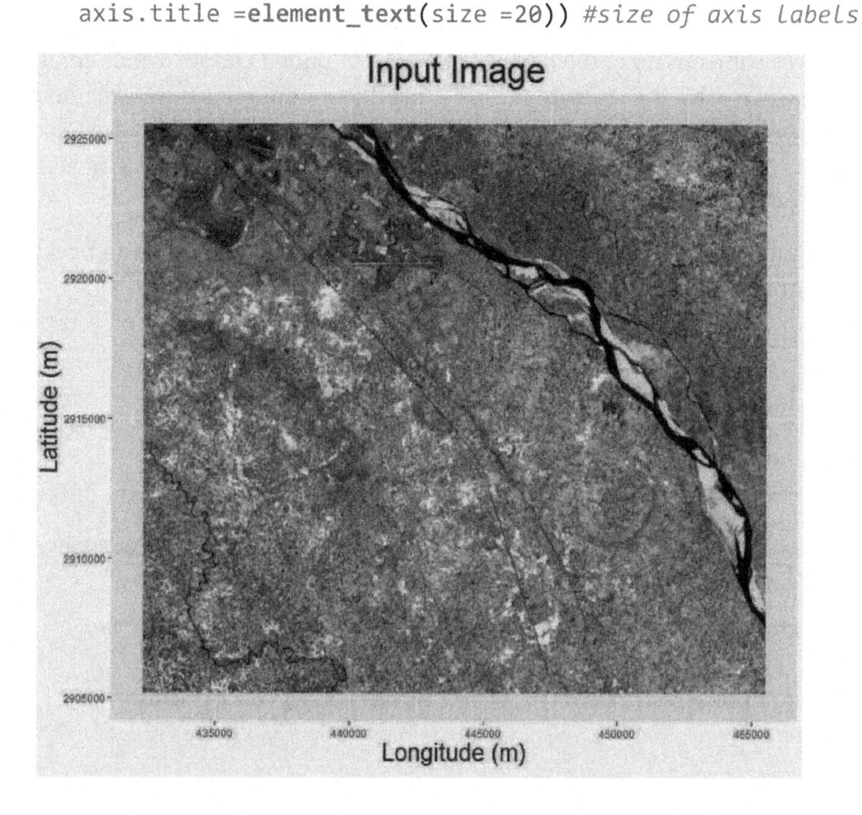

Now we will classify this input raster image in R by using following three methods of unsupervised classification.

1) K – means
2) CLARA
3) Random Forest (unsupervised)

The above stated methods and their implementation in R are discussed further in this chapter. The input raster data should be converted into matrix before applying these algorithms. This can be done by using **`getValues`** function of the `raster` package. No data values should also be omitted by using **`na.omit`** function as shown in the given R – code example.

```
# Convert raster into matrix and omit NA values
>raster_matrix =getValues(rast_img) #convert to matrix
>raster_matrix =na.omit(raster_matrix) #omit NA values
```

15.1. K-means method

In K-means algorithm, number of clusters (say k) are pre-defined by the user. The centroids (mean position of all points in the cluster)are arbitrarily assigned in the multispectral feature space for k number of clusters. The pixel data are grouped in these clusters based on the closest centroid. After partitioning of all pixels in the clusters, new centroid of the cluster is computed. This new centroid is then again used to classify the pixels in the clusters. This procedure is iterated until no significant difference is found in the successive cluster centroids or the pre-defined number of iterations are reached. In this way, this algorithm optimizes the position of the centroids and does the classification of pixels in the raster data. The drawback of K-means algorithm is that it is sensitive to outliers. This is the most commonly used method for unsupervised classification of raster provided the number of classes are known to the user.

The k-means clustering in R can be performed by using **`kmeans`** function. The number of clusters are defined in `centers` argument of **`kmeans`** function. Maximum number of iterations allowed are mentioned in `iter.max` argument. Algorithm to be used is mentioned in `algorithm` argument. Lloyd algorithm is most commonly used. The R – code example of computing k-means on matrix of raster data created earlier is depicted below.

```
#Perform k-means clustering on raster matrix
>kmn =kmeans(raster_matrix, centers=6, iter.max=1000,
algorithm=”Lloyd”)
>kmn                          #Display attributes of computed k-

means
```

```
## K-means clustering with 6 clusters of sizes 31246, 143710,
35233, 174305, 8310, 130612
##
## Cluster means:
##          Blue     Green       Red       NIR       SWIR     TIRS
## 1 0.1681479 0.1749835 0.1903507 0.2781202 0.31225222 114.5924
## 2 0.1596128 0.1648285 0.1792414 0.2786344 0.30477958 113.9377
## 3 0.1436890 0.1372336 0.1305691 0.2538766 0.20728759 112.4257
## 4 0.1549054 0.1571535 0.1671828 0.2718236 0.28528299 113.5080
## 5 0.1369125 0.1286484 0.1139124 0.1192478 0.08844418 110.5254
## 6 0.1526478 0.1511441 0.1549911 0.2614748 0.25317342 113.0405
##
## Clustering vector:
##     [1] 4 4 4 6 6 6 6 6 6 6 6 6 6 6 6 6 6 6 3 3 3 3 3 3 3 3 3 3 6 6
6 6 6 6 6
##    [35] 6 6 6 6 6 4 4 4 4 4 4 4 4 6 6 6 6 6 6 6 4 4 4 6 6 6 3 3 3
5 5 5 5 3
##    [69] 3 3 3 3 3 6 6 6 6 6 6 6 6 6 6 6 6 6 6 6 6 6 6 6 6 6 6 6 6
6 6 6 6 6
------------------------------------------------------------------------
##[99893] 4 6 6 6 3 3 3 3 3 3 3 3 3 3 3 3 3 3 3 3 3 3 3 3 3 6 6 6 6 4
4 4 4 4 2
##[99927] 2 2 2 2 2 2 2 2 2 4 4 4 4 2 2 2 2 4 4 6 6 3 3 3 3 6 6 6 4
6 6 6 3 3
##[99961] 3 3 3 5 5 5 5 5 5 5 5 5 5 5 5 5 5 5 5 5 5 5 5 5 3 3 3 5 5 5
5 5 5 5 5
##[99995] 5 3 3 6 4
##[ reached getOption("max.print") - omitted 423417 entries ]
##
## Within cluster sum of squares by cluster:
## [1] 3357.958 3317.029 3056.043 3221.284 2121.632 3328.175
##  (between_SS / total_SS =  91.8 %)
##
## Available components:
##
## [1] "cluster"      "centers"      "totss"        "withinss"
## [5] "tot.withinss" "betweenss"    "size"         "iter"
## [9] "ifault"
```

The output of **kmeans** function tells about the size of each cluster, mean value of all bands for each cluster, clustering vector *i.e.* cluster identity of each classified cell or pixel, within cluster sum of squares by cluster, and available components in the output data. The *'within cluster sum of squares by cluster'* provides the measure of total variance in the data which is explained by this algorithm. The results indicate that this algorithm captures 91.8% of variance in data.

The available components in the output of **kmeans** function with their respective meaning are depicted below.

```
#Display specific attribute of k-means clustering
>kmn$centers #Display cluster centres
```

```
##          Blue     Green       Red       NIR       SWIR     TIRS
## 1 0.1681479 0.1749835 0.1903507 0.2781202 0.31225222 114.5924
## 2 0.1596128 0.1648285 0.1792414 0.2786344 0.30477958 113.9377
## 3 0.1436890 0.1372336 0.1305691 0.2538766 0.20728759 112.4257
## 4 0.1549054 0.1571535 0.1671828 0.2718236 0.28528299 113.5080
## 5 0.1369125 0.1286484 0.1139124 0.1192478 0.08844418 110.5254
## 6 0.1526478 0.1511441 0.1549911 0.2614748 0.25317342 113.0405
```

```
>kmn$totss #Total sum of squares
```

```
## [1] 224992.7
```

```
>kmn$withinss #Within-cluster sum of squares
```

```
## [1] 3357.958 3317.029 3056.043 3221.284 2121.632 3328.175
```

```
>kmn$tot.withinss#Total of within-cluster sum of squares
```

```
## [1] 18402.12
```

```
>kmn$betweenss #Between-cluster sum of squares
```

```
## [1] 206590.6
```

```
>kmn$size #Number of points in each cluster
```

```
## [1]  31246 143710  35233 174305   8310 130612
```

```
>kmn$iter #Number of iterations done
```

```
## [1] 148
```

```
>kmn$ifault #Indicator of possible algorithm problem
```

```
## NULL
```

The result of k-means classification is in the form of matrix. Now we will convert the result from matrix to raster format. For this, we will first create an empty raster having attributes similar to the input raster as shown further.

```
# Get attributes of first raster layer without pixel values
>kmeans_raster =raster(rast_img)
>kmeans_raster #Display attributes of the created raster
```

```
## class      : RasterLayer
## dimensions : 678, 772, 523416  (nrow, ncol, ncell)
## resolution : 30, 30  (x, y)
```

```
## extent      : 432345, 455505, 2905185, 2925525  (xmin, xmax, ymin,
ymax)
## crs          : +proj=utm +zone=44 +datum=WGS84 +units=m +no_defs
+ellps=WGS84 +towgs84=0,0,0
```

```
>summary(kmeans_raster) #Summary of raster values
```

```
##          layer
## Min.        NA
## 1st Qu.     NA
## Median      NA
## 3rd Qu.     NA
## Max.        NA
## NA's        NA
```

The summary confirms that the created raster doesn't have any value *i.e.* it is empty. Now we will fill this empty raster with the values of the k-means classified matrix which is present in the *'cluster'* component of the output as depicted below.

```
#Fill K-means classified points in pixels of empty raster layer
>i =which(!is.na(raster_matrix))
>kmeans_raster[i] =kmn$cluster
>kmeans_raster #Display attributes of classified raster
```

```
## class       : RasterLayer
## dimensions : 678, 772, 523416  (nrow, ncol, ncell)
## resolution : 30, 30  (x, y)
## extent      : 432345, 455505, 2905185, 2925525  (xmin, xmax, ymin,
ymax)
## crs          : +proj=utm +zone=44 +datum=WGS84 +units=m +no_defs
+ellps=WGS84 +towgs84=0,0,0
## source      : memory
## names       : layer
## values      : 1, 6  (min, max)
```

The raster containing K-means classified cluster values is created. Now we will plot this raster and assign colour and class labels based on visual interpretation.

```
#Plot classified raster
>ggR(kmeans_raster, geom_raster =TRUE) +#Plot with ggplot2
       ggtitle('Unsupervised Classification by K-means') +#title
of plot
       labs(x='Longitude (m)', y='Latitude (m)')+#axis labels
       theme(plot.title =element_text(hjust =0.5, #Align title
in centre
       size =30), #size of title
       axis.title =element_text(size =20), #size of axis labels
       legend.key.size =unit(1, "cm"), #size of legend
```

```
      legend.title =element_text(size =20), #size of legend
title
      legend.text =element_text(size =15))+#size of legend text
      #Provide customized gradient colours
      scale_fill_gradientn(name ="Class",
      colours =c("wheat","white","skyblue",
"gray","forestgreen","burlywood"),
      labels =c('Riverine Sand','Salt-affected Soils',
'Water','Built-up','Vegetation',

'Fallow soil'))
```

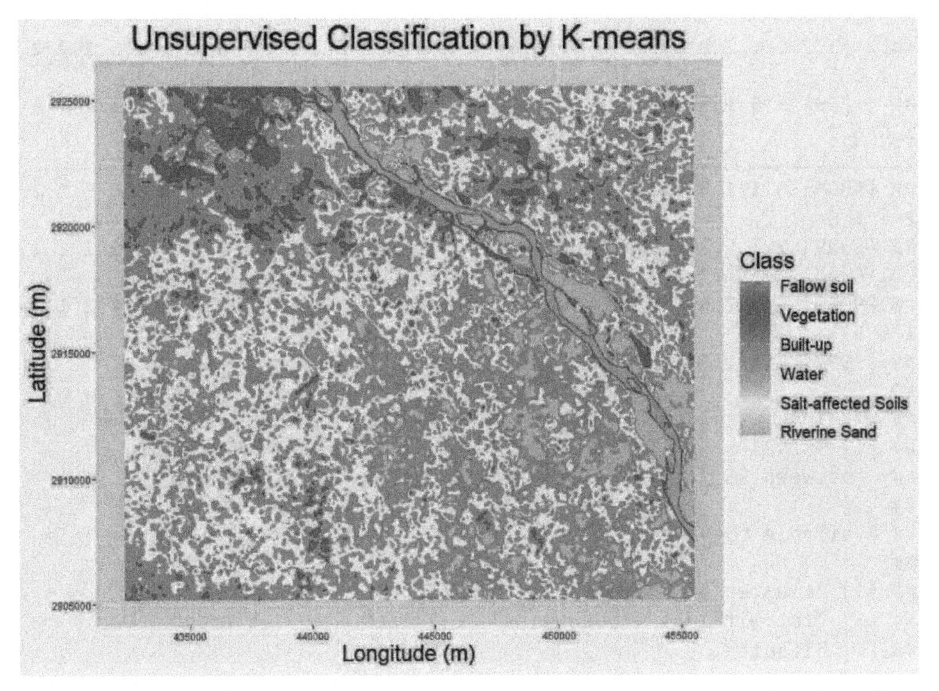

Classification by K-means method can be directly done on raster data by using `algorithm = "Lloyd"` argument in **`unsuperClass`** function of `RStoolbox` package as depicted below.

```
#K-means classification by 'RStoolbox' package
>kmeans_img =unsuperClass(rast_img, nSamples =10000, nClasses
=6,
nStarts =6, algorithm ="Lloyd")
>kmeans_img$model#Display parameters of K-means model

## K-means clustering with 6 clusters of sizes 3319, 2522, 553, 159,
2636, 811
```

```
##
## Cluster means:
##        Blue     Green       Red       NIR       SWIR     TIRS
## 1 0.1553994 0.1578821 0.1682475 0.2723771 0.28711713 113.5289
## 2 0.1527818 0.1516470 0.1559695 0.2627169 0.25620085 113.0738
## 3 0.1677470 0.1745891 0.1902789 0.2785207 0.31298487 114.6017
## 4 0.1368877 0.1285025 0.1137332 0.1218616 0.09006525 110.5187
## 5 0.1596732 0.1649426 0.1794117 0.2789785 0.30523895 113.9548
## 6 0.1447510 0.1386565 0.1329388 0.2557619 0.21209910 112.4797
##
## Clustering vector:
##    [1] 2 5 5 2 5 2 2 1 2 1 2 2 2 6 5 2 1 2 6 6 2 1 3 5 5 2 6 2 1 1
5 5 1 5
##   [35] 5 1 2 3 5 3 1 6 6 6 3 1 6 2 5 5 1 2 2 5 2 5 1 1 1 5 5 1 2 2
6 2 6 5
##   [69] 1 6 1 5 2 5 5 3 2 2 1 1 2 5 5 5 5 1 1 2 5 2 1 2 3 2 1 1 1 6
1 6 6 1
.........................................................................
## [9895] 5 1 1 2 4 1 1 2 5 5 5 2 3 1 2 2 2 2 1 5 2 6 1 5 1 6 6 5 1 2
2 1 1 6
## [9929] 2 1 5 2 1 1 1 2 6 1 6 1 3 2 1 3 1 1 5 2 2 1 2 3 3 5 2 1 1 2
2 1 6 2
## [9963] 5 5 2 5 1 5 5 5 2 1 2 5 5 3 5 1 3 1 5 1 1 1 1 1 5 5 1 2 1 1
1 6 5 3
## [9997] 1 1 1 6
##
## Within cluster sum of squares by cluster:
## [1] 58.92475 60.81556 58.29682 46.59328 60.73434 64.25072
##  (between_SS / total_SS =  91.8 %)
##
## Available components:
##
## [1] "cluster"      "centers"      "totss"        "withinss"
## [5] "tot.withinss" "betweenss"    "size"         "iter"
## [9] "ifault"
```

```
>kmeans_img$map#Display attributes of classified raster
```

```
## class      : RasterLayer
## dimensions : 678, 772, 523416  (nrow, ncol, ncell)
## resolution : 30, 30  (x, y)
## extent     : 432345, 455505, 2905185, 2925525  (xmin, xmax, ymin,
ymax)
## crs        : +proj=utm +zone=44 +datum=WGS84 +units=m +no_defs
+ellps=WGS84 +towgs84=0,0,0
## source     : memory
## names      : layer
## values     : 1, 6  (min, max)
```

In this way, unsupervised classification of raster data by using K-means method is done in R software.

15.2. CLARA method

CLARA is the abbreviation of Clustering LARge Application. This method was given by Kaufmann and Rousseeuw (1990). Unlike K-means, this method is more robust to outliers because it takes medoids rather than mean for clustering. Medoid is actual data point of a cluster having minimal dissimilarity with all points of the cluster. So, medoid is the most centrally located point in the cluster. CLARA derives multiple small samples from the dataset and applies PAM (Partitioning Around Medoids) algorithm to find the specified number of medoids. The number of medoids is specified by the user which is equal to the number of clusters. In PAM algorithm, initial random set of medoids are taken, and one of these medoids is iteratively replaced with one of non-medoids till lowest total swapping cost is achieved. In this way, optimum medoids are computed by the PAM algorithm. After getting optimum medoids, each observation of the entire dataset is assigned to its closest medoid. Dissimilarity of the observations to their closest medoid is computed. This process of sampling and clustering is iterated number of times and best cluster with minimal dissimilarity is selected as the final output.

CLARA can be applied in R by using `clara` function available in `cluster` package. This function is applied on matrix created from the raster data. Number of clusters is mentioned in the `k` argument of the `clara` function. The number of samples to be drawn from the dataset is mentioned in `samples` argument. The method for computing dissimilarities between points is specified in `metric` argument. The R – code example for application of `clara` function on raster data matrix for unsupervised classification into six classes is given below.

```
#Apply CLARA clustering on raster matrix
> clara_fun =clara(raster_matrix, k=6, samples=1000,
metric=”euclidean”)
> clara_fun #Display attributes of CLARA clusters
```

```
## Call:clara(x = raster_matrix, k=6, metric=”euclidean”, samples =
1000)
## Medoids:
##              Blue     Green       Red       NIR       SWIR     TIRS
## [1,] 0.1535965 0.1558692 0.1717594 0.2672927 0.29657111 113.5251
## [2,] 0.1472593 0.1454275 0.1526474 0.2682155 0.25667286 113.1988
## [3,] 0.1511470 0.1464991 0.1497557 0.2587401 0.23132218 112.7797
## [4,] 0.1351917 0.1241567 0.1108991 0.1080110 0.07986378 110.3465
## [5,] 0.1574617 0.1627999 0.1810217 0.2667751 0.30117640 114.2010
## [6,] 0.1596865 0.1667440 0.1853140 0.2818998 0.30808437 113.8368
## Objective function:   0.1416369
## Clustering vector:    int [1:523416] 1 1 2 2 2 2 2 2 2 3 3 3 3 3 3
3 3 3 ...
## Cluster sizes:            127461 111904 83323 8681 74113 117934
## Best sample:
```

```
## [1]    2189   9369  14648  15055  21557  30358  38464  39710  43600
53104
## [11]  63423  67440  76162  79532  84516  99954 109754 119721 123163
133219
## [21] 134185 135535 143010 149567 153113 193654 195794 199963 227190
240584
## [31] 245591 248411 261924 262475 273689 283975 300971 314037 320267
335178
## [41] 336256 375727 379185 407266 423351 434293 439069 465162 467653
478851
## [51] 482589 499448
##
## Available components:
##  [1] "sample"     "medoids"    "i.med"      "clustering" "objective"
##  [6] "clusinfo"   "diss"       "call"       "silinfo"    "data"
```

The output of `clara` function constitutes band values of each medoid, objective function for final clustering of entire dataset, clustering vector depicting the cluster identity of each classified pixel, size of each cluster, best sample *i.e.* pixel or cell numbers of best sample used by CLARA algorithm for final partitioning, and available components.

The available components in the output of `clara` function with their respective meaning are given below.

```
#Display specific attribute of CLARA
> clara_fun$sample #Pixel numbers of best sample used for final
partition

## [1]    2189   9369  14648  15055  21557  30358  38464  39710  43600
53104
## [11]  63423  67440  76162  79532  84516  99954 109754 119721 123163
133219
## [21] 134185 135535 143010 149567 153113 193654 195794 199963 227190
240584
## [31] 245591 248411 261924 262475 273689 283975 300971 314037 320267
335178
## [41] 336256 375727 379185 407266 423351 434293 439069 465162 467653
478851
## [51] 482589 499448


>clara_fun$medoids #Medoids of the clusters

##            Blue     Green       Red       NIR       SWIR     TIRS
## [1,] 0.1535965 0.1558692 0.1717594 0.2672927 0.29657111 113.5251
## [2,] 0.1472593 0.1454275 0.1526474 0.2682155 0.25667286 113.1988
## [3,] 0.1511470 0.1464991 0.1497557 0.2587401 0.23132218 112.7797
## [4,] 0.1351917 0.1241567 0.1108991 0.1080110 0.07986378 110.3465
## [5,] 0.1574617 0.1627999 0.1810217 0.2667751 0.30117640 114.2010
```

```
## [6,] 0.1596865 0.1667440 0.1853140 0.2818998 0.30808437 113.8368
>clara_fun$i.med    #Indices of the medoids
## [1] 320267  99954 135535 283975 314037 195794
#Objective function for final clustering of entire dataset
>clara_fun$objective
## [1] 0.1416369
>clara_fun$clusinfo #Provides information of each cluster
##         size  max_diss    av_diss isolation
## [1,] 127461 0.2790877 0.09382967 0.8920843
## [2,] 111904 0.3376254 0.10929160 1.0246613
## [3,]  83323 1.2360025 0.21577016 2.9428623
## [4,]   8681 1.2256004 0.47562794 0.5016811
## [5,]  74113 1.7011763 0.22498947 4.6663701
## [6,] 117934 0.2691394 0.09465486 0.8602853
```

The *'clusinfo'* component provides information about number of observations in each cluster *i.e.* 'size', maximum and average dissimilarity between the observations in the cluster and cluster's medoid *i.e.* 'max_diss' and 'av_diss', respectively, and the ratio of maximal dissimilarity between the observations in the cluster and the cluster's medoid to the minimal dissimilarity between the cluster's medoid and the medoid of any other cluster, *i.e.* 'isolation'.Smaller value of 'isolation' indicates that the cluster is well-separated from the other clusters. The dissimilarity matrix can be obtained from 'diss' component as shown below.

```
>clara_fun$diss       #Dissimilarity
## Dissimilarities :
##             1          2          3          4          5          6
## 2  0.54204907
## 3  0.47587117 0.06771780
## 4  1.03626131 0.49479175 0.56050548
## 5  0.40459070 0.14568060 0.07932855 0.63375257
## 6  0.44852067 0.13745419 0.10979396 0.60922325 0.13570142
## 7 0.45449529 0.12866611 0.08039216 0.59261765 0.06270881 0.17407837
## 8
.........................................................................
## 51 0.21627782 0.08775486 0.11942737 0.47204727 0.14262780 0.05236488
## 52 0.95903475 1.19536103 1.05417241 0.66631372 0.99312944 1.08843514
##            49         50         51
.........................................................................
## 50 0.74451951
## 51 1.02326322 0.28798023
## 52 0.11020674 0.85385381 1.13313451
```

```
##
## Metric :  euclidean
## Number of objects : 52
```

List with silhouette width information for the best sample is given by 'silinfo' component as shown below.

```
#List with silhouette width information for the best sample

> clara_fun$silinfo
## $widths
##        cluster neighbor   sil_width
## 320267       1        6  0.74751857
## 439069       1        6  0.73903243
## 193654       1        2  0.73666945
## 245591       1        6  0.71105302
## 43600        1        6  0.69289678
## 336256       1        2  0.62505284
## 423351       1        2  0.61968194
## 240584       1        6  0.60923227
## 379185       1        6  0.53812308
## 465162       1        2  0.53020748
## 335178       1        6  0.44215534
## 482589       1        2  0.30058936
## 21557        1        2  0.26290872
## 478851       1        6  0.06121407
## 99954        2        1  0.58203122
## 262475       2        3  0.52372244
## 9369         2        1  0.48265139
## 79532        2        3  0.44662766
## 273689       2        3  0.40482760
## 76162        2        3  0.32943416
## 261924       2        3  0.27889531
## 14648        2        1  0.20078945
## 407266       2        1  0.06825360
## 38464        2        1 -0.02973293
## 30358        2        1 -0.09008751
## 133219       3        2  0.78992799
## 135535       3        2  0.78983666
## 119721       3        2  0.77622792
## 109754       3        2  0.77231879
## 15055        3        2  0.75968998
## 123163       3        2  0.69763753
## 134185       3        2  0.68813658
## 143010       3        2  0.62044304
## 153113       3        2  0.31287475
## 283975       4        3  0.00000000
## 300971       5        6  0.60391453
## 149567       5        6  0.59191853
## 467653       5        6  0.57874503
```

```
## 314037       5        6  0.56956111
## 499448       5        6  0.51283169
## 39710        5        6 -0.02479348
## 67440        5        6 -0.19201290
## 195794       6        1  0.77872536
## 199963       6        1  0.77559458
## 227190       6        1  0.75155112
## 434293       6        1  0.74414231
## 375727       6        1  0.70171521
## 63423        6        1  0.69540719
## 2189         6        1  0.67158748
## 248411       6        5  0.54894976
## 84516        6        1  0.52838443
## 53104        6        5  0.33629056
##
## $clus.avg.widths
## [1] 0.5440240 0.2906739 0.6896770 0.0000000 0.3771664 0.6532348
##
## $avg.width
## [1] 0.5037183
```

As the output of CLARA classification is in the form of matrix, so, we will convert it into raster format following the same procedure as mentioned in the above K-means section. The R-code example is given below.

```
#Get attributes of first raster layer without pixel values

> clara_raster =raster(rast_img)

#Fill clara classified points in pixels of empty raster layer

> i =which(!is.na(raster_matrix))

> clara_raster[i] =clara_fun$clustering

> clara_raster #Display attributes of classified raster

## class      : RasterLayer
## dimensions : 678, 772, 523416  (nrow, ncol, ncell)
## resolution : 30, 30  (x, y)
## extent     : 432345, 455505, 2905185, 2925525  (xmin, xmax, ymin, ymax)
## crs        : +proj=utm +zone=44 +datum=WGS84 +units=m +no_defs +ellps=WGS84 +towgs84=0,0,0
## source     : memory
## names      : layer
## values     : 1, 6  (min, max)
```

After creating the classified raster file, we will plot the raster and assign the class labels to each cluster through visual interpretation. The plot of classified raster with their respective classes is given further.

```
#Plot classified raster
>ggR(clara_raster, geom_raster =TRUE) +#Plot with ggplot2
      ggtitle('Unsupervised Classification by Clara') +#Title
of plot
      labs(x='Longitude (m)', y='Latitude (m)')+#Axis labels
      theme(plot.title =element_text(hjust =0.5, #Align title
in centre
      size =30), #size of title
      axis.title =element_text(size =20), #size of axis labels
      legend.key.size =unit(1, "cm"), #size of legend
      legend.title =element_text(size =20), #size of legend
title
      legend.text =element_text(size =15))+#size of legend text
#Provide customized gradient colours
      scale_fill_gradientn(name ="Class",
colours =c("white", "forestgreen", "gray",
"skyblue", "wheat","burlywood"),
labels =c('Salt-affected Soils','Vegetation',
'Built-up','Water','Riverine Sand','Fallow soil'))
```

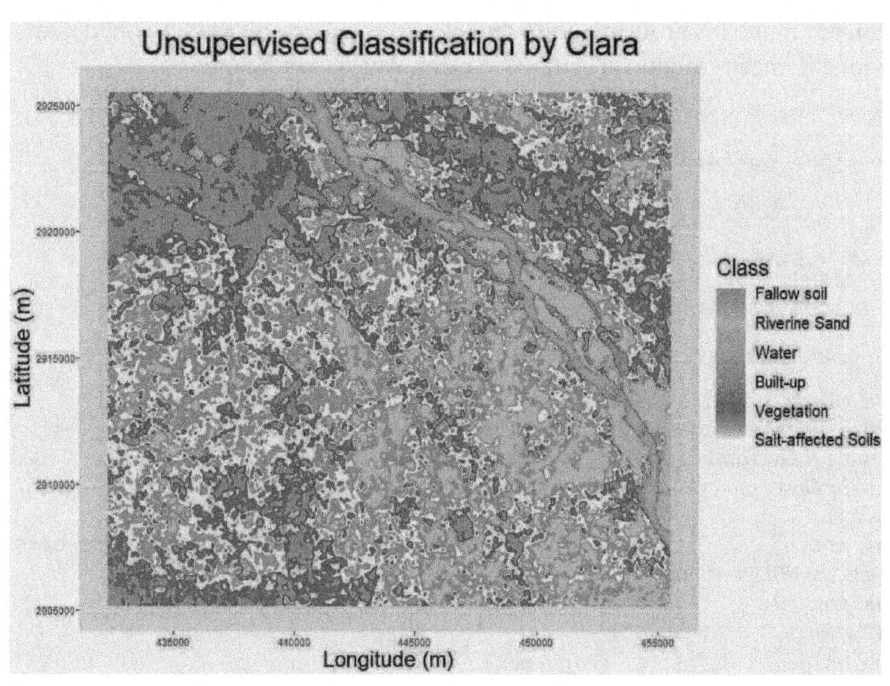

In this way, unsupervised classification of raster data using CLARA method is done in R software.

15.3. Random Forest (Unsupervised)

Random forest is a machine learning technique which is used for classification and regression. In random forest technique, multiple decision trees are constructed on training datasets. Mode of the outputs of individual decision trees is taken as final output of random forest for classification. Random forest is mainly used for supervised classification.The details of random forest for supervised classification is discussed in the next chapter. Random forest can be used for unsupervised classification in conjunction with K-means algorithm. K-means algorithm is used to create clusters on samples and these clusters are then used to train the random forest model. This trained model is used to predict the classes of raster dataset.

For performing unsupervised classification of raster data using random forest in R, random sample of specified size is taken from the raster data matrix as depicted below.

```
#Take sample of 2000 rows from the raster matrix
> samp =raster_matrix[sample(nrow(raster_matrix), 2000),]
>head(samp) #Display first few rows of sample

##              Blue     Green       Red       NIR      SWIR     TIRS
## [1,] 0.1518436 0.1524950 0.1593343 0.2963267 0.2788601 114.0162
## [2,] 0.1661135 0.1764790 0.1990267 0.2853658 0.3248700 113.7380
## [3,] 0.1554842 0.1552308 0.1638525 0.2703087 0.2844338 113.5491
## [4,] 0.1481132 0.1473426 0.1468189 0.2931532 0.2732218 112.9913
## [5,] 0.1488998 0.1447664 0.1469318 0.2674503 0.2609769 113.5926
## [6,] 0.1550122 0.1581946 0.1710591 0.2772633 0.2821311 113.4832
```

Now apply random forest on this sample to compute proximity values. Proximity is based on the frequency that pairs of data points are in the same terminal nodes. A variable name is assigned to these computed proximity values as depicted in following R – code. Random forest can be applied on matrix using `randomForest` function of `randomForest` package.

```
#Apply random forest on sample without training
> rf =randomForest(samp)


#Assign a variable to the matrix of proximity measures of random
forest

>rf_prox =rf$proximity
```

Apply K-means algorithm on these proximity values to compute clusters. In this R-code example, we are creating six clusters. These K-means classified clusters are further used for training the random forest model. The samples created earlier are used as training data. The number of trees to be constructed in random forest model is mentioned in the `ntree` argument of the `randomForest` function.

```
#Apply K-means clustering algorithm to the matrix of proximity measures
> kmn_rf =kmeans(rf_prox, centers =6, iter.max =10000)

#Train random forest with the clusters computed by K-means
> rf_train =randomForest(x = samp, y =as.factor(kmn_rf$cluster), ntree =2000)
>rf_train #Display the attributes of random forest model

##
## Call:
##  randomForest(x = samp, y = as.factor(kmn_rf$cluster), ntree = 2000)
##                Type of random forest: classification
##                      Number of trees: 2000
## No. of variables tried at each split: 2
##
##         OOB estimate of  error rate: 6.7%
## Confusion matrix:
##     1   2   3   4   5   6 class.error
## 1 186   5   0   0   0   8  0.06532663
## 2   7 187   0   0   0   0  0.03608247
## 3   0   0 162  10   0   0  0.05813953
## 4   0   0   8 164  21   0  0.15025907
## 5   1   1   3   7 831  24  0.04152249
## 6  15   4   0   0  20 336  0.10400000
```

The confusion matrix and OOB (Out-of-Bag) estimate of error rate indicates that the developed random forest has good accuracy. If the accuracy of developed random forest model is good, then this model can be used for the prediction of classes of the entire raster dataset using `predict` function of the `raster` package as shown in the following R-code example.

```
#Predict the classes for raster data using trained random forest model
> rf_raster =predict(object = rast_img, model = rf_train)
> rf_raster #Display attribute of classified raster

## class      : RasterLayer
## dimensions : 678, 772, 523416  (nrow, ncol, ncell)
## resolution : 30, 30  (x, y)
## extent     : 432345, 455505, 2905185, 2925525  (xmin, xmax, ymin, ymax)
## crs        : +proj=utm +zone=44 +datum=WGS84 +units=m +no_defs +ellps=WGS84 +towgs84=0,0,0
## source     : memory
## names      : layer
```

```
## values     : 1, 6  (min, max)
## attributes :
##        ID value
##  from:  1     1
##   to :  6     6
```

Now plot the classified raster and assign the class names to each cluster through visual interpretation as depicted below.

```
#Plot classified raster
>ggR(rf_raster, geom_raster =TRUE) +#Plot with ggplot2
ggtitle('Unsupervised Classification by Random Forest')+#Title of Plot
       labs(x='Longitude (m)', y='Latitude (m)')+#axis labels
       theme(plot.title =element_text(hjust =0.5, #Align title in centre
       size =30), #size of title
       axis.title =element_text(size =20), #size of axis labels
       legend.key.size =unit(1, "cm"), #size of legend
       legend.title =element_text(size =20), #size of legend title
       legend.text =element_text(size =15))+#size of legend text
       #Provide customized gradient colours
       scale_fill_manual(values =c("white","forestgreen","burlywood",
"gray","skyblue","wheat"),
       labels =c('Salt-affected Soils','Vegetation',
'Fallow soil','Built-up','Water','Riverine Sand'),
       name ='Class')
```

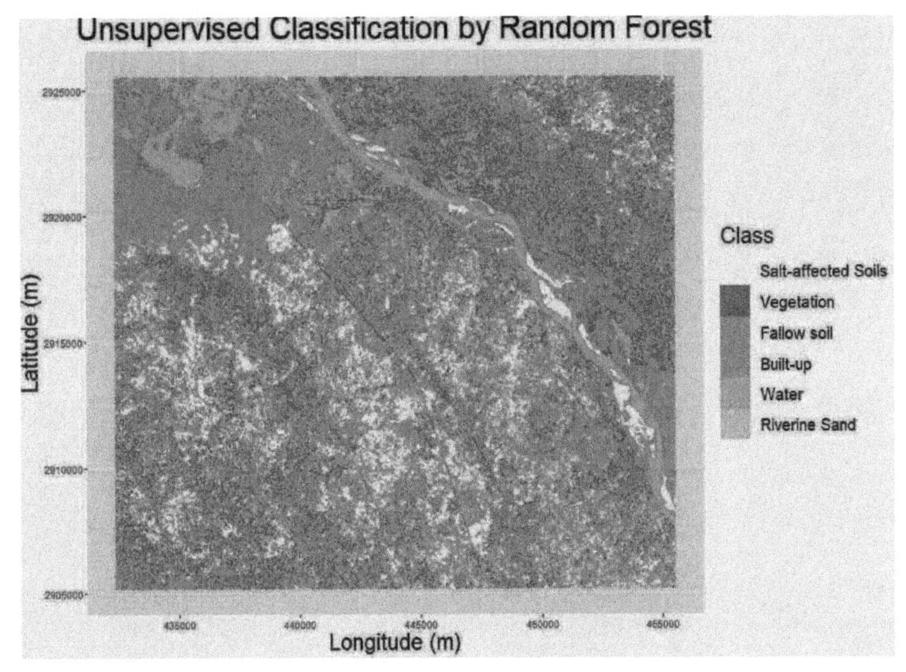

In this way, unsupervised classification of raster image can be performed in R by using K-means, CLARA and random forest methods.

Chapter 16

Supervised Classification

Supervised classification is the technique to classify raster image in which prior information about the classes in the form of training sites is provided by the user. Training site represents a homogeneous area in a particular class. The spectral response pattern of training sites is used to classify the raster image.The pixels of raster imagewhich have spectral response pattern similar to that of training sites of a particular class are grouped in that class. As there is some variability in the spectral response pattern of a particular class, therefore, multiple training sites per class are taken for supervised classification. Supervised classification of raster images can be performed by various methods. Conventional methods include maximum likelihood classifier, minimum distance to mean classifier, and parallelepiped classifier. Now a days, machine learning algorithms are also gaining importance in this context. These classifying algorithms are first trained or calibrated using the data of training sites.This trained model is then validated on the independent datasets for accuracy assessment. If the accuracy of the developed model is of acceptable level, then that model is used for the classification of raster data.

In this chapter, we are going to discuss the execution of maximum likelihood classifier and few machine learning techniques for supervised classification of raster image in R software. The training sites used in this exercise are in the form of points shapefile. The raster image is classified into six classes*i.e.* built-up, fallow soil, riverine sand, salt-affected soil, vegetation, and water. For performing this exercise, first of all,set the working directory and import the required packages.

```
# Set the working directory
>setwd('E:\\Chapter16\\data')
>getwd()    # Get the working directory

## [1] "E:/Chapter16/data"

#Import required packages
>library(rgdal)          #For importing and reprojection of
vector data
```

```
>library(raster)          #For importing and analysis of raster
data
>library(RStoolbox) #For visualization of raster data
>library(ggplot2)   #For visualization of raster data
>library(sp)        #For plotting vector data
```

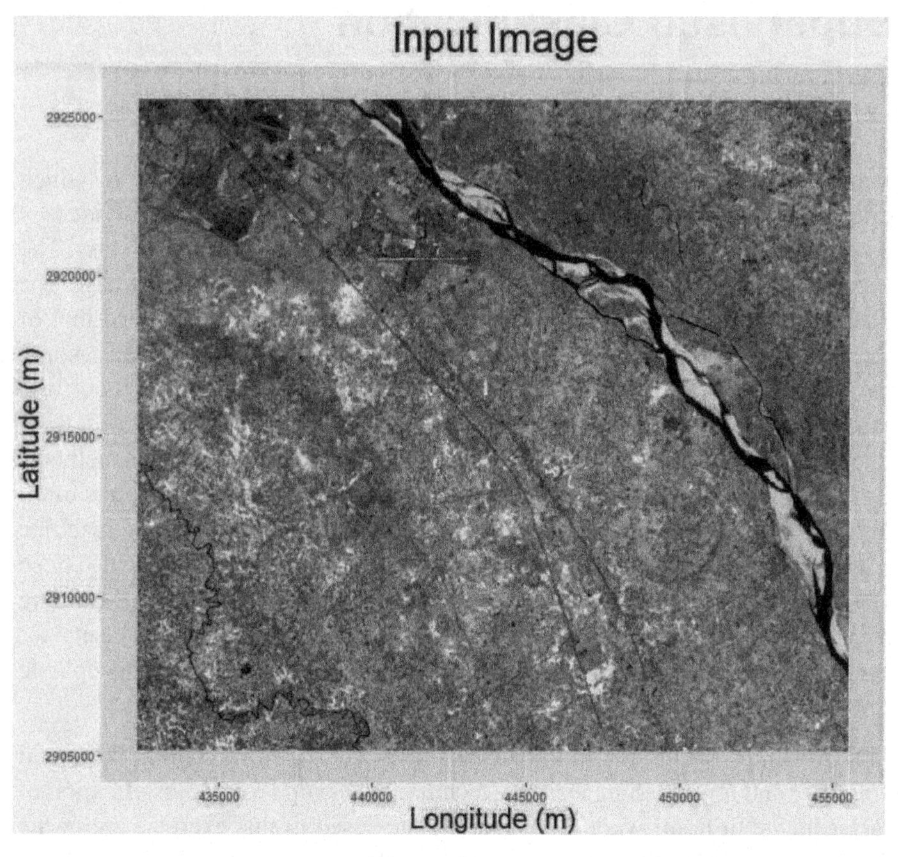

Now import the input multiband raster image which you want to classify and plot its False Colour Composite (FCC) by using following R-code.

```
#Import input raster image
>img =brick('input_img.tif')

#Set the names of bands in input raster
>names(img) =c('Coastal_aerosol', 'Blue', 'Green', 'Red',
'NIR', 'SWIR1', 'SWIR2','cirrus', 'TIRS1','TIRS2')
>img#Display attributes of imported raster

## class      : RasterBrick
## dimensions : 678, 772, 523416, 10  (nrow, ncol, ncell, nlayers)
## resolution : 30, 30  (x, y)
```

```
## extent     : 432345, 455505, 2905185, 2925525  (xmin, xmax, ymin,
ymax)
## crs        : +proj=utm +zone=44 +datum=WGS84 +units=m +no_defs
+ellps=WGS84 +towgs84=0,0,0
## source     : E:/................/Chapter16/data/input_img.tif
## names      : Coastal_aerosol, Blue, Green, Red, NIR, SWIR1, SWIR2,
cirrus, TIRS1, TIRS2
```

```
#Plot FCC of input raster
>ggRGB(img, r =5, g=4, b=3, stretch ='lin') +#plot with ggplot2
      ggtitle('Input Image') +#title of plot
      labs(x ='Longitude (m)', y ='Latitude (m)') +#axis labels
      theme(plot.title =element_text(hjust =0.5, #Align title
in centre
      size =30), #size of title
      axis.title =element_text(size =20)) #size of axis labels
```

Now import training sites and display its position with respect to its classes using following R-code.

```
#Import training data as spatial points
>trainData =readOGR(dsn ="training_points",
"training_points")

## OGR data source with driver: ESRI Shapefile
## Source: "E:\.......\Chapter16\data", layer: "training_points"
## with 400 features
## It has 2 fields
## Integer64 fields read as strings:  id

>trainData# Display attributes of training data

## class       : SpatialPointsDataFrame
## features    : 400
## extent      : 80.32199, 80.55325, 26.2663, 26.44891  (xmin, xmax,
ymin, ymax)
## crs         : +proj=longlat +datum=WGS84 +no_defs +ellps=WGS84
+towgs84=0,0,0
## variables   : 2
## names       : id,    Class
## min values  :  1, Built-up
## max values  :  6,    Water

#Plot training data
>spplot(trainData, zcol ="Class", #Attribute name
key.space ="right",  #Position of legend
scales=list(draw=TRUE),   #Scales of latitude and longitude
xlab ="Longitude", ylab ="Latitude",   #Axis labels
col.regions =c("grey", "burlywood", "wheat", "indianred",
      "darkgreen", "blue"), #Colours of classes
main ="Training Points")  #Title of plot
```

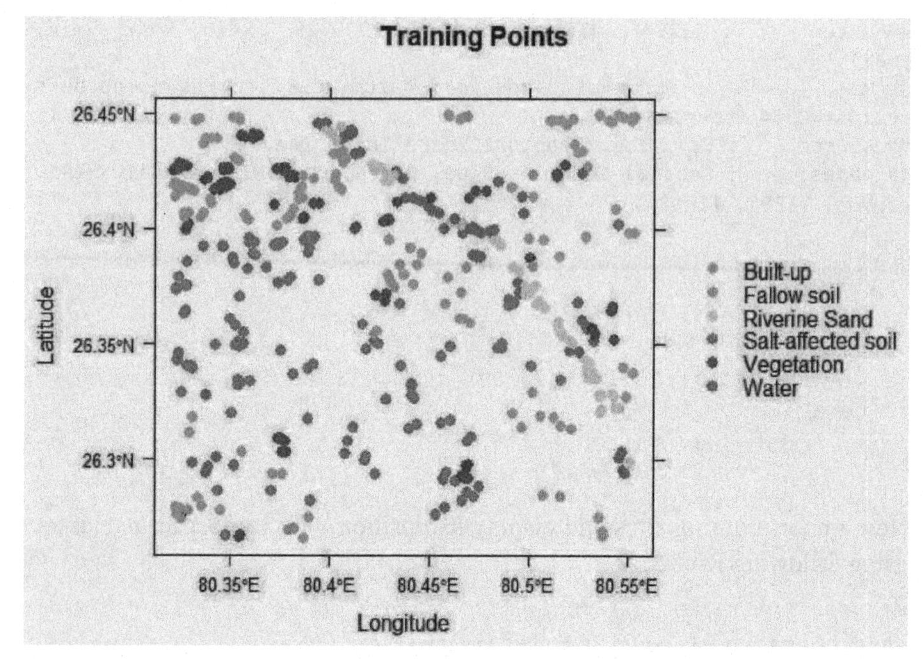

So, we have 400 training points with six classes as depicted in the plot. Compare the Coordinate Reference System (CRS) of both input image and training points shapefile.If the CRS of both these imported objects are different, then re-project them to the same CRS as shown in the following R-code.

```
# Compare CRS
>compareCRS(img, trainData)

## [1] FALSE

#Reproject shapefile of training points to the CRS of raster
>trainData =spTransform(trainData, CRS(projection(img)))
>compareCRS(img, trainData) # Compare CRS to check the reprojection

## [1] TRUE
```

After bringing both input raster image and points shapefile to the same projection, classification operation can be performed.

16.1. Conventional algorithms

Many conventional algorithms are used for the supervised classification of raster data like Minimum-distance-to-means classifier, parallelepiped classifier, and Maximum likelihood classifier. In this exercise, we are going to discuss only one algorithm *i.e.* maximum likelihood classifier and its execution in R software.

16.1.1. Maximum Likelihood Classification

Maximum likelihood classification technique classifies the unknown pixels based on statistical probability. This method assumes Gaussian or normal distribution of data for each class in each band. This method computes the probability density function for each class in each band, class mean vector and class covariance matrix. For classification of unknown pixel, the statistical probability or likelihood of pixel value to each class is computed, and the pixel will be assigned to the class having highest probability or maximum likelihood. In this way, this method considers both variance and covariance of classes while classifying unknown pixel. This method is computationally intensive.

For implementation of Maximum likelihood classification of raster data in R, `superClass` function of `RStoolbox` package is used, and `'mlc'` method is assigned in its `model` argument. The training data is provided in `trainData` argument. If separate validation data is not provided, then provided training data is partitioned into training and validation data. In the given R-code example, 0.8 value is assigned to `trainPartition` argument which denotes that 80% data will be used for training and 20% data for validation. The attribute name or column containing response variable or class labels is assigned in `responseCol` argument. The number of samples per land cover class is mentioned in `nSamples` argument. The purpose of model either for classification or regression is mentioned in `mode` argument. Number of cross-validation resamples during model tuning is mentioned in `kfold` argument. If `verbose` argument is set `TRUE`, then progress and statistics is printed during execution of code.

```
#Import required package
>library(RStoolbox) #For maximum likelihood classification
#Perform maximum likelihood classification of raster image

>img_mlc =superClass(img, trainData =trainData, responseCol
='Class',
nSamples =30, model ='mlc', mode ='classification',
trainPartition =0.8, kfold=5, verbose =TRUE)

## Begin sampling training data
## Starting to fit model
## Loading required package: lattice
## Starting spatial predict
##
```

```
  |
  |                                                                      |   0%
  |
  |=================                                                     |  25%
  |
  |===================================                                   |  50%
  |
  |====================================================                  |  75%
  |
  |======================================================================| 100%
##
## Begin validation
## ******************* Model summary *******************
## Maximum Likelihood Classification
##
## 324 samples
##  10 predictor
##   6 classes: 'Built-up', 'Fallow soil', 'Riverine Sand', 'Salt-
affected soil', 'Vegetation', 'Water'
##
## No pre-processing
## Resampling: Cross-Validated (5 fold)
## Summary of sample sizes: 258, 261, 260, 259, 258
## Resampling results:
##
##   Accuracy   Kappa
##   0.9784046  0.9729784
##
## [[1]]
##   TrainAccuracyTrainKappa method
## 1     0.9784046  0.9729784 custom
##
## [[2]]
## Cross-Validated (5 fold) Confusion Matrix
##
## (entries are average cell counts across resamples)
##
##                     Reference
## Prediction         Built-up Fallow soil Riverine Sand Salt-
affected soil
##   Built-up            7.0         0.0           0.2                 0.0
##   Fallow soil         0.0        13.2           0.0                 0.0
##   Riverine Sand       0.0         0.0           4.2                 0.0
##   Salt-affected soil  0.0         0.4           0.4                18.0
##   Vegetation          0.0         0.0           0.0                 0.0
##   Water               0.0         0.0           0.0                 0.0
##                     Reference
## Prediction           Vegetation Water
##   Built-up                  0.0   0.0
##   Fallow soil               0.0   0.0
##   Riverine Sand             0.0   0.0
```

```
##   Salt-affected soil        0.0   0.0
##   Vegetation               13.6   0.4
##   Water                     0.0   7.4
##
##  Accuracy (average) : 0.9784
## ******************** Validation summary ********************
## Confusion Matrix and Statistics
##
##                     Reference
## Prediction        Built-up Fallow soil Riverine Sand Salt-affected
soil
##   Built-up        8            0             0                  0
##   Fallow soil      0          16             0                  0
##   Riverine Sand    0           0             5                  0
##   Salt-affected soil 0         0             0                 22
##   Vegetation       0           0             0                  0
##   Water            0           0             0                  0
##                     Reference
## Prediction           Vegetation Water
##   Built-up                    0     1
##   Fallow soil                 0     0
##   Riverine Sand               0     0
##   Salt-affected soil          0     0
##   Vegetation                 16     0
##   Water                       0     8
##
## Overall Statistics
##
##                Accuracy : 0.9868
##                  95% CI : (0.9289, 0.9997)
##     No Information Rate : 0.2895
##     P-Value [Acc> NIR] : < 2.2e-16
##
##                   Kappa : 0.9835
##
##  Mcnemar’s Test P-Value : NA
##
## Statistics by Class:
##
##                      Class: Built-up Class: Fallow soil
## Sensitivity                   1.0000             1.0000
## Specificity                   0.9853             1.0000
## PosPred Value                0.8889             1.0000
## Neg Pred Value                1.0000             1.0000
## Prevalence                    0.1053             0.2105
## Detection Rate                0.1053             0.2105
## Detection Prevalence          0.1184             0.2105
## Balanced Accuracy             0.9926             1.0000
```

```
##                       Class: Riverine Sand Class: Salt-affected soil
## Sensitivity                        1.00000                    1.0000
## Specificity                        1.00000                    1.0000
## PosPred Value                      1.00000                    1.0000
## Neg Pred Value                     1.00000                    1.0000
## Prevalence                         0.06579                    0.2895
## Detection Rate                     0.06579                    0.2895
## Detection Prevalence               0.06579                    0.2895
## Balanced Accuracy                  1.00000                    1.0000
##                       Class: Vegetation Class: Water
## Sensitivity                      1.0000       0.8889
## Specificity                      1.0000       1.0000
## PosPred Value                    1.0000       1.0000
## Neg Pred Value                   1.0000       0.9853
## Prevalence                       0.2105       0.1184
## Detection Rate                   0.2105       0.1053
## Detection Prevalence             0.2105       0.1053
## Balanced Accuracy                1.0000       0.9444
```

The output of the given R-code is confusion matrix, accuracy and Kappa coefficient of both training and validation data, and statistics by class of validation data. Confusion matrix, accuracy and Kappa coefficient denotes that the classification accuracy is good. Statistics by class include sensitivity, specificity, positive predictive value, negative predictive value, prevalence, detection rate, detection prevalence and balanced accuracy. Sensitivity of a class is the ratio of predicted count to the actual count of samples. Specificity is the measure of not predicting the sample to a class if that sample belongs to a different class. Positive and Negative predictive values are proportions of positive and negative counts which are true positive and true negative counts, respectively. Prevalence is the proportion of actual counts in a particular class to the total counts of all classes. Detection rate is proportion of count of correct prediction of a particular class with respect to total counts of all classes. Detection prevalence is the percentage of prediction of a particular class in full sample of whole classes. Balanced accuracy is the average of sensitivity and specificity. The values of all the discussed parameters indicates that, the developed maximum likelihood classifier has classified the given raster image with accepted level of accuracy.

The attributes of maximum likelihood classified raster image can be displayed by using following R-code.

```
#Display attributes of maximum likelihood classified map
>img_mlc$map

## class      : RasterLayer
## dimensions : 678, 772, 523416  (nrow, ncol, ncell)
## resolution : 30, 30  (x, y)
```

```
## extent      : 432345, 455505, 2905185, 2925525  (xmin, xmax, ymin,
ymax)
## crs           : +proj=utm +zone=44 +datum=WGS84 +units=m +no_defs
+ellps=WGS84 +towgs84=0,0,0
## source      : memory
## names       : Class
## values      : 1, 6  (min, max)
## attributes :
##        ID    value
##  from:  1 Built-up
##    to :  6    Water
```

```
>img_mlc$classMapping# Display labels of each class
```

```
##   classID             class
## 1       1          Built-up
## 2       2       Fallow soil
## 3       3     Riverine Sand
## 4       4 Salt-affected soil
## 5       5        Vegetation
## 6       6             Water
```

As the classified raster image and respective labels are available, so, user need not to specify class based on visual interpretation unlike unsupervised classification. Now, we can plot the classified image and assign respective colours to the class labels as given in the output of above stated R-code.

```
#Plot classified raster
>ggR(img_mlc$map, geom_raster =TRUE) +#Plot with ggplot2
       ggtitle('Maximum Likelihood Classification') +#Title of
plot
       labs(x='Longitude (m)', y='Latitude (m)')+#Axis labels
       theme(plot.title =element_text(hjust =0.5, #Align title
in centre
       size =30), #size of title
       axis.title =element_text(size =20), #size of axis labels
       legend.key.size =unit(1, "cm"), #size of legend
       legend.title =element_text(size =20), #size of legend
title
       legend.text =element_text(size =15))+#size of legend text
       scale_fill_manual(values =c("grey", "burlywood", "wheat",
"white", "darkgreen", "skyblue"),
name ="Class")
```

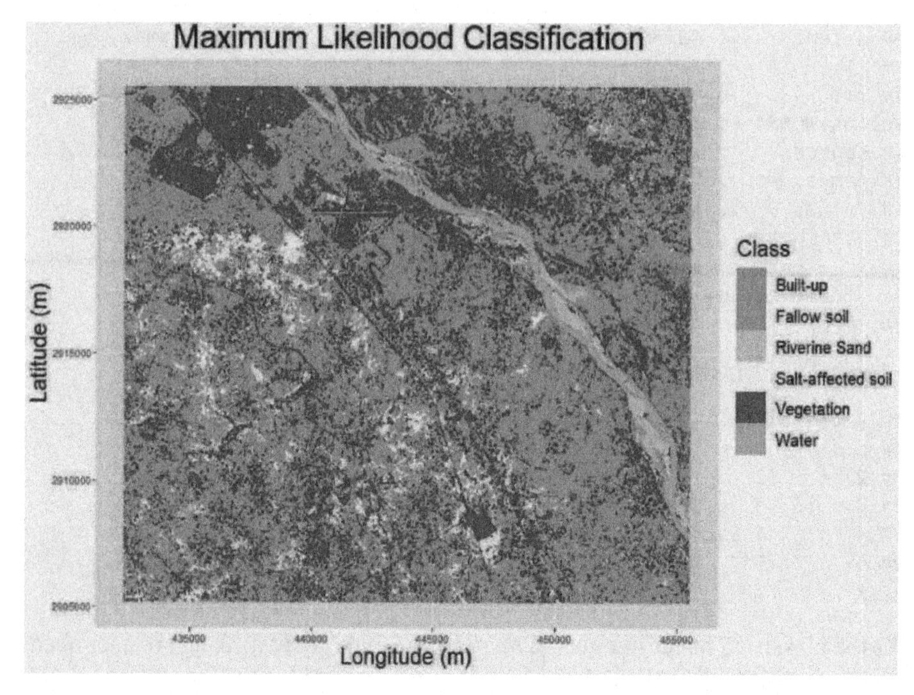

16.2. Machine learning algorithms

In todays world, machine learning algorithm has vast applicability in plethora of fields ranging from business, entertainment, software engineering, medicine, communication networks, computer security, fault diagnosis, speech recognition, web mining, and so on. Therefore, applications of these algorithms in remote sensing data analysis are obvious. According to Svensson and Söderberg(2008), 'Machine learning is concerned with design and development of algorithms and techniques that allow computers to "learn". The major focus of machine learning research is to extract information from data automatically, by computational and statistical methods. It is thus closely related to data mining and statistics'. Machine learning is a branch of Artificial Intelligence. Therefore, machine learning algorithms don't require pre-determined programmed model equation, but they build algorithms automatically from input data and provide the output. The output gets updated when new set of input data is added by the user. Many machine learning algorithms can be used for supervised classification of raster data. Some of these below listed algorithms and its implementation in R software is discussed in this chapter.

1) Binomial logistic regression
2) Multiple logistic regression
3) Support Vector Machine

4) Decision Tree
5) Random Forest

For implementation of above-listed algorithms in R software, the input raster data imported earlier is used. The sample data for training and validation is present in the '.csv' file.

16.2.1. Binomial logistic regression

Binomial logistic regression is expression of ordinary linear regression in logarithmic terms which predict the probability of occurrence of an observation in the dichotomous response variable *i.e.* dependent variable having two categories. The independent variables can be continuous, categorical or both. This method considers both mean and variance for classification of data.

In this exercise, we are going to classify the pixels of raster image into presence (TRUE) or absence (FALSE) of salt-affected soils representing dichotomous variable. First of all, import the '.csv' file which contains sample data for training and validation through following R-code.

```
# Read the file containing sample data for training and validation
>samp_data =read.csv('binomial_logistic_input.csv', header =TRUE)
>head(samp_data) #Display first few rows of sample data
```

```
## Salt.affectedCoastal_aerosol       Blue      Green       Red       NIR
## 1          FALSE           0.1402806 0.1322478 0.1230623 0.1045736
0.09900824
## 2          FALSE           0.1399093 0.1319781 0.1212157 0.1040089
0.09855811
## 3          FALSE           0.1387720 0.1311691 0.1212841 0.1033537
0.09932334
## 4          FALSE           0.1413483 0.1343827 0.1266189 0.1091596
0.10341961
## 5          FALSE           0.1412090 0.1333040 0.1234955 0.1053643
0.10155153
## 6          FALSE           0.1422767 0.1354613 0.1274396 0.1098373
0.10344211

##      SWIR1      SWIR2      cirrus     TIRS1    TIRS2
## 1 0.06813533 0.04542350 0.001279298 109.8694 240.4842
## 2 0.06641371 0.04414189 0.001416180 109.7238 240.2817
## 3 0.06729604 0.04447805 0.001142416 109.7752 240.4096
## 4 0.07237478 0.04882711 0.001324925 110.3856 241.1747
## 5 0.06906069 0.04668410 0.001028348 109.8828 240.5764
## 6 0.07099750 0.04840691 0.001484621 110.3582 241.0991
```

From the first few rows of the imported sample file, we can see that first column shows the presence or absence of salt-affected soils in terms of TRUE or FALSE, respectively whereas remaining columns contain the values of reflectance in their respective bands. The first column is a categorical variable. In R software, the categories are represented in terms of levels. So, check the levels of first column.

```
#Check levels of 'Salt.affected' column
>levels(samp_data$Salt.affected)
```

```
## NULL
```

The output shows that first column of sample data *i.e.* 'Salt.affected' column doesn't have levels. Therefore, R is not considering it as categorical variable. Convert this column to levels by using `factor` function as shown below.

```
#Convert 'Salt.affected' column to factor
>samp_data$Salt.affected =factor(samp_data$Salt.affected)
```

```
#Check whether 'Salt.affected' column converted to factor or
not
>levels(samp_data$Salt.affected)
```

```
## [1] "FALSE" "TRUE"
```

From the output of `levels` function, we can see that now R considers two levels of 'Salt.affected' column *i.e.* TRUE and FASLE representing it as dichotomous dependent variable.

Now we will partition the sample data into two parts such that 70% of data will be used for training and 30% of data will be used for testing of the binomial logistic regression algorithm. This partitioning can be done by taking random samples from both the categories in the given proportion. This can be done by using `sample` function in which number of sample sets (2 in this case, for training and testing) is mentioned in first argument and size of sample in second argument. If you want to do sampling with or without replacement, then it is mentioned in the `replace` argument. The proportion of samples for training and testing, respectively is assigned in `prob` argument. The R-code example for partitioning of sample data and for assigning separate variable names to the partitioned training and testing data is given further. The minimum, maximum, 1st and 3rd quartiles, median and mean of all columns in randomly selected training and testing data can be displayed by using `summary` function.

```
# Partitioning of sample data into train and test data
>samp =sample(2, nrow(samp_data), replace =TRUE, prob =c(0.7,
0.3))
>train =samp_data[samp==1,]#Assigning variable name to training
data
>test =samp_data[samp==2,]      #Assigning variable name to
testing data
```

```
>summary(train) #Summary of training dataset
```

```
## Salt.affectedCoastal_aerosol      Blue            Green
## FALSE:242      Min.   :0.1354  Min.   :0.1260  Min.   :0.1123
## TRUE : 78      1st Qu.:0.1436  1st Qu.:0.1374  1st Qu.:0.1284
##                Median :0.1643  Median :0.1616  Median :0.1619
##                Mean   :0.1675  Mean   :0.1665  Mean   :0.1700
##                3rd Qu.:0.1887  3rd Qu.:0.1945  3rd Qu.:0.2129
##                Max.   :0.2204  Max.   :0.2265  Max.   :0.2457
##       Red              NIR              SWIR1            SWIR2
## Min.   :0.09104  Min.   :0.09856  Min.   :0.06641  Min.   :0.04414
## 1st Qu.:0.11851  1st Qu.:0.22726  1st Qu.:0.18693  1st Qu.:0.12718
## Median :0.17363  Median :0.27702  Median :0.28268  Median :0.20615
## Mean   :0.17723  Mean   :0.26307  Mean   :0.26261  Mean   :0.20700
## 3rd Qu.:0.23317  3rd Qu.:0.31696  3rd Qu.:0.34214  3rd Qu.:0.29245
## Max.   :0.28266  Max.   :0.38802  Max.   :0.42438  Max.   :0.36333
##     cirrus               TIRS1           TIRS2
## Min.   :0.0005493  Min.   :109.7  Min.   :240.3
## 1st Qu.:0.0010968  1st Qu.:112.8  1st Qu.:245.0
## Median :0.0012337  Median :113.4  Median :246.0
## Mean   :0.0012483  Mean   :113.3  Mean   :245.7
## 3rd Qu.:0.0013706  3rd Qu.:114.0  3rd Qu.:246.7
## Max.   :0.0020321  Max.   :115.8  Max.   :249.6
```

```
>summary(test) #Summary of testing dataset
```

```
## Salt.affectedCoastal_aerosol      Blue            Green
## FALSE:104      Min.   :0.1366  Min.   :0.1280  Min.   :0.1134
## TRUE : 34      1st Qu.:0.1445  1st Qu.:0.1376  1st Qu.:0.1280
##                Median :0.1654  Median :0.1629  Median :0.1612
##                Mean   :0.1691  Mean   :0.1682  Mean   :0.1719
##                3rd Qu.:0.1878  3rd Qu.:0.1942  3rd Qu.:0.2169
##                Max.   :0.2173  Max.   :0.2245  Max.   :0.2426
##       Red              NIR              SWIR1            SWIR2
## Min.   :0.09463  Min.   :0.09948  Min.   :0.06549  Min.   :0.04507
## 1st Qu.:0.11654  1st Qu.:0.22605  1st Qu.:0.18547  1st Qu.:0.12460
## Median :0.17397  Median :0.28136  Median :0.29119  Median :0.21769
## Mean   :0.17923  Mean   :0.26620  Mean   :0.26563  Mean   :0.21007
## 3rd Qu.:0.23751  3rd Qu.:0.31641  3rd Qu.:0.35079  3rd Qu.:0.30002
## Max.   :0.28049  Max.   :0.36880  Max.   :0.40650  Max.   :0.35364
##     cirrus               TIRS1           TIRS2
## Min.   :0.0005493  Min.   :109.7  Min.   :240.3
## 1st Qu.:0.0010797  1st Qu.:112.8  1st Qu.:245.0
## Median :0.0012337  Median :113.5  Median :246.0
## Mean   :0.0012438  Mean   :113.4  Mean   :245.9
## 3rd Qu.:0.0013877  3rd Qu.:114.1  3rd Qu.:246.8
## Max.   :0.0018496  Max.   :115.8  Max.   :249.6
```

Now, develop binomial logistic regression algorithm on training dataset by using `glm` function and assigning `binomial'` parameter in its `family` argument.

```
#Apply Binomial Logistic regression on training dataset
>blr =glm(Salt.affected~., data = train, family ='binomial')
```

In this R-code, the model is developed by taking 'Salt.affected' column as dependent variable and remaining columns as independent variables represented as `Salt.affected~.` where'`.`' denotes all remaining columns in the data frame. Now predict the probabilities of occurrence of observations in training data with respect to dependent variable in order to check the accuracy of developed binomial logistic regression model. This can be done by using `predict` function and assigning model name in its first argument, name of dataset in the second argument and type of prediction as `'response'` in `type` argument as shown below.

```
#Prediction for training dataset
>blr_train_prob =predict(blr, train, type ='response')
```

If the predicted probability is more than 50%, then it would be salt-affected soil, otherwise not. So, assign the values in the predicted training dataset as 'TRUE' if their corresponding probability is more than 0.5, otherwise 'FALSE' by using `ifelse` control statement as shown below.

```
#Assign True to data having probability > 0.5, otherwise False
>blr_train =ifelse(blr_train_prob>0.5, "TRUE", "FALSE")
```

Create the confusion matrix of actual and predicted class of the training dataset to check the accuracy of developed binomial logistic regression model by using `table` function as depicted below.

```
#Confusion Matrix for training dataset
>table(Predicted =blr_train, Actual =train$Salt.affected)
##           Actual
## Predicted FALSE TRUE
##      FALSE   242    0
##      TRUE      0   78
```

From the confusion matrix of training dataset, it is clear that the counts of both actual and predicted values are same for both the classes which shows that the developed model has good accuracy which can be further validated on testing datasct. So, apply this developed model on independent testing dataset to predict the probabilities of dependent variable and create its confusion matrix and compute accuracy using following R-codes.

```
#Prediction for testing dataset
>blr_test_prob =predict(blr, test, type ='response')

#Assign True to data having probability > 0.5, otherwise False
>blr_test =ifelse(blr_test_prob>0.5, "TRUE", "FALSE")

#Confusion Matrix for testing dataset
```

```
>test_tab =table(Predicted =blr_test, Actual =test$Salt.affected)
>test_tab#Display table

##          Actual
## Predicted FALSE TRUE
##     FALSE   101    2
##     TRUE      3   32

#Accuracy of classification
>sum(diag(test_tab))/sum(test_tab)

## [1] 0.9637681
```

The confusion matrix and accuracy of classification of testing data shows that the developed binomial logistic regression model has good accuracy, so, it can be used for classification of raster dataset. As we don't have pixelwise observed classification corresponding to raster data, therefore, large number of point samples or polygon samples are required for training and testing of the machine learning algorithms. The classification results of raster data can be validated through ground-truth observations.

The developed binomial logistic regression algorithm can be applied on raster data by using `clusterR` function of `raster` package in parallel processing mode to speed up the execution process of R-code. For parallel processing, `doParallel` package is required. For applying any model on raster object through `clusterR` function, R-code is executed through following three steps:

1) Create a cluster object by using beginCluster() function which detects the number of cores available in the computer system and uses them for parallel processing.
2) Apply the model on raster data by using **clusterR** function in which raster data name and **predict** function from the raster package are assigned as arguments. Its args argument contains list of parameters for the **predict** function in which model name and prediction type is assigned.
3) The last step is closure of cluster object through **endCluster**() function.

The R-code example is given below.

```
# Import required packages
>library(raster)   # For raster processing
>library(doParallel)   # For parallel processing

#Applying binomial logistic regression on raster image

>beginCluster()    #Creates cluster object

## 8 cores detected, using 7
```

```
#Apply algorithm on raster data for prediction of response variable
>blr_raster_prob =clusterR(img, raster::predict,
args =list(model =blr, type ='response'))
>endCluster()        #Close the cluster object
```

The output of the above R-code is a raster file containing pixel-wise predicted probabilities for the dependent variable. The attributes of the predicted raster can be displayed as follows.

```
# Display attributes of predicted raster
>blr_raster_prob

## class      : RasterLayer
## dimensions : 678, 772, 523416  (nrow, ncol, ncell)
## resolution : 30, 30  (x, y)
## extent     : 432345, 455505, 2905185, 2925525  (xmin, xmax, ymin, ymax)
## crs        : +proj=utm +zone=44 +datum=WGS84 +units=m +no_defs +ellps=WGS84 +towgs84=0,0,0
## source     : memory
## names      : layer
## values     : 2.220446e-16, 1  (min, max)
```

Now we need to reclassify these probabilities values into two classes *i.e.* 'TRUE' if the probability value is more than 0.5, otherwise 'FASLE'. Here, we are denoting 'TRUE' as '1' and 'FALSE' as '0' in the raster. For performing this step, we need to convert output raster into matrix, and then reclassify them based on their corresponding probability value using `ifelse` control statement as depicted below.

```
#Convert the predicted raster to matrix
>blr_raster_matrix =getValues(blr_raster_prob)

>summary(blr_raster_matrix) #Display summary of created matrix

##    Min. 1st Qu.  Median    Mean 3rd Qu.    Max.
## 0.00000 0.00000 0.00000 0.02159 0.00000 1.00000

>head(blr_raster_matrix) #Display first few values of matrix

## [1] 2.220446e-16 2.220446e-162.220446e-162.220446e-162.220446e-16
## [6] 2.220446e-16

# Assign 1 to the value having probability > 0.5, otherwise 0
# Here, 1 denotes salt-affected soils whereas 0 denotes other categories

>blr_raster_TF =ifelse(blr_raster_matrix>0.5, 1, 0)

>head(blr_raster_TF) #Display first few values of matrix
```

```
## [1] 0 0 0 0 0 0
```

So, it is clear from the output that the raster matrix data has been converted into '0' and '1'. Now we will convert this matrix into raster format by using the procedure described in section 15.1 of this book. The R-codes for this conversion process are depicted below.

```
#Get attributes of first raster layer without pixel values
>blr_raster =raster(img)

>blr_raster#Display attributes of the created raster

## class      : RasterLayer
## dimensions : 678, 772, 523416  (nrow, ncol, ncell)
## resolution : 30, 30  (x, y)
## extent     : 432345, 455505, 2905185, 2925525  (xmin, xmax, ymin,
ymax)
## crs         : +proj=utm +zone=44 +datum=WGS84 +units=m +no_defs
+ellps=WGS84 +towgs84=0,0,0

>summary(blr_raster)  #Summary of raster denotes that it is
empty

##          layer
## Min.        NA
## 1st Qu.     NA
## Median      NA
## 3rd Qu.     NA
## Max.        NA
## NA's        NA

#Fill Binomial logistic regression classified pixels in empty
raster layer
>i =which(!is.na(blr_raster_TF))
>blr_raster[i] =blr_raster_TF

>blr_raster#Display attributes of classified raster

## class      : RasterLayer
## dimensions : 678, 772, 523416  (nrow, ncol, ncell)
## resolution : 30, 30  (x, y)
## extent     : 432345, 455505, 2905185, 2925525  (xmin, xmax, ymin,
ymax)
## crs         : +proj=utm +zone=44 +datum=WGS84 +units=m +no_defs
+ellps=WGS84 +towgs84=0,0,0
## source     : memory
## names      : layer
## values     : 0, 1  (min, max)
```

So, the classified raster containing two values *i.e.* '0' and '1' is created in which '1' represents presence of salt-affected soils whereas '0' represents its absence. Now we will plot this raster to visualize the results. The legends assigned to this plot are 'TRUE' and 'FALSE' denoting presence and absence of salt-affected soils, respectively.

```
#Plot classified Image
>ggR(blr_raster, geom_raster =TRUE) +#Plot with ggplot2
      ggtitle('Binomial Logistic Regression for identification
of salt-affected soils')+#Title of plot
      labs(x='Longitude (m)', y='Latitude (m)')+#Axis labels
      theme(plot.title =element_text(hjust =0.5, #Align title
in centre
      size =24), #size of title
      axis.title =element_text(size =20), #size of axis labels
      legend.key.size =unit(1, "cm"), #size of legend
      legend.title =element_text(size =20), #size of legend
title
      legend.text =element_text(size =15)) +#size of legend
text
      scale_fill_gradient(low ="gray48", high ="ivory",
name ="Salt-affected soils",
labels =c("False", "True"), breaks =c(0,1))
```

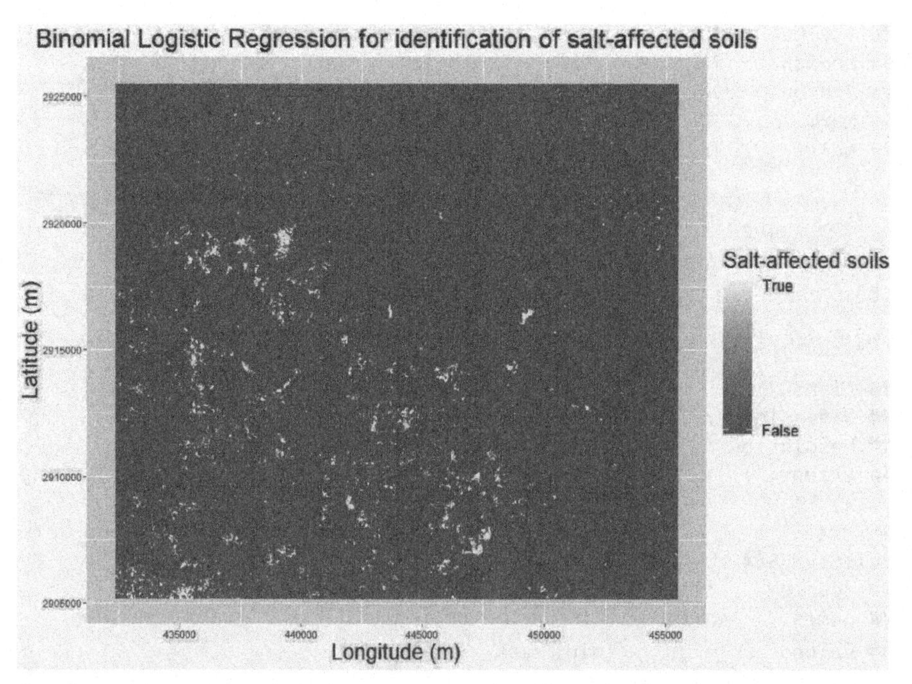

So, the white coloured pixels represent salt-affected soils whereas remaining pixels are grey in colour. In this way, binomial logistic regression technique can be used for classification of raster data in two categories.

16.2.2. Multiple logistic regression

Multiple logistic regression is similar to binomial logistic regression except that there are more than two categories of response or dependent variable. This algorithm can be used for supervised classification of raster data in multiple categories. In this section, we are going to execute this algorithm in R software for supervised classification of input raster data imported earlier. For implementation of multiple linear regression in R, `nnet` package is required. The remaining packages are same which were imported in the previous sections.

```
#Import required packages
>library(raster)   # For raster processing
>library(RStoolbox)   # For visualization of raster with ggplot2
>library(ggplot2)      # For visualization of raster
>library(doParallel)   # For Parallel processing
>library(nnet)    # For multiple logistic regression
```

Now we will import the '.csv' files containing sample data for training and testing as shown below.

```
# Read the file containing sample data for training and
validation
>samp_data =read.csv('samples_data.csv', header =TRUE)
>head(samp_data) #Display first few rows of sample data

##    Class Coastal_aerosol       Blue      Green        Red
NIR
## 1 Water       0.1402806 0.1322478 0.1230623 0.1045736 0.09900824
## 2 Water       0.1399093 0.1319781 0.1212157 0.1040089 0.09855811
## 3 Water       0.1387720 0.1311691 0.1212841 0.1033537 0.09932334
## 4 Water       0.1413483 0.1343827 0.1266189 0.1091596 0.10341961
## 5 Water       0.1412090 0.1333040 0.1234955 0.1053643 0.10155153
## 6 Water       0.1422767 0.1354613 0.1274396 0.1098373 0.10344211
##         SWIR1       SWIR2       cirrus     TIRS1    TIRS2
## 1 0.06813533 0.04542350 0.001279298 109.8694 240.4842
## 2 0.06641371 0.04414189 0.001416180 109.7238 240.2817
## 3 0.06729604 0.04447805 0.001142416 109.7752 240.4096
## 4 0.07237478 0.04882711 0.001324925 110.3856 241.1747
## 5 0.06906069 0.04668410 0.001028348 109.8828 240.5764
## 6 0.07099750 0.04840691 0.001484621 110.3582 241.0991
```

So, the first column *i.e.* 'Class' is the dependent variable and remaining columns containing reflectance values in respective bands are independent variables. Now we will partition the sample data into two parts for training and testing using the procedure described in section 16.2.1. The R-code example is given below.

```
#partitioning of sample data into train and test data
>samp =sample(2, nrow(samp_data), replace =TRUE, prob =c(0.7,
0.3))
> train =samp_data[samp==1,]
> test =samp_data[samp==2,]
```

```
>summary(train) #Summary of training dataset
```

```
##                  Class      Coastal_aerosol          Blue
##  Built-up           :39    Min.   :0.1354    Min.   :0.1260
##  Fallow soil        :61    1st Qu.:0.1440    1st Qu.:0.1371
##  Riverine Sand      :46    Median :0.1639    Median :0.1613
##  Salt-affected soil:76     Mean   :0.1675    Mean   :0.1664
##  Vegetation         :53    3rd Qu.:0.1877    3rd Qu.:0.1939
##  Water              :49    Max.   :0.2204    Max.   :0.2265
##      Green               Red                 NIR                SWIR1
## Min.   :0.1123   Min.   :0.09104   Min.   :0.09856   Min.   :0.06549
## 1st Qu.:0.1277   1st Qu.:0.11608   1st Qu.:0.22125   1st Qu.:0.18172
## Median :0.1590   Median :0.17214   Median :0.27275   Median :0.27672
## Mean   :0.1694   Mean   :0.17639   Mean   :0.25876   Mean   :0.25925
## 3rd Qu.:0.2135   3rd Qu.:0.23331   3rd Qu.:0.31572   3rd Qu.:0.34261
## Max.   :0.2457   Max.   :0.28266   Max.   :0.38802   Max.   :0.42438
##      SWIR2               cirrus                TIRS1             TIRS2
## Min.   :0.04414   Min.   :0.0005493   Min.   :109.7   Min.   :240.3
## 1st Qu.:0.12029   1st Qu.:0.0010968   1st Qu.:112.8   1st Qu.:245.0
## Median :0.20594   Median :0.0012337   Median :113.4   Median :246.0
## Mean   :0.20469   Mean   :0.0012456   Mean   :113.3   Mean   :245.7
## 3rd Qu.:0.29398   3rd Qu.:0.0013706   3rd Qu.:113.9   3rd Qu.:246.7
## Max.   :0.36333   Max.   :0.0019865   Max.   :115.8   Max.   :249.6
```

```
>summary(test) #Summary of testing dataset
```

```
##                  Class      Coastal_aerosol          Blue
##  Built-up           :14    Min.   :0.1367    Min.   :0.1276
##  Fallow soil        :23    1st Qu.:0.1440    1st Qu.:0.1382
##  Riverine Sand      :21    Median :0.1688    Median :0.1683
##  Salt-affected soil:36     Mean   :0.1692    Mean   :0.1686
##  Vegetation         :31    3rd Qu.:0.1892    3rd Qu.:0.1967
##  Water              : 9    Max.   :0.2143    Max.   :0.2188
##      Green              Red                NIR                SWIR1
## Min.   :0.1136   Min.   :0.0945   Min.   :0.09989   Min.   :0.07242
## 1st Qu.:0.1301   1st Qu.:0.1209   1st Qu.:0.24340   1st Qu.:0.19883
## Median :0.1707   Median :0.1798   Median :0.28802   Median :0.29598
## Mean   :0.1734   Mean   :0.1813   Mean   :0.27672   Mean   :0.27386
## 3rd Qu.:0.2133   3rd Qu.:0.2365   3rd Qu.:0.32112   3rd Qu.:0.35777
## Max.   :0.2407   Max.   :0.2733   Max.   :0.36749   Max.   :0.40247
##      SWIR2               cirrus                TIRS1             TIRS2
## Min.   :0.05009   Min.   :0.0005493   Min.   :109.9   Min.   :240.6
## 1st Qu.:0.13950   1st Qu.:0.0011196   1st Qu.:112.9   1st Qu.:245.1
## Median :0.23314   Median :0.0012337   Median :113.4   Median :245.9
## Mean   :0.21574   Mean   :0.0012502   Mean   :113.4   Mean   :245.9
```

```
## 3rd Qu.:0.30195   3rd Qu.:0.0013477   3rd Qu.:114.1   3rd Qu.:247.0
## Max.   :0.34229   Max.   :0.0020321   Max.   :115.7   Max.   :249.2
```

It is clear from the summary of training and testing data that samples are taken from each of the six classes. Now we will develop multiple linear regression model on training dataset using the `multinom` function of `nnet` package as shown below where 'Class' is dependent variable and remaining columns represented by '`.`' are independent variables.

```
#Multiple Logistic Regression on Training dataset
>mlr =multinom(Class~., data = train)
```

Now predict the classes of samples in training dataset using the developed multiple logistic regression model using `predict` function and `"class"` parameter in `type` argument.Compare the predicted class with actual class through confusion matrix and compute its accuracy.

```
#Predict for training dataset
>mlr_train =predict(mlr, train, type ="class")

#Confusion Matrix of training dataset
>train_table =table(Predicted =mlr_train, Actual =train$Class)
>train_table #Display confusion matrix

##                       Actual
## Predicted            Built-up Fallow soil Riverine Sand Salt-affected
soil
##   Built-up                39           0             0             0
##   Fallow soil              0          61             0             0
##   Riverine Sand            0           0            46             0
##   Salt-affected soil       0           0             0            76
##   Vegetation               0           0             0             0
##   Water                    0           0             0             0
##                       Actual
## Predicted              Vegetation Water
##   Built-up                      0     0
##   Fallow soil                   0     0
##   Riverine Sand                 0     0
##   Salt-affected soil            0     0
##   Vegetation                   53     0
##   Water                         0    49

#Accuracy of classification
>sum(diag(train_table))/sum(train_table)

## [1] 1
```

Accuracy of 1 denotes 100% accuracy for training dataset. Now validate the performance of developed multiple logistic regression model on independent set of testing dataset using following R-codes.

```
#Predict for testing dataset
>mlr_test =predict(mlr, test, type =”class”)

#Confusion Matrix of testing dataset
>test_table =table(Predicted =mlr_test, Actual =test$Class)
>test_table  # Display confusion matrix

##                        Actual
## Predicted             Built-up Fallow soil Riverine Sand Salt-affected
soil
##    Built-up             14          0            0               0
##    Fallow soil           0         22            0               0
##    Riverine Sand         0          0           21               0
##    Salt-affected soil    0          1            0              36
##    Vegetation            0          0            0               0
##    Water                 0          0            0               0
##                        Actual
## Predicted             Vegetation Water
##    Built-up                    0     0
##    Fallow soil                 1     0
##    Riverine Sand               0     0
##    Salt-affected soil          0     0
##    Vegetation                 30     0
##    Water                       0     9

#Accuracy of classification
>sum(diag(test_table))/sum(test_table)

## [1] 0.9850746
```

The accuracy of 98.5% on testing data depicts that the developed model has very good accuracy. Therefore, it can be further used for the classification of raster image. The procedure for classification of raster using `clusterR` algorithm is discussed in the section 16.2.1. The R-code example is given below.

```
#Apply multiple logistic regression on raster image
>beginCluster()     #Creates cluster object

## 8 cores detected, using 7

#Apply algorithm on raster data for prediction of response
variable
>mlr_raster =clusterR(img, raster::predict, args =list(model
=mlr))
>endCluster()        #Close the cluster object

>mlr_raster#Display the attributes of classified raster

## class      : RasterLayer
## dimensions : 678, 772, 523416  (nrow, ncol, ncell)
## resolution : 30, 30  (x, y)
```

```
## extent     : 432345, 455505, 2905185, 2925525  (xmin, xmax, ymin,
ymax)
## crs          : +proj=utm +zone=44 +datum=WGS84 +units=m +no_defs
+ellps=WGS84 +towgs84=0,0,0
## source     : memory
## names      : layer
## values     : 1, 6  (min, max)
```

The classified output raster has six classes corresponding to the levels of the 'Class' column of the sample data. Now we can plot the classified raster image to visualize the results.

```
#PLot cLassified Image
>ggR(mlr_raster, geom_raster =TRUE) +#PLot with ggpLot2
       ggtitle('Classification by Multiple Logistic
Regression')+#TitLe
       labs(x='Longitude (m)', y='Latitude (m)')+#Axis LabeLs
       theme(plot.title =element_text(hjust =0.5, #ALign titLe
in centre
       size =30), #size of titLe
       axis.title =element_text(size =20), #size of axis LabeLs
       legend.key.size =unit(1, "cm"), #size of Legend
       legend.title =element_text(size =20), #size of Legend
titLe
       legend.text =element_text(size =15))+#size of Legend text
       scale_fill_gradientn(colours =c("grey", "burlywood",
"wheat",
       "white", "darkgreen", "skyblue"),
       name ="Class", labels =levels(samp_data$Class))
```

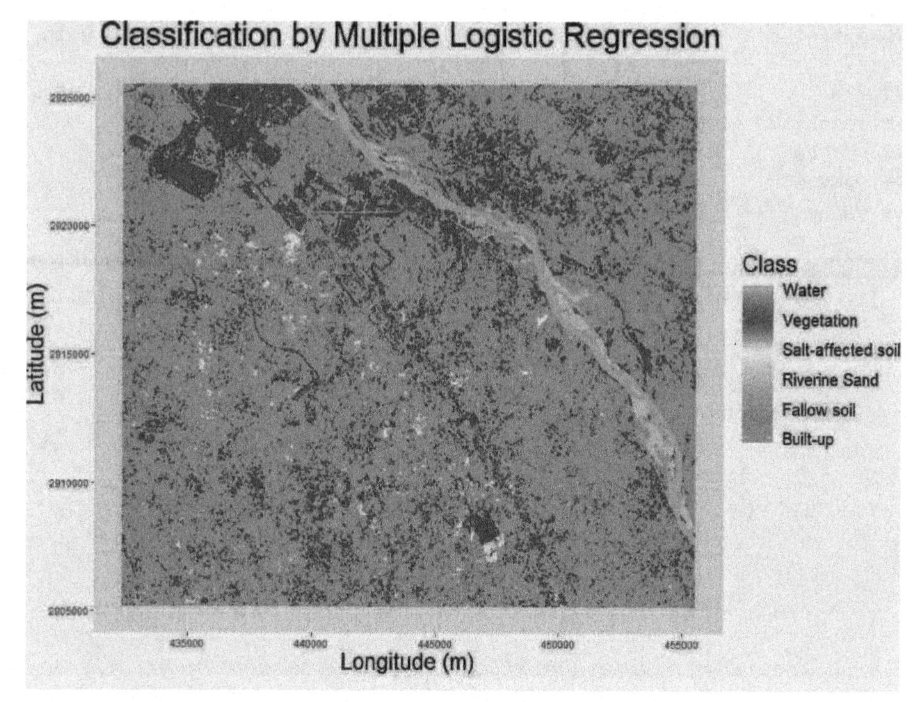

16.2.3. Support Vector Machine

Support Vector Machine (SVM) is supervised machine learning algorithm for both classification and regression. However, it is mostly used for classification. In Support Vector Machine, the data is plotted as points in n-dimensional space where each dimension represent a variable. The basic idea of this algorithm is to find a hyperplane that best divides a dataset into two classes.

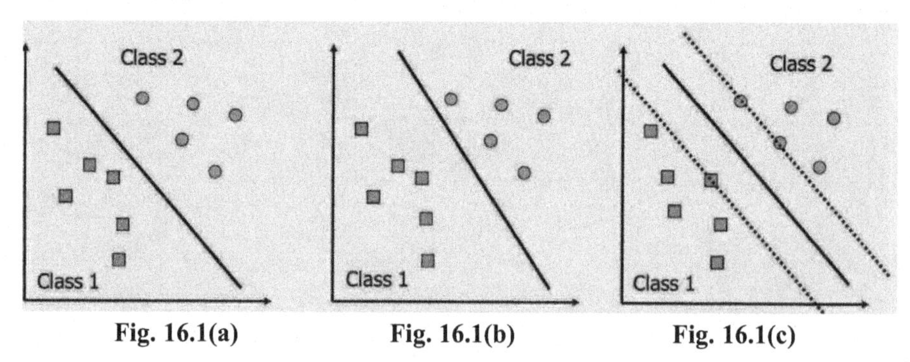

Fig. 16.1(a) **Fig. 16.1(b)** **Fig. 16.1(c)**

In the above figures, figure 16.1(a) and figure 16.1(b) represents bad decision because they are separating two classes with unequal margin which increases the chances of misclassification. However, Figure 16.1(c) correctly classifies the two classes with the hyperplane having equal margin. So, the process of finding out the correct hyperplane for classification with the help of support vectors,*i.e.* the points of each class present at the margin, is the basic principle of support vector machines. The Figure 16.2 depicts all components of Support Vector Machine (SVM).

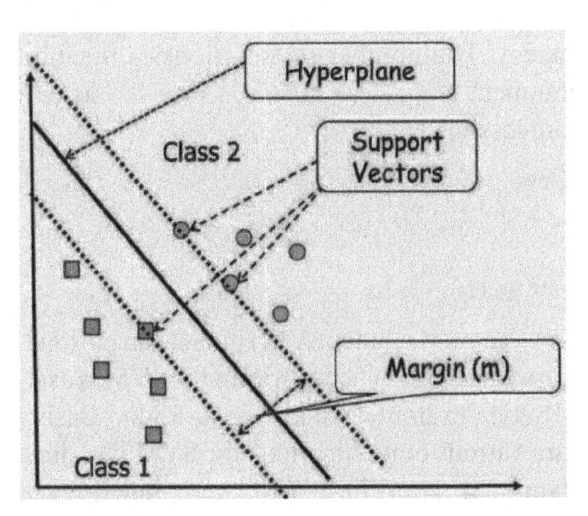

Fig. 16.2: Components of Support Vector Machine (SVM)

For performing support vector machine in R software, 'caret' and 'kernlab' packages are required along with other packages for raster analysis, parallel processing and raster visualization.

```
#Import required packages
>library(caret)   # For classification training
>library(kernlab)   # For support vector machine
>library(rgdal)   # For spatial data processing
>library(raster)   # For raster processing
>library(RStoolbox)   # For visualization of raster with ggplot2
>library(ggplot2)      # For visualization of raster
>library(doParallel)   # For parallel processing
```

In this exercise, we will use the input multiband raster data and '.csv' file containing sample data which was imported in the previous section (section 16.2.2). We will partition the sample data into testing and training set as explained in the previous section. After partitioning of sample data, we will develop the Support vector machine algorithm for classification on training dataset. We will execute this R-code in parallel processing mode.

```
#Parallel processing
>mc =makeCluster(detectCores())#make cluster of detected cores
>registerDoParallel(mc)#register parallel backend with 'foreach'
package
```

We will tune the parameters for training of model by using `trainControl` function of `caret` package. The resampling method is assigned in its `method` argument. We have used "repeatedcv" method as it repeatedly splits the training data into k-folds for cross-validation. The number of k-folds is provided in `number` argument. The number of iterations is mentionedin `repeats` argument. The argument `allowParallel` is set `TRUE` as we are running this code in parallel processing.

```
#Tuning Parameters for training
>tune_par =trainControl(method="repeatedcv", number=10,
repeats=5,
allowParallel=TRUE)
```

Training of Support Vector Machine (SVM) model can be done by using `train` function of the `caret` package. The algorithm for SVM is assigned in `method` function. In this R-code example, we are using Radial Basis Kernel function for SVM. There are various other functions for SVM like linear, least squares, exponential, polynomial, etc. The list of these functions along with their respective character string names for `method` argument are available on the website https://topepo.github.io/caret/train-models-by-tag.html#Support_Vector_Machines . You can develop SVM model using multiple functions and select the best function depending upon their accuracy. The summary metric to be used for selection of optimum model is assigned in `metric` method. We are using "Accuracy" method for this purpose in the given R-code example. Assign the tuning parameters in `trControl` argument. After training of model, parallel processing can be stopped by using `stopCluster` function.

```
#Train SVM model
>svm_model =train(Class ~., data=train, method ="svmRadial",
metric="Accuracy", trControl =tune_par)

>svm_model   #Display attributes of trained SVM model

## Support Vector Machines with Radial Basis Function Kernel
##
## 306 samples
##  10 predictor
##    6 classes: 'Built-up', 'Fallow soil', 'Riverine Sand',
'Salt-affected soil', 'Vegetation', 'Water'
##
## No pre-processing
## Resampling: Cross-Validated (10 fold, repeated 5 times)
## Summary of sample sizes: 276, 274, 275, 275, 277, 276, ...
## Resampling results across tuning parameters:
```

```
##
##    C      Accuracy   Kappa
##    0.25   0.9826620  0.9788198
##    0.50   0.9890203  0.9865966
##    1.00   0.9928560  0.9912658
##
## Tuning parameter 'sigma' was held constant at a value of 0.2022912
## Accuracy was used to select the optimal model using the largest
value.
## The final values used for the model were sigma = 0.2022912 and C =
1.
```

```
#Stop parallel processing
>stopCluster(mc)
```

The attributes of the developed SVM model shows the summary of model and values of cost *i.e.* C and sigma parameters. The cost parameter denotes the cost of constraints violation. Sigma parameter indicates the inverse kernel width for radial basis function.

Now we will predict the classified values using developed model for training dataset for accuracy assessment. The `confusionMatrix` function of `caret` package is used for accuracy assessment as it provides confusion matrix, overall statistics and statistics by class for evaluation of the performance of SVM model.

```
#Prediction of classes for training data
>pred_train =predict(svm_model, train)
```

```
#Accuracy assessment of SVM model
>confusionMatrix(pred_train, train$Class)
```

```
## Confusion Matrix and Statistics
##
##                    Reference
## Prediction         Built-up Fallow soil Riverine Sand Salt-affected
soil
##   Built-up             31           0             0             0
##   Fallow soil           0          56             0             0
##   Riverine Sand         0           0            46             0
##   Salt-affected soil    0           0             0            75
##   Vegetation            0           0             0             0
##   Water                 0           0             0             0
##                    Reference
## Prediction           Vegetation Water
##   Built-up                    0     0
##   Fallow soil                 0     0
##   Riverine Sand               0     0
##   Salt-affected soil          0     0
```

```
##   Vegetation                 57     0
##   Water                       0    41
##
## Overall Statistics
##                   Accuracy : 1
##                     95% CI : (0.988, 1)
##      No Information Rate : 0.2451
##      P-Value [Acc> NIR] : < 2.2e-16
##                      Kappa : 1
##
##  Mcnemar's Test P-Value : NA
##
## Statistics by Class:
##
##                      Class: Built-up Class: Fallow soil
## Sensitivity                   1.0000              1.000
## Specificity                   1.0000              1.000
## PosPred Value                1.0000              1.000
## Neg Pred Value                1.0000              1.000
## Prevalence                    0.1013              0.183
## Detection Rate                0.1013              0.183
## Detection Prevalence          0.1013              0.183
## Balanced Accuracy             1.0000              1.000
##                     Class: Riverine Sand Class: Salt-affected soil
## Sensitivity                        1.0000                    1.0000
## Specificity                        1.0000                    1.0000
## PosPred Value                     1.0000                    1.0000
## Neg Pred Value                     1.0000                    1.0000
## Prevalence                         0.1503                    0.2451
## Detection Rate                     0.1503                    0.2451
## Detection Prevalence               0.1503                    0.2451
## Balanced Accuracy                  1.0000                    1.0000
##                     Class: Vegetation Class: Water
## Sensitivity                     1.0000        1.000
## Specificity                     1.0000        1.000
## PosPred Value                  1.0000        1.000
## Neg Pred Value                  1.0000        1.000
## Prevalence                      0.1863        0.134
## Detection Rate                  0.1863        0.134
## Detection Prevalence            0.1863        0.134
## Balanced Accuracy               1.0000        1.000
```

Now validate the SVM model on testing dataset by using following R-codes.

```
#Prediction of classes for testing dataset
>pred_test =predict(svm_model, test)

#Accuracy assessment of SVM model on testing dataset
>confusionMatrix(pred_test, test$Class)
```

```
## Confusion Matrix and Statistics
##
##                     Reference
## Prediction           Built-up Fallow soil Riverine Sand Salt-affected
soil
##   Built-up                22           0             0             0
##   Fallow soil              0          25             0             0
##   Riverine Sand            0           0            21             0
##   Salt-affected soil       0           3             0            37
##   Vegetation               0           0             0             0
##   Water                    0           0             0             0
##                     Reference
## Prediction           Vegetation Water
##   Built-up                    0     0
##   Fallow soil                 0     0
##   Riverine Sand               0     0
##   Salt-affected soil          0     0
##   Vegetation                 27     0
##   Water                       0    17
##
## Overall Statistics
##
##                Accuracy : 0.9803
##                  95% CI : (0.9434, 0.9959)
##     No Information Rate : 0.2434
##     P-Value [Acc> NIR] : < 2.2e-16
##
##                   Kappa : 0.976
##
##  Mcnemar's Test P-Value : NA
##
## Statistics by Class:
##
##                      Class: Built-up Class: Fallow soil
## Sensitivity                   1.0000             0.8929
## Specificity                   1.0000             1.0000
## PosPred Value                 1.0000             1.0000
## Neg Pred Value                1.0000             0.9764
## Prevalence                    0.1447             0.1842
## Detection Rate                0.1447             0.1645
## Detection Prevalence          0.1447             0.1645
## Balanced Accuracy             1.0000             0.9464
##                      Class: Riverine Sand Class: Salt-affected soil
## Sensitivity                        1.0000                    1.0000
## Specificity                        1.0000                    0.9739
```

```
## PosPred Value                 1.0000                  0.9250
## Neg Pred Value                1.0000                  1.0000
## Prevalence                    0.1382                  0.2434
## Detection Rate                0.1382                  0.2434
## Detection Prevalence          0.1382                  0.2632
## Balanced Accuracy             1.0000                  0.9870
##                        Class: Vegetation Class: Water
## Sensitivity                      1.0000       1.0000
## Specificity                      1.0000       1.0000
## PosPred Value                    1.0000       1.0000
## Neg Pred Value                   1.0000       1.0000
## Prevalence                       0.1776       0.1118
## Detection Rate                   0.1776       0.1118
## Detection Prevalence             0.1776       0.1118
## Balanced Accuracy                1.0000       1.0000
```

The output shows that the accuracy of developed SVM model is quite good. Now this model can be used for classification of the raster data as follows.

```
#Apply support vector machine algorithm on raster image
>beginCluster()     #Creates cluster object

## 8 cores detected, using 7

#Apply algorithm on raster data for prediction of classes
>svm_raster = clusterR(img, raster::predict, args=
list(model=svm_model))
>endCluster()       #Close the cluster object

>svm_raster#Display the attributes of classified raster

## class      : RasterLayer
## dimensions : 678, 772, 523416  (nrow, ncol, ncell)
## resolution : 30, 30  (x, y)
## extent     : 432345, 455505, 2905185, 2925525  (xmin, xmax, ymin,
ymax)
## crs        : +proj=utm +zone=44 +datum=WGS84 +units=m +no_defs
+ellps=WGS84 +towgs84=0,0,0
## source     : memory
## names      : layer
## values     : 1, 6  (min, max)
```

Now we can plot the classified raster to visualize the classification results using following R-code.

```
#Plot classified Image
>ggR(svm_raster, geom_raster =TRUE) +#Plot with ggplot2
      ggtitle('Classification by Support Vector Machine', #Title
      subtitle ="Radial basis function") +#Subtitle
      labs(x='Longitude (m)', y='Latitude (m)')+#Axis labels
      theme(plot.title =element_text(hjust =0.5, #Align title
in centre
      size =30), #size of title
      plot.subtitle =element_text(hjust =0.5, #Subtitle in centre
      size =20), #size of subtitle
      axis.title =element_text(size =20), #size of axis labels
      legend.key.size =unit(1, "cm"), #size of legend
      legend.title =element_text(size =20), #size of legend
title
      legend.text =element_text(size =15))+#size of legend text
      scale_fill_gradientn(colours =c("grey", "burlywood",
"wheat",
      "white", "darkgreen", "skyblue"),
name ="Class", labels =levels(samp_data$Class))
```

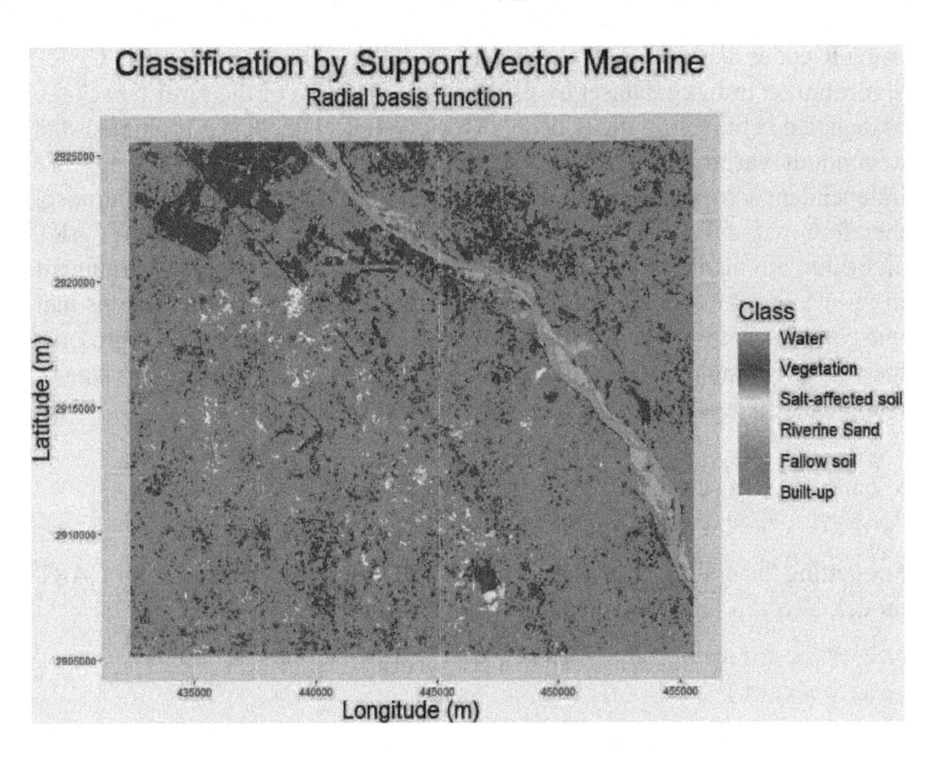

In this way, supervised classification of raster image through Support Vector Machine can be done in R.

16.2.4. Classification and Regression Tree

Classification and Regression Tree (CART) is a machine learning technique which constructs a tree-like graphical structure from the data containing multiple decision rules at their root node, branches and leaf nodes for classification or regression. For classification, CART takes categorical discrete values as the dependent variable whereas for regression, the dependent variable is continuous.CART is a predictive model which can be easily interpreted by the users and can handle non-parametric and non-linear data efficiently.

For implementing CART in R software for supervised classification of raster data, `rpart` package is required.`rpart.plot` package is also required for plotting the 'rpart' models.

```
#Import required packages
>library(rpart)  #For Classification and Regression Trees (CART)
>library(rpart.plot)   #For plotting 'rpart' models
```

The input multiband raster data andthe '.csv' file containing samples data which were imported and used in the previous section are also being used in this section. The sample data can be partitioned into training and testing datasets using R-codes discussed being in the section 16.2.1. Now we will apply CART algorithm on training dataset by using `rpart` function of the `rpart` package as depicted in the following R-code. In the given R-code, 'Class' column is the dependent variable whereas remaining columns represented as '`.`' are independent variables. As we are using CART for classification purpose, therefore, `method ='class'` is assigned. The control parameters for the CART algorithm are mentioned in the`control`argument. The `minsplit` argument in `rpart.control` function denotes minimum number of observations that must be present in a node while attempting a split. The argument `cp` represents the value of complexity parameter which determines the split. Higher value of complexity parameter results in smaller tree and vice-versa.

```
#Apply CART on training dataset
> cart =rpart(Class ~., data = train, method ='class',
control =rpart.control(cp =0.0, minsplit =2))
```

For getting the optimum number of splits and complexity parameter for CART model, `plotcp` function is used which graphically depicts the error plot.

```
#Plot to get the optimum number of splits and cp value
>plotcp(cart)
```

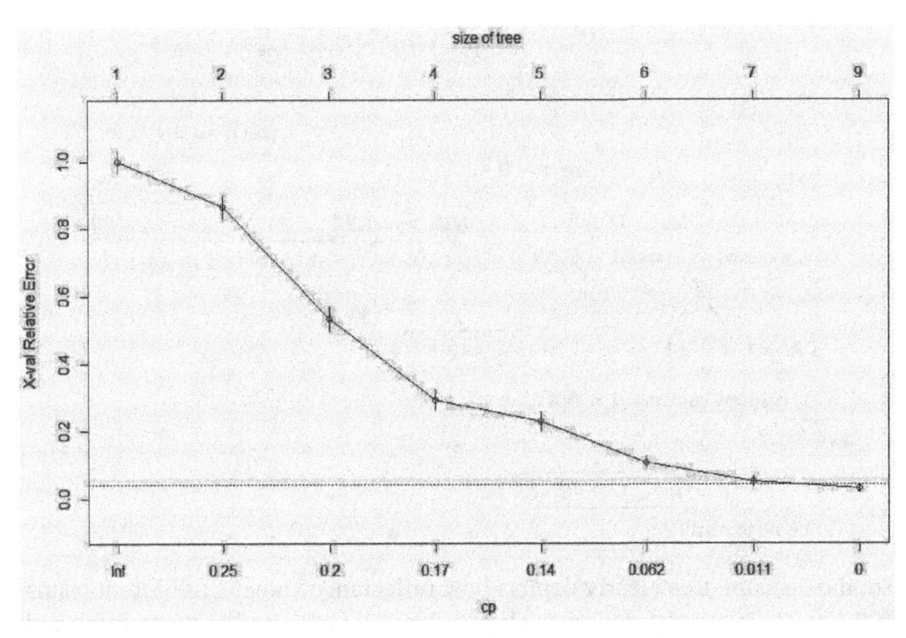

From the error plot, we can see that the relative error is lowest at cp value of '0' and size of tree as '9'. So, now will prune the earlier fully-grown tree using these parameters.

```
# Prune the CART
>cart_mod =rpart(Class ~., data = train, method ='class',
control =rpart.control(cp =0.0, minsplit =9))
```

Plot the decision tree to graphically visualize its structure using `prp` function of `rpart.plot` package as shown below. The argument `varlen =0` represents full variable names in the plot. The `cex` argument determines size of text in the plot.

```
# Plot CART
>prp(cart_mod, varlen =0, yesno =TRUE, cex =0.8)
```

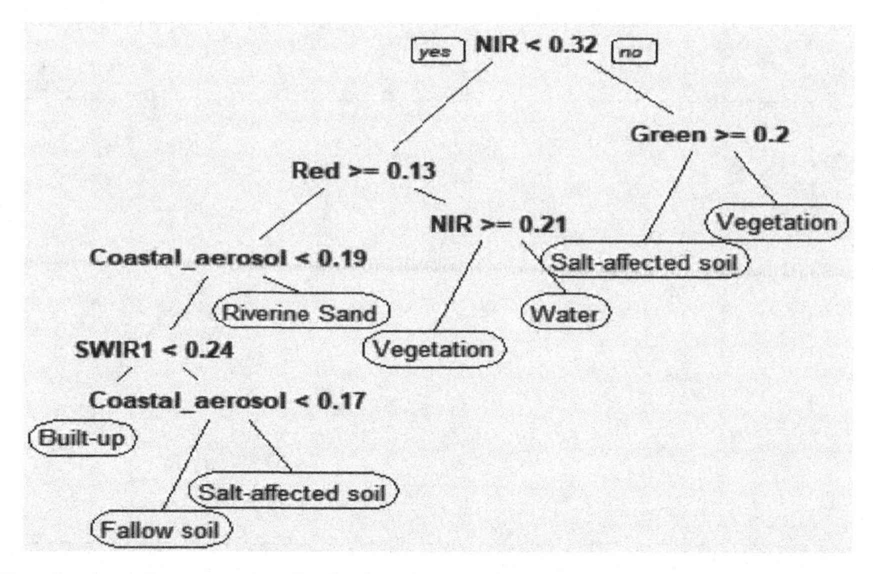

So, the decision tree clearly depicts how reflectance values of different bands in the raster image determine its classes. Now we will predict the training and testing dataset to compare the predicted class with actual class. The accuracy of model can be determined by computation of confusion matrix and accuracy by using following R-codes.

```
#Apply CART for predicting training dataset
>cart_train =predict(cart_mod, train, type =”class”)

#Confusion matrix of training dataset
> tr =table(actual = train[,1], predictions =cart_train)
> tr #Display confusion matrix
```

```
##                          predictions
## actual               Built-up Fallow soil Riverine Sand Salt-affected
soil
##    Built-up              39           0             0             0
##    Fallow soil            0          58             0             0
##    Riverine Sand          0           0            43             0
##    Salt-affected soil     0           0             0            79
##    Vegetation             0           1             0             0
##    Water                  0           0             0             0
##                       predictions
## actual                 Vegetation Water
##    Built-up                     0     0
##    Fallow soil                  2     0
##    Riverine Sand                0     0
##    Salt-affected soil           0     0
##    Vegetation                  56     0
##    Water                        0    43
```

```
>sum(diag(tr))/sum(tr) #Accuracy of classification for training
dataset

## [1] 0.9906542

#Apply CART for predicting testing dataset
>cart_test =predict(cart_mod, test, type =”class”)

#Confusion matrix of testing dataset
>ts =table(actual = test[,1], predictions =cart_test)
>ts#Display confusion matrix

##                     predictions
## actual         Built-up Fallow soil Riverine Sand Salt-affected soil
##   Built-up           14           0             0                  0
##   Fallow soil         0          24             0                  0
##   Riverine Sand       0           0            24                  0
##   Salt-affected soil  0           0             0                 33
##   Vegetation          0           1             0                  0
##   Water               0           0             0                  0
##                     predictions
## actual               Vegetation Water
##   Built-up                    0     0
##   Fallow soil                 0     0
##   Riverine Sand               0     0
##   Salt-affected soil          0     0
##   Vegetation                 26     0
##   Water                       0    15

>sum(diag(ts))/sum(ts)#Accuracy of classification for testing
dataset

## [1] 0.9927007
```

As we can see that the accuracy of developed CART model is quite good. So, we can use this model for classification of raster image by using following R-code.

```
# Apply CART model on raster data for classification
>raster_cart =predict(img, cart_mod, type=’class’)
>raster_cart #Display attributes of classified raster image

## class      : RasterLayer
## dimensions : 678, 772, 523416  (nrow, ncol, ncell)
## resolution : 30, 30  (x, y)
## extent     : 432345, 455505, 2905185, 2925525  (xmin, xmax, ymin,
ymax)
## crs         : +proj=utm +zone=44 +datum=WGS84 +units=m +no_defs
+ellps=WGS84 +towgs84=0,0,0
## source     : memory
## names      : layer
## values     : 1, 6  (min, max)
```

```
## attributes :
##          ID    value
##   from:   1 Built-up
##    to :   6    Water
```

Now we can plot this classified raster to visualize the results.

```
#Plot classified raster
>ggR(raster_cart, geom_raster =TRUE) +#Plot with ggplot2
       ggtitle('Classification and Regression Tree') +#Title of
plot
       labs(x='Longitude (m)', y='Latitude (m)')+#axis labels
       theme(plot.title =element_text(hjust =0.5, #Align title
in centre
       size =30), #size of title
       axis.title =element_text(size =20), #size of axis labels
       legend.key.size =unit(1, "cm"), #size of legend
       legend.title =element_text(size =20), #size of legend
title
       legend.text =element_text(size =15))+#size of legend text
       scale_fill_manual(values =c("grey", "burlywood", "wheat",
       "white", "darkgreen", "skyblue"),
       name ="Class")
```

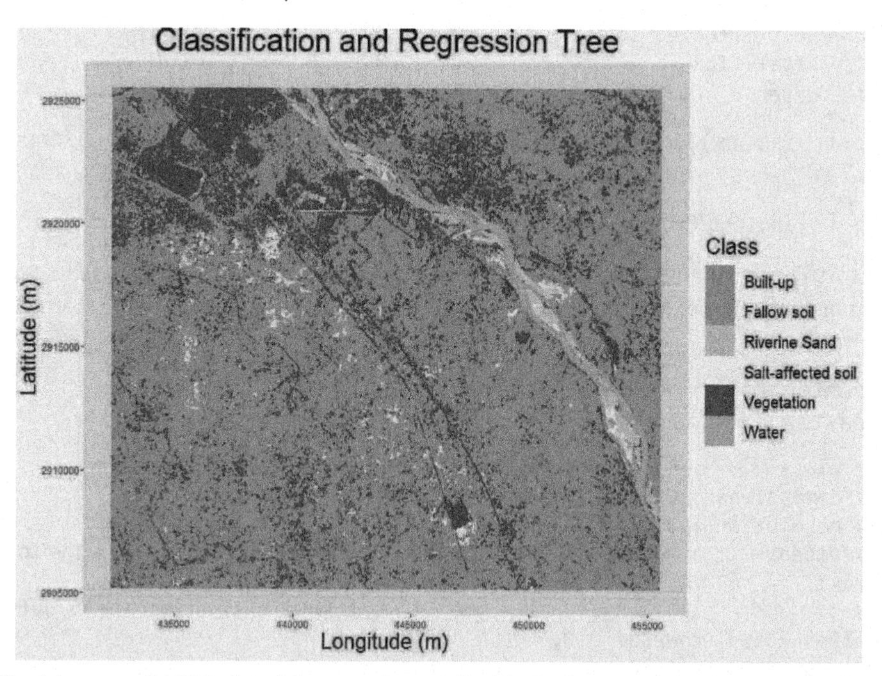

In this way, CART algorithm can be applied in R for supervised classification of the raster data.

16.2.5. Random Forest

Random forest is a machine learning technique which constructs multiple random uncorrelated decision trees from the subsets of training dataset and ensemble their outputs to provide the final decision or output. Random forest is used for classification and regression. Mode and mean prediction of the outputs of the individual decision trees is taken as the final output of the random forest in case of classification and regression, respectively. Random forest avoids the problem of over-fitting.

In R software, random forest technique can be applied in two ways. First way is by using built-in function in `RStoolbox` package which directly applies random forest on raster data. Second way is conversion of raster data into matrix, and then apply random forest technique using `randomForest` package. We will discuss both the methods in this section. First of all, lets import all packages required for performing this exercise.

```
#Import required packages
>library(rgdal)   #For vector data processing
>library(raster)   #For raster processing
>library(RStoolbox) #For random forest and visualization of
raster
>library(ggplot2)  #For visualization of raster with 'RStoolbox'
>library(randomForest) #For Random Forest
>library(caret)     #For creating confusion matrix
```

We will use the same multiband raster as input which was imported in the earlier section for performing this exercise.

16.2.5.1. Random forest by 'RStoolbox' package:

For applying random forest technique using `RStoolbox` package, import the shapefile containing training sites and re-project it to the Coordinate Reference System (CRS) of raster data using following R-codes.

```
#Import training data as spatial points
>trainData =readOGR(dsn =”training_points”,
“training_points”)

## OGR data source with driver: ESRI Shapefile
## Source: “E:\alka_PAT\Manual on
R\Chapter16\data\training_points”, layer: “training_points”
## with 458 features
## It has 2 fields
## Integer64 fields read as strings:  id
```

```
>trainData#Display attributes of training data
## class       : SpatialPointsDataFrame
## features    : 458
## extent      : 80.32199, 80.55325, 26.2663, 26.44891
(xmin, xmax, ymin, ymax)
## crs         : +proj=longlat +datum=WGS84 +no_defs
+ellps=WGS84 +towgs84=0,0,0
## variables   : 2
## names       : id,    Class
## min values  :  1, Built-up
## max values  :  6,    Water
```

```
>compareCRS(img, trainData) #compare CRS
## [1] FALSE
```

```
#Reproject shapefile of training points to the CRS of raster
>trainData =spTransform(trainData, CRS(projection(img)))
>compareCRS(img, trainData) # compare CRS to check the
reprojection
```

```
## [1] TRUE
```

Assign `model ='rf'` argument in the **`superClass`** function as depicted below. Use 70% data for training and 30% data for testing by assigning `trainPartition =0.7` argument in it. The description of other parameters is given in section 16.1.1. *i.e.* R-code for maximum likelihood classification.

```
#Supervised classification by Random Forest using 'RStoolbox'
package
>img_rf =superClass(img, trainData =trainData, responseCol
='Class',
nSamples =50, model ='rf', mode ='classification',
trainPartition =0.7, tuneLength =3, kfold =10)
```

```
## ******************* Model summary *******************
## Random Forest
##
## 323 samples
##  10 predictor
##   6 classes: 'Built-up', 'Fallow soil', 'Riverine Sand', 'Salt-
affected soil', 'Vegetation', 'Water'
##
## No pre-processing
## Resampling: Cross-Validated (10 fold)
## Summary of sample sizes: 291, 291, 291, 292, 291, 291, ...
## Resampling results across tuning parameters:
##
##   mtry  Accuracy   Kappa
```

```
##    2    0.9845644  0.9812244
##    6    0.9937500  0.9923628
##   10    0.9873992  0.9846151
##
## Accuracy was used to select the optimal model using the largest
value.
## The final value used for the model was mtry = 6.
## [[1]]
##   TrainAccuracyTrainKappa method
## 1       0.99375  0.9923628     rf
##
## [[2]]
## Cross-Validated (10 fold) Confusion Matrix
##
## (entries are average cell counts across resamples)
##
##                     Reference
## Prediction        Built-up Fallow soil Riverine Sand Salt-affected
soil
##   Built-up            3.8         0.0           0.0               0.0
##   Fallow soil         0.0         5.8           0.0               0.0
##   Riverine Sand       0.0         0.0           4.7               0.0
##   Salt-affected soil  0.0         0.1           0.0               7.9
##   Vegetation          0.0         0.0           0.0               0.0
##   Water               0.0         0.0           0.0               0.0
##                     Reference
## Prediction           Vegetation Water
##   Built-up                  0.0   0.0
##   Fallow soil               0.1   0.0
##   Riverine Sand             0.0   0.0
##   Salt-affected soil        0.0   0.0
##   Vegetation                5.8   0.0
##   Water                     0.0   4.1
##
##  Accuracy (average) : 0.9938
## ******************* Validation summary *******************
## Confusion Matrix and Statistics
##

##                     Reference
## Prediction         Built-up Fallow soil Riverine Sand Salt-affected
soil
##   Built-up             15          0             0                0
##   Fallow soil           0         22             0                0
##   Riverine Sand         0          0            19                0
##   Salt-affected soil    0          1             1               33
##   Vegetation            0          1             0                0
##   Water                 0          1             0                0
```

```
##                         Reference
## Prediction               Vegetation Water
##    Built-up                       1     0
##    Fallow soil                    1     0
##    Riverine Sand                  0     0
##    Salt-affected soil             0     0
##    Vegetation                    21     0
##    Water                          0    13
##
## Overall Statistics
##
##                 Accuracy : 0.9535
##                   95% CI : (0.9015, 0.9827)
##      No Information Rate : 0.2558
##      P-Value [Acc> NIR] : < 2.2e-16
##
##                    Kappa : 0.9431
##
##  Mcnemar's Test P-Value : NA
##
## Statistics by Class:
##
##                      Class: Built-up Class: Fallow soil
## Sensitivity                   1.0000             0.8800
## Specificity                   0.9912             0.9904
## PosPred Value                0.9375             0.9565
## Neg Pred Value                1.0000             0.9717
## Prevalence                    0.1163             0.1938
## Detection Rate                0.1163             0.1705
## Detection Prevalence          0.1240             0.1783
## Balanced Accuracy             0.9956             0.9352
##                     Class: Riverine Sand Class: Salt-affected soil
## Sensitivity                       0.9500                    1.0000
## Specificity                       1.0000                    0.9792
## PosPred Value                    1.0000                    0.9429
## Neg Pred Value                    0.9909                    1.0000
## Prevalence                        0.1550                    0.2558
## Detection Rate                    0.1473                    0.2558
## Detection Prevalence              0.1473                    0.2713
## Balanced Accuracy                 0.9750                    0.9896
##                      Class: Vegetation Class: Water
## Sensitivity                     0.9130       1.0000
## Specificity                     0.9906       0.9914
## PosPred Value                  0.9545       0.9286
## Neg Pred Value                  0.9813       1.0000
## Prevalence                      0.1783       0.1008
## Detection Rate                  0.1628       0.1008
## Detection Prevalence            0.1705       0.1085
## Balanced Accuracy               0.9518       0.9957
```

The output of the R-code depicts that the developed random forest model has very good accuracy. We can see the attributes of classified raster and labels for each class by using the given R-codes.

```
#Display attributes of Random Forest classified raster
>img_rf$map

## class       : RasterLayer
## dimensions : 678, 772, 523416  (nrow, ncol, ncell)
## resolution : 30, 30  (x, y)
## extent     : 432345, 455505, 2905185, 2925525  (xmin, xmax,
ymin, ymax)
## crs        : +proj=utm +zone=44 +datum=WGS84 +units=m +no_defs
+ellps=WGS84 +towgs84=0,0,0
## source     : memory
## names      : Class
## values     : 1, 6  (min, max)
## attributes :
##         ID    value
##  from:  1 Built-up
##    to :  6    Water

>img_rf$classMapping#Display labels of each class

##   classID              class
## 1       1           Built-up
## 2       2        Fallow soil
## 3       3     Riverine Sand
## 4       4 Salt-affected soil
## 5       5         Vegetation
## 6       6              Water
```

In this way, direct classification of raster by random forest technique can be done using `RStoolbox` package.

16.2.5.2. Random forest by 'randomForest' package:

For performing classification of raster data using `randomForest` package, raster data should be converted into matrix by using following R-codes.

```
#convert raster into matrix and omit no data values
>img_matrix =getValues(img) #convert to matrix
>img_matrix =na.omit(img_matrix) #omit NA values
```

Import the '.csv' file containing samples data and partition it into training and testing datasets using the given R-codes. 70% of data is used for training and 30% for testing.

```
# Read the file containing sample data for training and
validation
>samp_data =read.csv('samples_data.csv', header =TRUE)
#Partitioning of sample data into train and test data
>samp =sample(2, nrow(samp_data), replace =TRUE, prob =c(0.7,
0.3))
> train =samp_data[samp==1,]
> test =samp_data[samp==2,]
```

Now we will apply the random forest model on training dataset with say 2000 decision trees mentioned in the `ntree` argument of `randomForest` function. As the purpose of this exercise is classification, therefore, `'Class'` method is assigned in `strata` argument. 'Class' column is the dependent variable whereas remaining columns are independent variables. The R-code example is given below.

```
#Apply random forest technique on training dataset
>rf =randomForest(Class ~., data=train, ntree =2000, strata
='Class'
>rf#Display attributes of random forest model
##
## Call:
##   randomForest(formula = Class ~ ., data = train, ntree =
2000,      strata = "Class")
##                Type of random forest: classification
##                      Number of trees: 2000
## No. of variables tried at each split: 3
##
##         OOB estimate of  error rate: 1.6%
## Confusion matrix:
##                      Built-up Fallow soil Riverine Sand Salt-
affected soil
## Built-up               39         0           0              0
## Fallow soil             0        49           0              1
## Riverine Sand           0         0          48              1
## Salt-affected soil       0         0           0             70
## Vegetation              0         2           0              0
## Water                  0         0           0              0
##                       Vegetation Water class.error
## Built-up                      0     0  0.00000000
## Fallow soil                   0     0  0.02000000
## Riverine Sand                 0     0  0.02040816
## Salt-affected soil            0     0  0.00000000
## Vegetation                   64     1  0.04477612
## Water                         0    38  0.00000000
```

The output of `randomForest` function provides OOB estimate of error rate and confusion matrix. OOB or Out-Of-Bag estimate of error provides misclassification rate. The OOB error estimate and confusion matrix depicts that the accuracy of constructed random forest model is good.We can further improve the performance of random forest model by selecting the optimum number of trees. This can be obtained from the error plot of random forest model as depicted below.

```
#Error rate of random forest
>plot(rf, main ='Error Plot')
```

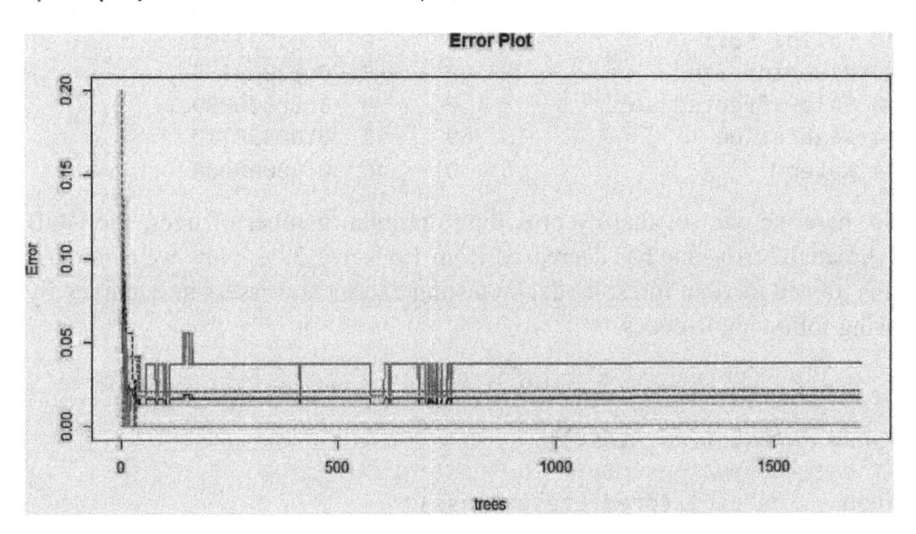

The error plot of random forest shows that error rate is constant at approximately 700 trees. Therefore, we will improve the existing random forest model by altering the number of trees as depicted below.

```
#Improve random forest model
>rf =randomForest(Class ~., data = train, ntree =700,
importance =TRUE, proximity =TRUE)

>rf   #Display attributes of random forest model

##
## Call:
##  randomForest(formula = Class ~ ., data = train, ntree =
1700,      importance = TRUE, proximity = TRUE)
##                Type of random forest: classification
##                      Number of trees: 700
## No. of variables tried at each split: 3
##
##         OOB estimate of  error rate: 1.29%
```

```
## Confusion matrix:
##                    Built-up Fallow soil Riverine Sand Salt-
affected soil
## Built-up               37       0            0            0
## Fallow soil             0      62            0            1
## Riverine Sand           0       0           47            1
## Salt-affected soil      0       0            0           81
## Vegetation              0       2            0            0
## Water                   0       0            0            0
##                     Vegetation Water class.error
## Built-up                     0     0  0.00000000
## Fallow soil                  0     0  0.01587302
## Riverine Sand                0     0  0.02083333
## Salt-affected soil           0     0  0.00000000
## Vegetation                  59     1  0.04838710
## Water                        0    40  0.00000000
```

So, here we can see that by providing optimum number of trees, the OOB estimate of error rate has decreased from 1.6% to 1.29%. Now we can apply this trained random forest model on testing dataset and assess its accuracy by using following R-codes.

```
#Apply random forest model for prediction of classes in testing
dataset
>pred =predict(rf, test)
#Create Confusion matrix for testing dataset
>confusionMatrix(pred, test$Class)
```

```
## Confusion Matrix and Statistics
##
##                       Reference
## Prediction           Built-up Fallow soil Riverine Sand Salt-
affected soil
##   Built-up               16       0            0            0
##   Fallow soil             0      21            0            0
##   Riverine Sand           0       0           19            0
##   Salt-affected soil      0       0            0           31
##   Vegetation              0       0            0            0
##   Water                   0       0            0            0
##                       Reference
## Prediction             Vegetation Water
##   Built-up                      0     0
##   Fallow soil                   0     0
##   Riverine Sand                 0     0
##   Salt-affected soil            0     0
##   Vegetation                   22     0
##   Water                         0    18
```

```
##
## Overall Statistics
##
##                Accuracy : 1
##                  95% CI : (0.9714, 1)
##     No Information Rate : 0.2441
##     P-Value [Acc> NIR] : < 2.2e-16
##
##                   Kappa : 1
##
##  Mcnemar's Test P-Value : NA
##
## Statistics by Class:
##                      Class: Built-up Class: Fallow soil
## Sensitivity                    1.000             1.0000
## Specificity                    1.000             1.0000
## PosPred Value                 1.000             1.0000
## Neg Pred Value                 1.000             1.0000
## Prevalence                     0.126             0.1654
## Detection Rate                 0.126             0.1654
## Detection Prevalence           0.126             0.1654
## Balanced Accuracy              1.000             1.0000
##                   Class: Riverine Sand Class: Salt-affected
soil
## Sensitivity                      1.0000                 1.0000
## Specificity                      1.0000                 1.0000
## PosPred Value                   1.0000                 1.0000
## Neg Pred Value                   1.0000                 1.0000
## Prevalence                       0.1496                 0.2441
## Detection Rate                   0.1496                 0.2441
## Detection Prevalence              0.1496                 0.2441
## Balanced Accuracy                1.0000                 1.0000
##                      Class: Vegetation Class: Water
## Sensitivity                      1.0000       1.0000
## Specificity                      1.0000       1.0000
## PosPred Value                   1.0000       1.0000
## Neg Pred Value                   1.0000       1.0000
## Prevalence                       0.1732       0.1417
## Detection Rate                   0.1732       0.1417
## Detection Prevalence             0.1732       0.1417
## Balanced Accuracy                1.0000       1.0000
```

The output of the `confusionMatrix` function clearly shows that the classification accuracy of developed random forest model on testing dataset is very good.

The importance of variables in classification through developed random forest model can be obtained by plotting variable importance plot using following R-code.

```
#Variable Importance Plot
>varImpPlot(rf)
```

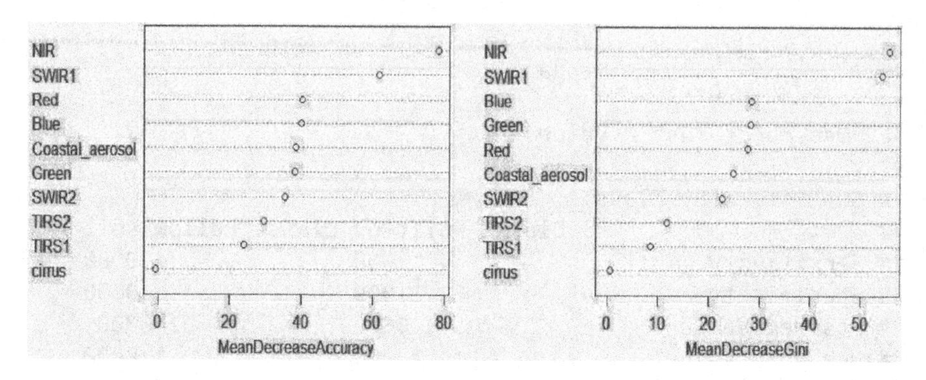

In the variable importance plot, Mean Decrease Accuracy represents the amount of decrease in model accuracy if that particular variable is omitted. Similarly, Mean Decrease Gini measure variable importance based on Gini impurity index which is used for calculation of splits in trees. Higher value of Mean Decrease Accuracy or Mean Decrease Gini indicates higher importance of that variable in the random forest model. The above variable importance plot shows that reflectance in NIR and SWIR-1 bands have highest importance as compared to reflectance in other bands in the developed random forest model.

The matrix of importance measure of each variable in each class can be obtained from the `importance` function as depicted below.

```
# Obtain matrix of importance measure of each variable in each class
>importance(rf)

##                  Built-up Fallow soil Riverine Sand Salt-affected soil
## Coastal_aerosol 28.588309   30.272909    28.5145611         18.1321073
## Blue            30.900934   30.679791    29.7626022         20.5330276
## Green           32.528095   34.175397    29.6279976         32.5265062
## Red             35.278150   39.472643    23.8579429         21.8139154
## NIR             29.038827   25.489521    37.5631990         44.9489977
## SWIR1           56.088847   51.912133    25.7842469         29.1249236
## SWIR2           22.503948   30.167992    10.3804539         15.2529870
## cirrus          -2.117376    2.376518    -0.7620208          0.5627397
## TIRS1           10.623363   14.356289    12.2272799         12.4039435
## TIRS2           10.824186   16.604938    15.1486662         14.9325334
##                Vegetation    Water MeanDecreaseAccuracyMeanDecreaseGini
## Coastal_aerosol23.429524 18.496408            39.387181        24.08140
## Blue            28.150763 23.699790            39.080133        26.64552
```

```
## Green        28.894709 29.705712       40.575634      28.61178
## Red          39.301826 35.096587       43.570876      33.30808
## NIR          42.645428 53.364627       71.329569      54.15635
## SWIR1        29.909423 45.018097       64.313230      61.34729
## SWIR2        23.926953 22.468689       34.882454      22.11152
## cirrus       -3.305374 -1.812400       -1.452145       0.59101
## TIRS1        15.727884  3.438327       25.785764       8.20811
## TIRS2        16.291908  5.171917       28.065399      11.85949
```

The number of times each variable used in the random forest model can be obtained from `varUsed` function as depicted below.

```
#Number of times each variable used in random forest
>varUsed(rf)

##  [1] 2472 2397 2152 2450 3691 3721 2280  508 1484 1636
```

If you want to know the effect of reflectance values in a particular band in predicting a specific class in random forest model, then partial dependence plot can be created. Partial dependence plot provides graphical representation of the marginal effect of a variable on the class probability. Suppose you want to see the effect of variation in NIR reflectance values in predicting the vegetation class in random forest model, then partial dependence plot can be created using `partialPlot` function as depicted below.

```
#Partial dependence plot of class 'Vegetation' on NIR
variable

>partialPlot(rf, train, NIR, "Vegetation")
```

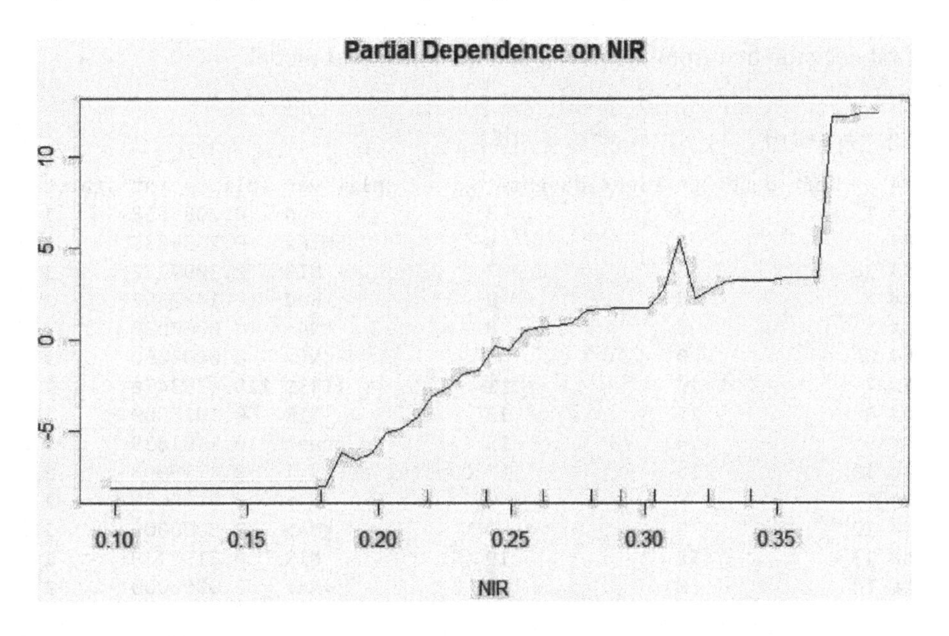

In this plot, we can see that when NIR reflectance value is more than 0.35, then there is highest probability of prediction of 'Vegetation' class. Similarly, partial dependence plot of class 'Water' on Red band can be created using following R-code.

```
#Partial dependence plot of class 'Water' on Red variable
>partialPlot(rf, train, Red, "Water")
```

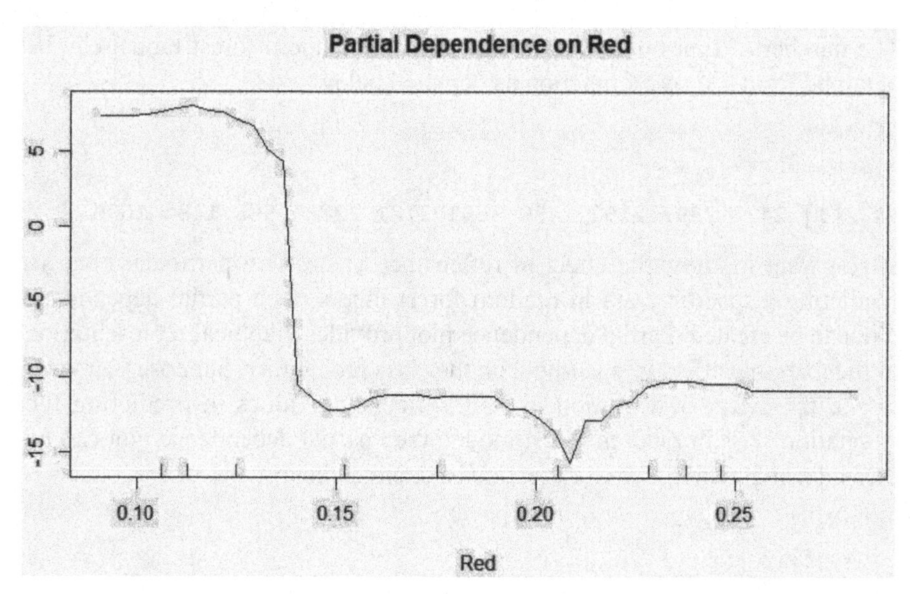

If you want to get a single decision tree of the random forest model, then `getTree` function is used. In the following R-code example, we are extracting first decision tree from the developed random forest model.

```
#Extract Single tree from the random forest model
>getTree(rf, 1, labelVar =TRUE)
```

```
##    left daughter right daughter       split var split point status
## 1              2              3             Red   0.2083568      1
## 2              4              5           SWIR1   0.2639036      1
## 3              6              7             NIR   0.3099772      1
## 4              8              9             Red   0.1437352      1
## 5              0              0            <NA>   0.0000000     -1
## 6              0              0            <NA>   0.0000000     -1
## 7             10             11           TIRS1 115.0353470      1
## 8             12             13             NIR   0.1925809      1
## 9             14             15           Green   0.1401839      1
## 10            16             17 Coastal_aerosol   0.1715098      1
## 11             0              0            <NA>   0.0000000     -1
## 12             0              0            <NA>   0.0000000     -1
## 13            18             19             NIR   0.2153129      1
## 14             0              0            <NA>   0.0000000     -1
```

```
## 15          0          0      <NA>   0.0000000    -1
## 16          0          0      <NA>   0.0000000    -1
## 17          0          0      <NA>   0.0000000    -1
## 18         20         21     TIRS2 245.5340882     1
## 19         22         23     TIRS2 246.1467819     1
## 20          0          0      <NA>   0.0000000    -1
## 21          0          0      <NA>   0.0000000    -1
## 22         24         25     Green   0.1370149     1
## 23          0          0      <NA>   0.0000000    -1
## 24          0          0      <NA>   0.0000000    -1
## 25         26         27     Green   0.1375849     1
## 26          0          0      <NA>   0.0000000    -1
## 27          0          0      <NA>   0.0000000    -1
##            prediction
## 1                <NA>
## 2                <NA>
## 3                <NA>
## 4                <NA>
## 5         Fallow soil
## 6       Riverine Sand
## 7                <NA>
## 8                <NA>
## 9                <NA>
## 10               <NA>
## 11      Riverine Sand
## 12              Water
## 13               <NA>
## 14        Fallow soil
## 15           Built-up
## 16        Fallow soil
## 17 Salt-affected soil
## 18               <NA>
## 19               <NA>
## 20        Fallow soil
## 21           Built-up
## 22               <NA>
## 23        Fallow soil
## 24         Vegetation
## 25               <NA>
## 26        Fallow soil
## 27         Vegetation
```

In this way, various parameters of developed random forest model can be derived. Now we will use this developed random forest for classification of raster matrix using following R-codes.

```
#Apply random forest to raster matrix
>matrix_rf =predict(rf, img_matrix)
```

Now convert this classified matrix into raster format using the procedure explained in section 16.2.1. Binomial logistic regression. The R-codes are given below.

```
#Get attributes of first raster layer without pixel values
>raster_rf =raster(img)
>raster_rf   #Display attributes of the created raster

## class      : RasterLayer
## dimensions : 678, 772, 523416  (nrow, ncol, ncell)
## resolution : 30, 30  (x, y)
## extent     : 432345, 455505, 2905185, 2925525  (xmin, xmax, ymin,
ymax)
## crs         : +proj=utm +zone=44 +datum=WGS84 +units=m +no_defs
+ellps=WGS84 +towgs84=0,0,0

>summary(raster_rf) #Summary of raster denotes that it is empty

##          layer
## Min.        NA
## 1st Qu.     NA
## Median      NA
## 3rd Qu.     NA
## Max.        NA
## NA's        NA

#Fill Random Forest classified pixels in empty raster layer
>i =which(!is.na(img_matrix))
>raster_rf[i] =matrix_rf
>raster_rf#Display attributes of classified raster

## class      : RasterLayer
## dimensions : 678, 772, 523416  (nrow, ncol, ncell)
## resolution : 30, 30  (x, y)
## extent     : 432345, 455505, 2905185, 2925525  (xmin, xmax, ymin,
ymax)
## crs         : +proj=utm +zone=44 +datum=WGS84 +units=m +no_defs
+ellps=WGS84 +towgs84=0,0,0
## source     : memory
## names      : layer
## values     : 1, 6  (min, max)
```

The raster layer classified by random forest technique can be plotted using following R-code to visualize the results.

```
#Plot classified raster
>ggR(raster_rf, geom_raster =TRUE) +#Plot with ggplot2
      ggtitle('Random Forest Classification') +#Title ofplot
      labs(x='Longitude (m)', y='Latitude (m)')+#axis labels
      theme(plot.title =element_text(hjust =0.5, #Align title
in centre
```

```
      size =30), #size of title
      axis.title =element_text(size =20), #size of axis labels
      legend.key.size =unit(1, "cm"), #size of legend
      legend.title =element_text(size =20), #size of legend
title
      legend.text =element_text(size =15))+#size of legend text
      scale_fill_gradientn(colors =c("grey", "burlywood",
"wheat",
      "white", "darkgreen", "skyblue"),
      name ="Class", labels =levels(samp_data$Class))
```

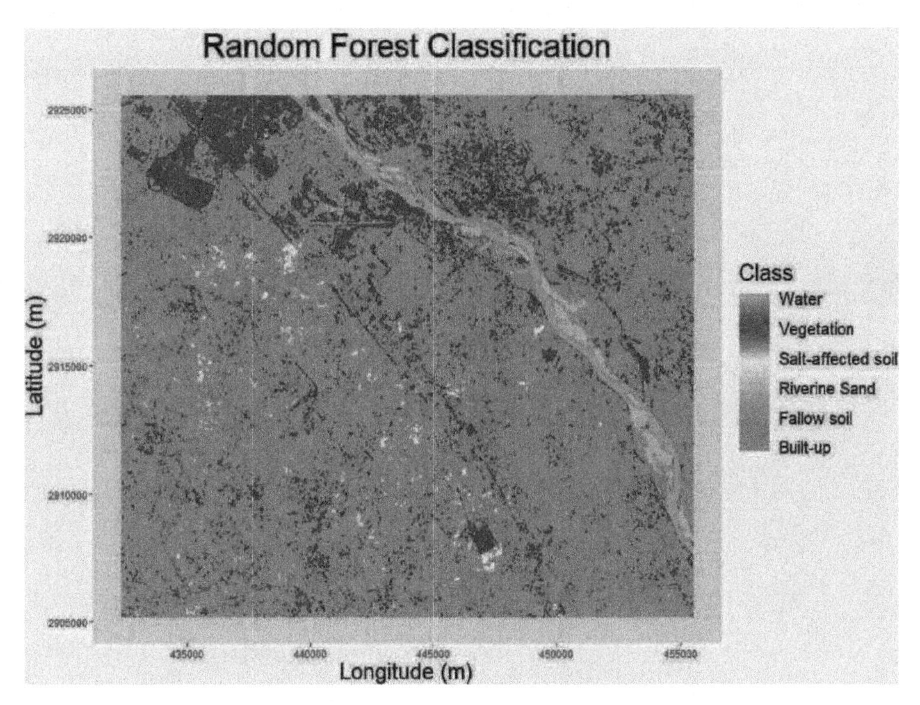

In this way, random forest algorithm can be used for classification of raster data.

Chapter 17

Digital Terrain Analysis

Elevation or height data of land surface above a defined vertical datum is stored in the form of raster grids which is known as Digital Elevation Model (DEM). The DEM data can be processed to derive many useful parameters like slope, aspect, curvature, roughness, etc. This process of quantitatively describing terrain obtained from DEM data through computation of many variables is known as Digital Terrain Analysis. Digital terrain analysis has applications in the various fields like hydrology, ecology, geomorphology, etc. DEM data for digital terrain analysis are available from the satellites like ASTER (Advanced Spaceborne Thermal Emission and Reflection Radiometer), Cartosat, ALOS (Advanced Land Observation Satellite), and SRTM (Shuttle Radar Topography Mission).

In this exercise, we are going to discuss the derivation of some primary and secondary terrain attributes from DEM data in R software. For performing digital terrain analysis in R, following packages are required.

```
#Set the working directory
> setwd('E:\\...............\\Chapter17\\data')
#Import required packages
>library(RSAGA)       #SAGA geoprocessing and terrain analysis
>library(raster)      #For raster data processing
>library(RStoolbox)  #For visualization of raster with 'ggplot2'
>library(rasterVis)  #For 3D visualization of raster
>library(ggplot2)     #For visualization of raster
>library(rgdal)       #For spatial data processing
```

In this exercise, we are going to perform digital terrain analysis on the DEM data derived from SRTM having spatial resolution of about 30 m. So, let's import the DEM data using the following R-code.

```
#Import DEM image
>dem =raster('DEM_SRTM.tif')
>dem #Display attributes of DEM layer
```

```
## class      : RasterLayer
## dimensions : 3115, 3484, 10852660  (nrow, ncol, ncell)
## resolution : 0.0002777778, 0.0002777778  (x, y)
## extent     : 76.02069, 76.98847, 23.09736, 23.96264  (xmin, xmax,
ymin, ymax)
## crs         : +proj=longlat +datum=WGS84 +no_defs +ellps=WGS84
+towgs84=0,0,0
## source     : E:/................................/Chapter17/data/DEM_SRTM.tif
## names      : DEM_SRTM
## values     : -32768, 32767  (min, max)
```

We can see from the attributes of DEM raster data that the pixels are present in angular units. We need to convert the pixels from angular to linear units for performing digital terrain analysis in R software. Therefore, DEM raster data is re-projected to UTM Coordinate Reference System (CRS) having linear units using following R-codes.

```
#Reproject DEM raster to UTM
>utm_proj =CRS('+proj=utm +zone=43 +datum=WGS84 +units=m +no_defs
+ellps=WGS84 +towgs84=0,0,0 ') #Define CRS object
>dem_utm =projectRaster(dem, crs = utm_proj) #Re-project to UTM
>dem_utm  #Display attributes of re-projected DEM
```

```
## class      : RasterLayer
## dimensions : 3155, 3525, 11121375  (nrow, ncol, ncell)
## resolution : 28.4, 30.8  (x, y)
## extent     : 603707.6, 703817.6, 2554497, 2651671  (xmin, xmax, ymin,
ymax)
## crs         : +proj=utm +zone=43 +datum=WGS84 +units=m +no_defs
+ellps=WGS84 +towgs84=0,0,0
## source     : memory
## names      : DEM_SRTM
## values     : 351.0788, 602.3245  (min, max)
```

We can see from the attributes of re-projected DEM that it is now having units in meter (m). We can plot the DEM using following R-code.

```
#2-D plot of DEM
>ggR(dem_utm, geom_raster =TRUE) +#Plot with ggplot2
      ggtitle('Digital Elevation Model Image')+#Title
      labs(x='Longitude (m)', y='Latitude (m)')+#Axis labels
      theme(plot.title =element_text(hjust =0.5, #Align title in
centre
      size =24), #size of title
      axis.title =element_text(size =20), #size of axis labels
      legend.key.size =unit(1, "cm"), #size of legend
      legend.title =element_text(size =20), #size of legend title
      legend.text =element_text(size =15)) +#size of legend text
      #Define colours
scale_fill_gradientn(colours =terrain.colors(10), name
="Elevation")
```

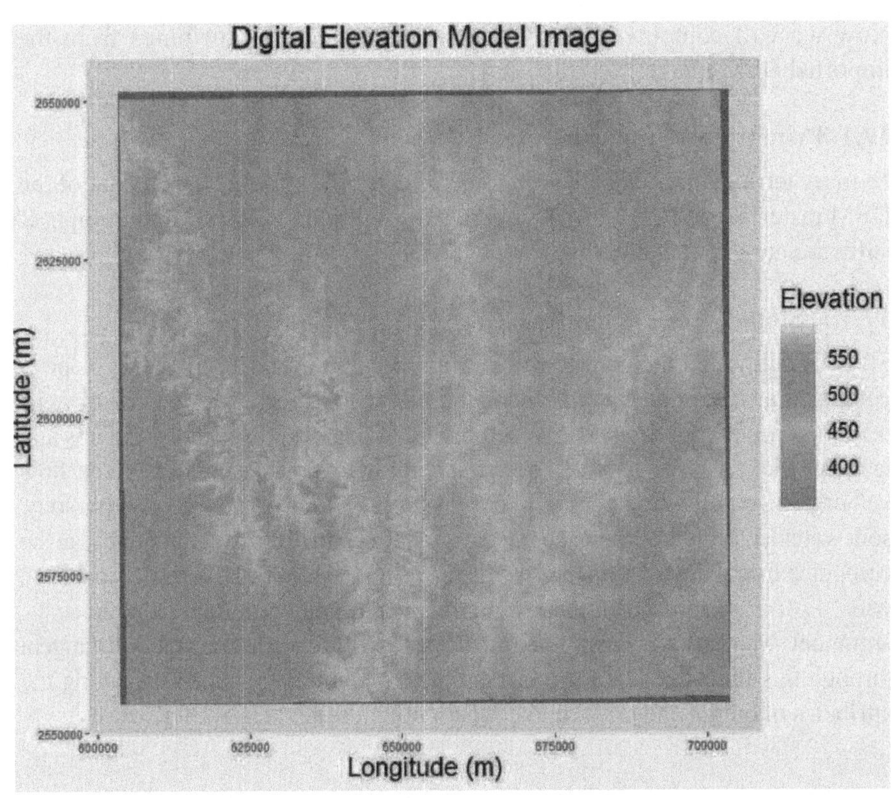

If you want to view the interactive three-dimensional plot of the DEM raster, then you can use the `plot3D` function of `rasterVis` package. The interactive 3-D plot will be displayed in the viewer window of R software.

```
#Interactive 3-D plot of DEM
>plot3D(dem_utm)
```

Now we will compute primary and secondary terrain attributes from the imported DEM image.

17.1. Primary terrain attributes

Primary terrain attributes are computed directly from the elevation data of the DEM raster. Some of the primary terrain attributes and their computation in R software are explained below.

17.1.1. Slope

Slope is defined as the change in elevation per unit distance. In DEM, slope is computed as maximum rate of change in elevation values from one pixel to its neighbouring pixels. Slope can be expressed in degrees, percentage, radians and tangent. Slope can be used for determination of overland and subsurface flow velocity and runoff rate, precipitation intensity, vegetation density, geomorphology, soil water content, and land cover classification. In R software, slope can be computed from DEM raster image by using `terrain` function of `raster` package with `'slope'` parameter in the `opt` argument and units of output slope in `unit` argument. Slope can be computed in the units of degrees, radians and tangent through this function. The R-code example for computing slope, displaying the attributes of output and creating the plot of output raster are given below.

```
#Slope
>slope =terrain(dem_utm, opt='slope', unit='degrees')
>slope #Display attributes of slope layer

## class      : RasterLayer
## dimensions : 3155, 3525, 11121375  (nrow, ncol, ncell)
## resolution : 28.4, 30.8  (x, y)
## extent     : 603707.6, 703817.6, 2554497, 2651671  (xmin, xmax, ymin,
ymax)
## crs          : +proj=utm +zone=43 +datum=WGS84 +units=m +no_defs
+ellps=WGS84 +towgs84=0,0,0
## source     : memory
## names      : slope
## values     : 0, 27.37814  (min, max)

#Plot Slope
>ggR(slope, geom_raster =TRUE) +#Plot with ggplot2
       ggtitle('Slope')+#Title
       labs(x='Longitude (m)', y='Latitude (m)')+#axis labels
       theme(plot.title =element_text(hjust =0.5, #Align title
in centre
       size =24), #size of title
       axis.title =element_text(size =20), #size of axis labels
       legend.key.size =unit(1, "cm"), #size of legend
```

```
        legend.title =element_text(size =20), #size of Legend
title
        legend.text =element_text(size =15)) +#size of Legend
text
#Define colours
        scale_fill_gradientn(colours =terrain.colors(5), name
="Slope (degrees)")
```

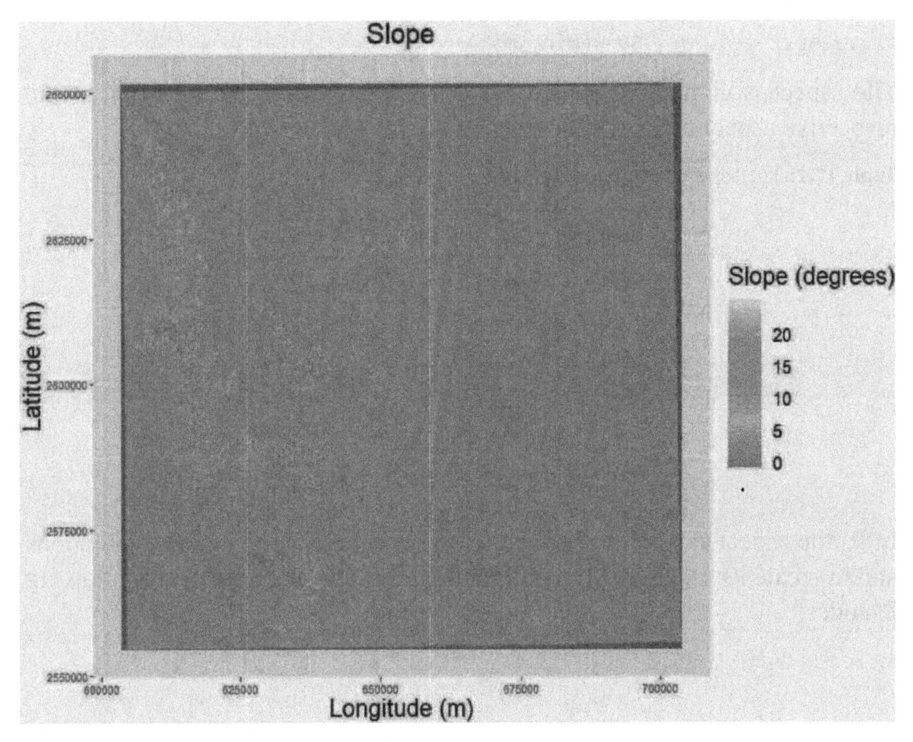

You can create the output raster of this slope by using **writeRaster** function as explained in the section 3.7 of Chapter 3.

17.1.2. Aspect

Aspect is the cardinal direction of slope and is expressed in terms of degrees. It can be used for determining solar insolation, evapotranspiration, soil moisture, distribution and abundance of flora and fauna, fire management, etc. It can be computed in R by assigning `opt='aspect'` argument in **terrain** function as depicted below.

```
#Aspect
> aspect =terrain(dem_utm, opt='aspect',unit='degrees')
> aspect #Display attributes of aspect Layer
```

```
## class       : RasterLayer
## dimensions : 3155, 3525, 11121375  (nrow, ncol, ncell)
## resolution : 28.4, 30.8  (x, y)
## extent    : 603707.6, 703817.6, 2554497, 2651671 (xmin, xmax, ymin,
ymax)
## crs          : +proj=utm +zone=43 +datum=WGS84 +units=m +no_defs
+ellps=WGS84 +towgs84=0,0,0
## source      : memory
## names       : aspect
## values      : 0, 360  (min, max)
```

The aspect is computed in terms of 0 to 360 degrees. The direction of the respective aspect in degrees is given in the following table 17.1.

Table 17.1: The aspects and their respective directions

Aspect in degrees	Direction
0 – 22.5& 337.5 - 360	North
22.5 – 67.5	Northeast
67.5 – 112.5	East
112.5 – 157.5	Southeast
157.5 – 202.5	South
202.5 – 247.5	Southwest
247.5 – 292.5	West
292.5 – 337.5	Northwest

In R, the aspect can be reclassified to assign the above-mentioned directions and to create its plot with proper direction labels by using the set of following R-codes.

```
#Reclassify the aspect

#Define reclassification matrix
>aspect_matrix =matrix(c(0, 22.5, 1, 22.5, 67.5, 2, 67.5, 112.5,
3,
112.5, 157.5, 4, 157.5, 202.5, 5, 202.5, 247.5,
6, 247.5, 292.5, 7, 292.5, 337.5, 8, 337.5,
360, 1), ncol =3, byrow =TRUE)

>aspect_matrix #Display matrix

##          [,1]  [,2] [,3]
##   [1,]    0.0  22.5    1
##   [2,]   22.5  67.5    2
##   [3,]   67.5 112.5    3
##   [4,] 112.5 157.5    4
##   [5,] 157.5 202.5    5
##   [6,] 202.5 247.5    6
##   [7,] 247.5 292.5    7
```

```
##  [8,] 292.5 337.5    8
##  [9,] 337.5 360.0    1

#Reclassify aspect layer according to matrix
>aspect_dir =reclassify(aspect, aspect_matrix)

>aspect_dir #Display attributes of classified aspect layer

## class      : RasterLayer
## dimensions : 3155, 3525, 11121375  (nrow, ncol, ncell)
## resolution : 28.4, 30.8  (x, y)
## extent     : 603707.6, 703817.6, 2554497, 2651671  (xmin, xmax, ymin,
ymax)
## crs          : +proj=utm +zone=43 +datum=WGS84 +units=m +no_defs
+ellps=WGS84 +towgs84=0,0,0
## source     : memory
## names      : aspect
## values     : 1, 8  (min, max)

#Define name of labels for aspect plot
>lab =c('North', 'Northeast', 'East', 'Southeast', 'South',
'Southwest', 'West', 'Northwest')

#Plot the aspect
>ggR(aspect_dir, geom_raster =TRUE) +#Plot with ggplot2
      ggtitle('Aspect')+#Title
      labs(x='Longitude (m)', y='Latitude (m)')+#Axis labels
      theme(plot.title =element_text(hjust =0.5, #Align title
in centre
      size =24), #size of title
      axis.title =element_text(size =20), #size of axis labels
      legend.key.size =unit(1, "cm"), #size of legend
      legend.title =element_text(size =20), #size of legend
title
      legend.text =element_text(size =15)) +#size of legend
text
      #Define customized colours
      scale_fill_gradientn(colours =c('red', 'orange', 'yellow',
'green',
'skyblue','blue','darkblue','magenta'),
      name ="Aspect Direction", labels = lab,
      breaks =c(1,2,3,4,5,6,7,8))
```

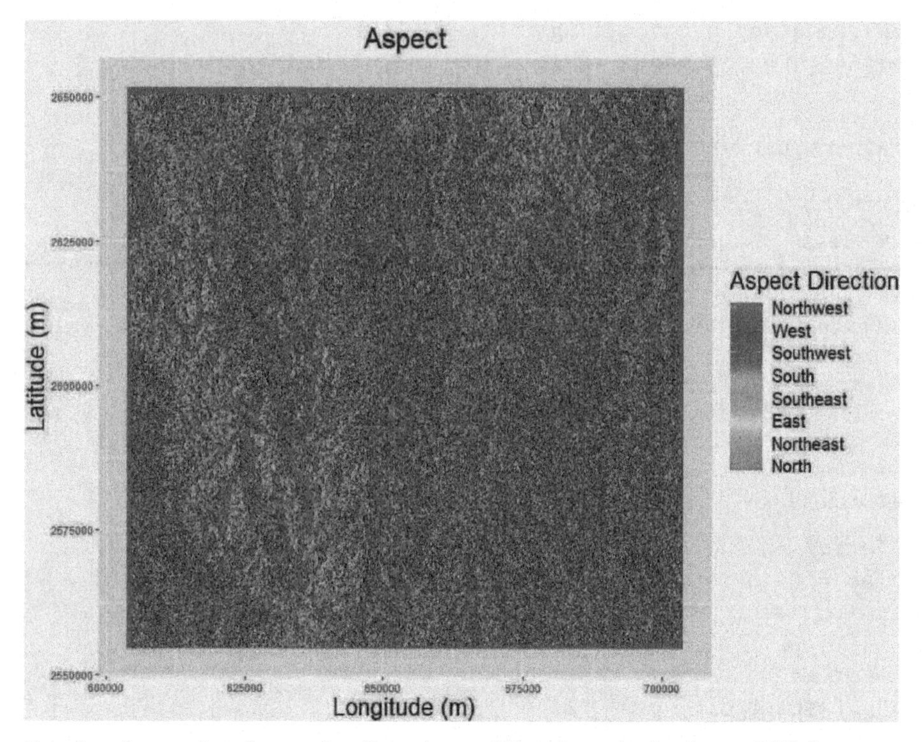

So, the above plot shows the directions of the slope in the input DEM raster.

17.1.3. Curvature

Curvature tells about the shape of the slope. It is second derivative of the elevation surface. Curvatures are mainly of following three types:

1) *Planform curvature:* Planform curvature is measured perpendicular to the direction of steepest or maximum slope. It denotes change of aspect. Its positive, negative and zero value indicates that the surface is sidewardly convex, sidewardly concave and linear, respectively. It is also known as plan curvature or contour curvature. Planform curvature has significance in describing convergence or divergence of flow, soil water content and soil characteristics.

2) *Profile curvature:* Profile curvature is measured parallel to the direction of steepest slope *i.e.* it represents the direction of maximum slope. It denotes change of slope. Its positive, negative and zero value indicates that the surface is upwardly concave, upwardly convex and linear, respectively. It has significance in measuring flow acceleration or deceleration rate, erosion or deposition rate, and in geomorphological studies.

3) *Total curvature:* Total curvature combines both planiform and profile curvature. Therefore, it helps in understanding the water flow over surface more accurately.

In R software, curvature can be computed by `spatialEco` package.

```
>library(spatialEco) #For computation of curvature
```

The planform curvature of DEM raster can be computed by assigning"`planform`"method in the `type` argument of **`curvature`** function as depicted in the following R-code.

```
#Compute Planform Curvature
>plan_cur =curvature(dem_utm, type ="planform")

>plan_cur #Display attributes of planform curvature layer

## class      : RasterLayer
## dimensions : 3155, 3525, 11121375  (nrow, ncol, ncell)
## resolution : 28.4, 30.8  (x, y)
## extent    : 603707.6, 703817.6, 2554497, 2651671  (xmin, xmax, ymin,
ymax)
## crs          : +proj=utm +zone=43 +datum=WGS84 +units=m +no_defs
+ellps=WGS84 +towgs84=0,0,0
## source     : memory
## names      : layer
## values     : -0.018893, 0.010966  (min, max)

#Plot Planform curvature layer

>ggR(plan_cur, geom_raster =TRUE) +#Plot with ggplot2
      ggtitle('Planform Curvature')+#Title
      labs(x='Longitude (m)', y='Latitude (m)')+#axis labels
      theme(plot.title =element_text(hjust =0.5, #Align title
in centre
      size =24), #size of title
      axis.title =element_text(size =20), #size of axis labels
      legend.key.size =unit(1, "cm"), #size of legend
      legend.title =element_text(size =20), #size of legend
title
      legend.text =element_text(size =15)) +#size of legend
text
      scale_fill_gradientn(colours =terrain.colors(5),
      name ="Planform Curvature") #Define colours
```

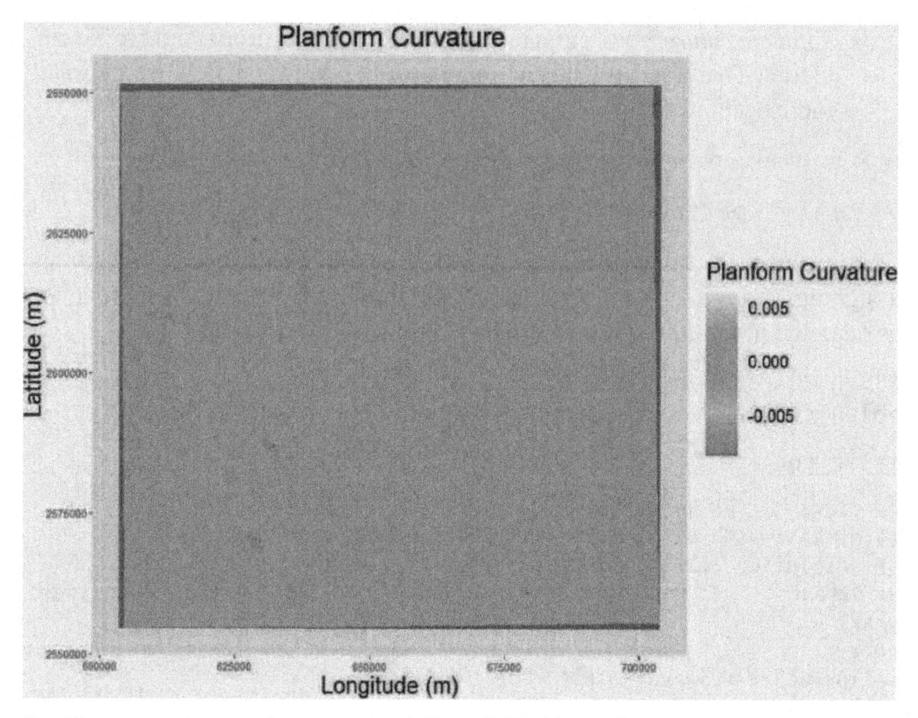

Profile curvature can be computed from DEM by using `type ="profile"` argument in `curvature` function and can be plotted by using following R-codes.

```
# Compute Profile Curvature
> prof_cur =curvature(dem_utm, type ="profile")

> prof_cur  #Display attributes of profile curvature layer

## class      : RasterLayer
## dimensions : 3155, 3525, 11121375  (nrow, ncol, ncell)
## resolution : 28.4, 30.8  (x, y)
## extent     : 603707.6, 703817.6, 2554497, 2651671  (xmin, xmax, ymin, ymax)
## crs        : +proj=utm +zone=43 +datum=WGS84 +units=m +no_defs +ellps=WGS84 +towgs84=0,0,0
## source     : memory
## names      : layer
## values     : -0.011942, 0.012639  (min, max)

#Plot Profile curvature layer
>ggR(prof_cur, geom_raster =TRUE) +#Plot with ggplot2
       ggtitle('Profile Curvature')+#Title
       labs(x='Longitude (m)', y='Latitude (m)')+#axis labels
       theme(plot.title =element_text(hjust =0.5, #Align title in centre
       size =24), #size of title
```

```
      axis.title =element_text(size =20), #size of axis labels
      legend.key.size =unit(1, "cm"), #size of legend
      legend.title =element_text(size =20), #size of legend title
      legend.text =element_text(size =15)) +#size of legend text
      scale_fill_gradientn(colours =terrain.colors(5),
      name ="Profile Curvature") #Define colours
```

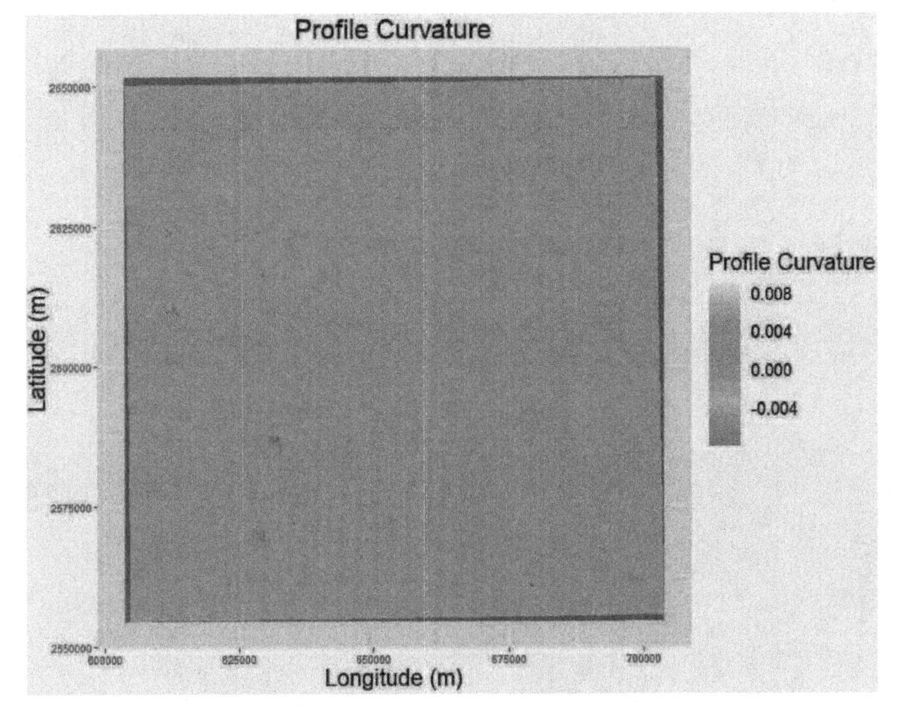

Similarly, total curvature can be computed from the DEM by assigning `type ="total"` argument in `curvature` function.

```
#Compute Total Curvature
> total_cur =curvature(dem_utm, type ="total")
> total_cur #Display attributes of total curvature layer

## class      : RasterLayer
## dimensions : 3155, 3525, 11121375  (nrow, ncol, ncell)
## resolution : 28.4, 30.8  (x, y)
## extent     : 603707.6, 703817.6, 2554497, 2651671  (xmin, xmax, ymin, ymax)
## crs        : +proj=utm +zone=43 +datum=WGS84 +units=m +no_defs +ellps=WGS84 +towgs84=0,0,0
## source     : memory
## names      : layer
## values     : -0.024782, 0.019868  (min, max)
```

```
#Plot Total curvature layer
>ggR(total_cur, geom_raster =TRUE) +#Plot with ggplot2
      ggtitle('Total Curvature')+#Title
      labs(x='Longitude (m)', y='Latitude (m)')+#axis labels
      theme(plot.title =element_text(hjust =0.5, #Align title
in centre
      size =24), #size of title
      axis.title =element_text(size =20), #size of axis labels
      legend.key.size =unit(1, "cm"), #size of legend
      legend.title =element_text(size =20), #size of legend
title
      legend.text =element_text(size =15)) +#size of legend
text
      scale_fill_gradientn(colours =terrain.colors(5),
      name ="Total Curvature")  #Define colours
```

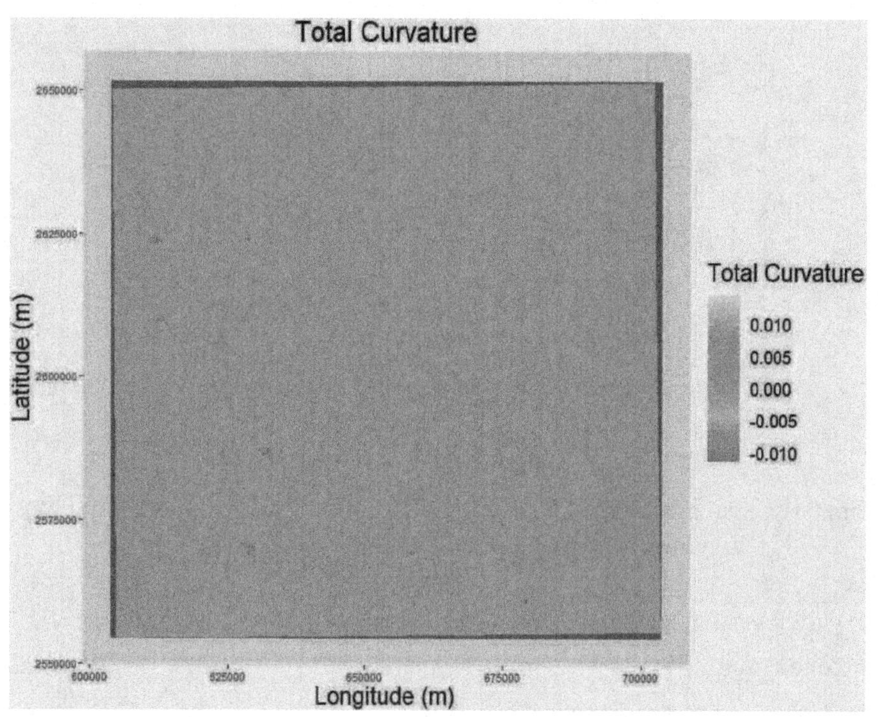

In this way, planform, profile and total curvatures are computed from DEM in R software.

17.1.4. Hillshades

Hillshades creates the shaded relief raster for enhancing the grayscale three-dimensional visualization of DEM by taking into account the relative position of sun in creating the shaded effect. The relative position of sun is computed by using its azimuth and elevation angle. Hillshades can be computed in R by using `hillShade` function of `raster` package. This function takes slope and aspect values in radians, as well as value of elevation angle in `angle` argument and azimuth angle in `direction` argument expressed in degree. For computation of hillshades in R, we will first compute slope and aspect in radians, and then compute the hillshades as depicted in the following R-code example.

```
# Hill-shades

#Compute slope in radians
>slope_rad =terrain(dem_utm, opt='slope', unit='radians')

#Compute aspect in radians
>aspect_rad =terrain(dem_utm, opt='aspect',unit='radians')

#Compute hillshades
>hill =hillShade(slope=slope_rad, aspect=aspect_rad, angle=45,
direction=0)

>hill #Display attributes of Hill-shade layer

## class      : RasterLayer
## dimensions : 3155, 3525, 11121375  (nrow, ncol, ncell)
## resolution : 28.4, 30.8  (x, y)
## extent     : 603707.6, 703817.6, 2554497, 2651671  (xmin, xmax, ymin,
ymax)
## crs        : +proj=utm +zone=43 +datum=WGS84 +units=m +no_defs
+ellps=WGS84 +towgs84=0,0,0
## source     : memory
## names      : layer
## values     : 0.3974948, 0.9226059  (min, max)
```

Now this computed hillshade layer can be plotted in grayscale by using following R-code.

```
#Plot Hill shade
>ggR(hill, geom_raster =TRUE) +#Plot with ggplot2
       ggtitle('Hill-shades')+#Title
       labs(x='Longitude (m)', y='Latitude (m)')+#axis labels
       theme(plot.title =element_text(hjust =0.5, #Align title
in centre
       size =24), #size of title
       axis.title =element_text(size =20), #size of axis labels
       legend.key.size =unit(1, "cm"), #size of legend
       legend.title =element_text(size =20), #size of legend
title
       legend.text =element_text(size =15)) +#size of legend
text
#Define colours
       scale_fill_gradientn(colours =c('azure2', 'grey',
'black'),    name ="Hill-shades") #legend title
```

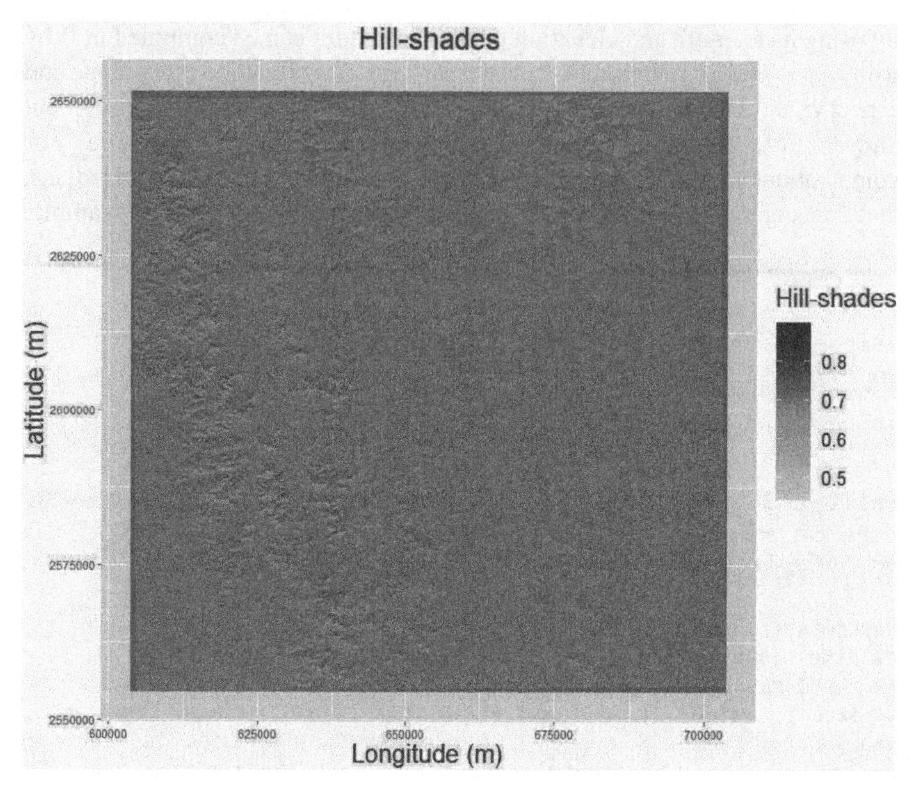

In the above plot, we can visualize the terrain properties better by using hillshade function.

17.1.5. Flow direction

Flow direction describes the direction from each pixel in DEM to its neighbouring steepest downslope pixelor neighbouring pixels having maximum drop in elevation. It tells about the directionin which water will flow. It is useful in hydrological applications. In R, it can be computed by assigning `opt ='flowdir'` argument in `terrain` function of `raster` package. The number of neighbouring pixels to be used for its calculation is assigned in `neighbors` argument. The R-code example is given below.

```
#Compute Flow direction
>flow_direction =terrain(dem_utm, opt ='flowdir', neighbors =8)

>flow_direction #Display attributes of flow direction layer

## class      : RasterLayer
## dimensions : 3155, 3525, 11121375  (nrow, ncol, ncell)
## resolution : 28.4, 30.8  (x, y)
## extent     : 603707.6, 703817.6, 2554497, 2651671  (xmin, xmax, ymin, ymax)
```

```
## crs          : +proj=utm +zone=43 +datum=WGS84 +units=m +no_defs
+ellps=WGS84 +towgs84=0,0,0
## source      : memory
## names       : flowdir
## values      : 1, 128  (min, max)
```

The flow direction is coded as numbers given in the following table 17.2.

Table 17.2: Flow direction and their code

Code	Flow Direction
1	East
2	Southeast
4	South
8	Southwest
16	West
32	Northwest
64	North
128	Northeast

The output flow direction raster can be reclassified in order to assign direction to the output plot. The R-codes for reclassification and for creating plot of the flow direction is given below.

```
#Define reclassication matrix
>fd_matrix =matrix(c(0, 1, 1, 1, 2, 2, 2, 4, 3, 4, 8, 4, 8, 16,
5,
16,32, 6, 32, 64, 7, 64, 128, 8),
ncol =3, byrow =TRUE)

>fd_matrix #Display matrix

##        [,1] [,2] [,3]
## [1,]      01    1
## [2,]      1    2    2
## [3,]  2    4    3
## [4,]  4    8    4
## [5,]  8   16    5
## [6,] 16   32    6
## [7,] 32   64    7
## [8,] 64  128    8

#Reclassify aspect layer according to matrix
>fd_dir =reclassify(flow_direction, fd_matrix)

#Define name of labels for flow direction plot
>fd_lab =c('East', 'Southeast', 'South', 'Southwest',
'West', 'Northwest', 'North', 'Northeast')
```

```
#Plot Flow direction layer
>ggR(fd_dir, geom_raster =TRUE) +#Plot with ggplot2
      ggtitle('Flow Direction')+#Title
      labs(x='Longitude (m)', y='Latitude (m)')+#axis labels
      theme(plot.title =element_text(hjust =0.5, #Align title
in centre
      size =24), #size of title
      axis.title =element_text(size =20), #size of axis labels
      legend.key.size =unit(1, "cm"), #size of legend
      legend.title =element_text(size =20), #size of legend
title
      legend.text =element_text(size =15)) +#size of legend
text
#Define customized colours
      scale_fill_gradientn(colours =c('red', 'orange', 'yellow',
'green',
'skyblue','blue','darkblue','magenta'),
name ="Flow Direction", labels = fd_lab,
breaks =c(1,2,3,4,5,6,7,8))
```

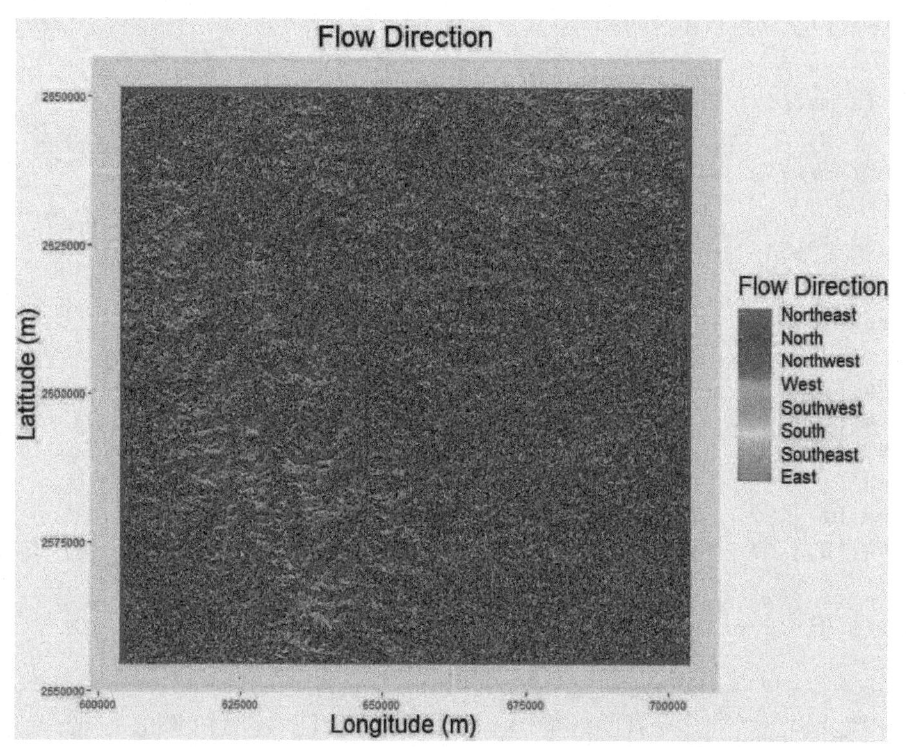

In this way, flow direction can be computed and plotted in R software.

17.1.6. Roughness

Roughness is the measure of degree of irregularity of the surface. It is calculated by computing the difference between maximum and minimum value of a pixel and its neighbouring 8 pixels in the DEM raster.Roughness can be computed in R by assigning `opt=’roughness’` argument in `terrain` function as depicted below. The R-code for creating plot of the output raster layer is also given below.

```
#Compute Roughness
>roughness =terrain(dem_utm, opt =’roughness’)
>roughness #Display attributes of roughness layer

## class       : RasterLayer
## dimensions : 3155, 3525, 11121375  (nrow, ncol, ncell)
## resolution : 28.4, 30.8  (x, y)
## extent     : 603707.6, 703817.6, 2554497, 2651671  (xmin, xmax, ymin,
ymax)
## crs          : +proj=utm +zone=43 +datum=WGS84 +units=m +no_defs
+ellps=WGS84 +towgs84=0,0,0
## source      : memory
## names       : roughness
## values      : 0, 37.82228  (min, max)

#Plot Roughness layer
>ggR(roughness, geom_raster =TRUE) +#Plot with ggplot2
      ggtitle(‘Roughness’)+#Title
      labs(x=’Longitude (m)’, y=’Latitude (m)’)+#axis labels
      theme(plot.title =element_text(hjust =0.5, #Align title
in centre
      size =24), #size of title
      axis.title =element_text(size =20), #size of axis labels
      legend.key.size =unit(1, “cm”), #size of legend
      legend.title =element_text(size =20), #size of legend
title
      legend.text =element_text(size =15)) +#size of legend
text
      scale_fill_gradientn(colours =terrain.colors(5), #define
colours
      name =”Roughness”) #Legend title
```

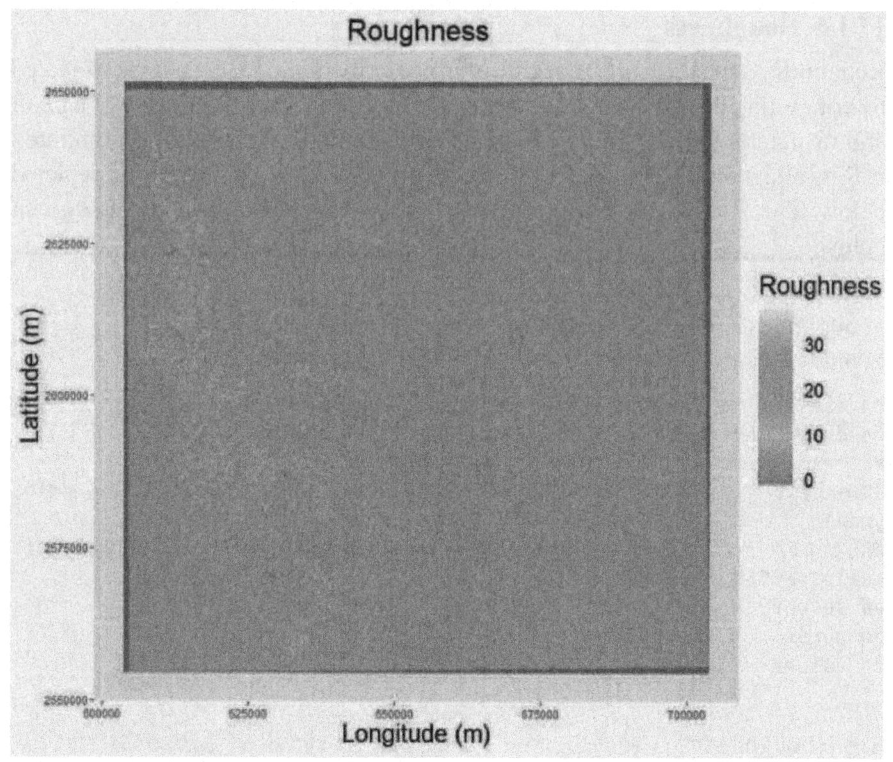

17.2. Secondary terrain attributes

Secondary terrain attributes are computed from the primary terrain attributes for describing any phenomenon or process. Some of the secondary terrain attributes along with their computation in R software are explained further.

17.2.1. Topographic Position Index (TPI)

Topographic Position Index (TPI) is the difference between elevation value of a central pixel in DEM and mean of its surrounding pixels. Positive TPI value indicates ridge, negative value indicates valley and zero denotes flat areas or areas having constant slope. So, TPI can be used for identification of landforms. TPI can be computed in R by assigning `opt='tpi'` argument in **`terrain`** function of `raster` package. In this function, mean of neighbouring 8 pixels are taken. The R-code example for computation of TPI and plotting it is given below.

```
#Compute Topographic Position Index
>tpi =terrain(dem_utm, opt='tpi')
>tpi #Display attributes of TPI layer
## class      : RasterLayer
## dimensions : 3155, 3525, 11121375  (nrow, ncol, ncell)
## resolution : 28.4, 30.8  (x, y)
```

```
## extent    : 603707.6, 703817.6, 2554497, 2651671  (xmin, xmax, ymin,
ymax)
## crs         : +proj=utm +zone=43 +datum=WGS84 +units=m +no_defs
+ellps=WGS84 +towgs84=0,0,0
## source     : memory
## names      : tpi
## values     : -14.4472, 10.37761  (min, max)
#Plot TPI layer
>ggR(tpi, geom_raster =TRUE) +#Plot with ggplot2
      ggtitle('Topographic Position Index')+#Title
      labs(x='Longitude (m)', y='Latitude (m)')+#axis labels
      theme(plot.title =element_text(hjust =0.5, #Align title
in centre
      size =24), #size of title
      axis.title =element_text(size =20), #size of axis labels
      legend.key.size =unit(1, "cm"), #size of legend
      legend.title  =element_text(size  =20),  #size  of  legend
title
      legend.text  =element_text(size  =15))  +#size  of  legend
text
      scale_fill_gradientn(colours  =c('azure',  'grey28'),
#define colours
      name ="TPI") #legend title
```

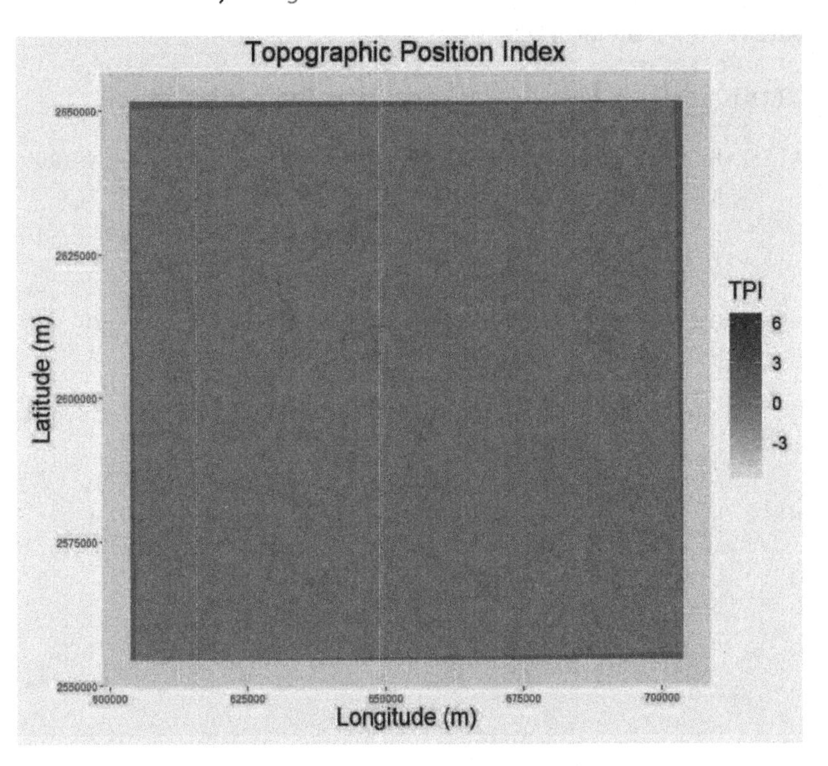

17.2.2. Terrain Ruggedness Index (TRI)

Terrain Ruggedness Index (TRI) is the mean of the absolute differences between elevation value of a central pixel in DEM raster and its immediately surrounding pixels. For computation of TRI, first of all, the differences between central pixel and its eight adjacent pixels are calculated. After that, average of the square of these differences are computed. The square root of that average value is the Terrain Ruggedness Index. This Terrain Ruggedness Index quantifies the topographic heterogeneity. TRI can be computed in R by assigning `opt='tri'` argument in `terrain` function of the `raster` package. The R-code example for its computation and for creating plot is depicted below.

```
#Compute Terrain Ruggedness Index (TRI)
>tri =terrain(dem_utm, opt='tri')
>tri #Display attributes of TRI layer
## class      : RasterLayer
## dimensions : 3155, 3525, 11121375  (nrow, ncol, ncell)
## resolution : 28.4, 30.8  (x, y)
## extent     : 603707.6, 703817.6, 2554497, 2651671  (xmin, xmax, ymin,
ymax)
## crs          : +proj=utm +zone=43 +datum=WGS84 +units=m +no_defs
+ellps=WGS84 +towgs84=0,0,0
## source     : memory
## names      : tri
## values     : 0, 14.4472  (min, max)
#Plot TRI layer
>ggR(tri, geom_raster =TRUE) +#Plot with ggplot2
       ggtitle('Terrain Ruggedness Index')+#Title
       labs(x='Longitude (m)', y='Latitude (m)')+#axis labels
       theme(plot.title =element_text(hjust =0.5, #Align title
in centre
       size =24), #size of title
       axis.title =element_text(size =20), #size of axis labels
       legend.key.size =unit(1, "cm"), #size of legend
       legend.title =element_text(size =20), #size of legend
title
       legend.text =element_text(size =15)) +#size of legend
text
       scale_fill_gradientn(colours =c('gray', 'black'), #define
colours
       name ="TRI") #legend title
```

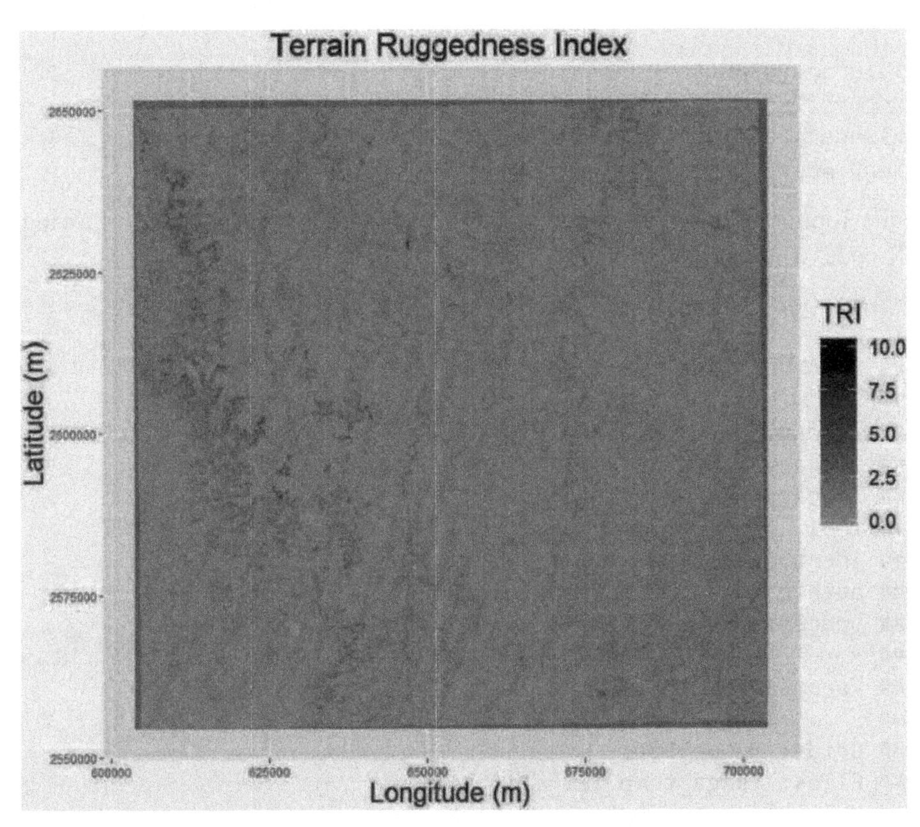

17.2.3. Topographic Wetness Index (TWI)

Topographic Wetness Index (TWI) is computed from the local upslope contributing area or flow accumulation and slope.

Topographic Wetness Index (TWI) = $\ln(\frac{A}{tan\beta})$

where, A is the ratio of upslope area to flow width, and â is the slope.

Topographic Wetness Index is also known as Compound Topographic Index (CTI). TWI measures the potential wetness of any portion in the landscape. Higher TWI value indicates higher potential wetness of that area and vice-versa.It is used for quantifying topographic control on hydrological processes like soil moisture conditions, runoff generation potential, identification of areas for water collection or pond formation, etc.TWI can be computed in R by using SAGA-GIS environment and `envirem` package. First of all, import `RSAGA` and `envirem` packages and define the RSAGA environment using following R-codes.

```
#Import packages
>library(RSAGA)          #For SAGA geoprocessing and terrain
analysis
>library(envirem)  #For computation of Topographic Wetness Index
>env =rsaga.env()  #Setting the environment of RSAGA
```

The Topographic Wetness Index (TWI) can be computed by using following R-code.

```
>twi =topoWetnessIndex(dem = dem_utm, sagaEnv = env)#Compute
TWI
## SAGA Version: 6.3.0 (64 bit)
##
## library path: C:\SAGA-GIS\tools\
## library name: io_gdal
## library     : io_gdal
## tool        : Import Raster
## identifier  : 0
## author      : O.Conrad (c) 2007 (A.Ringeler)
## processors  : 8 [8]
##
## Parameters
##
## Grids: No objects
## Files: "saga_temp_twi_gtif.tif"
## Multiple Bands Output: automatic
## Select from Multiple Bands:
## Transformation: yes
## Resampling: Nearest Neighbour
##
## loading: saga_temp_twi_gtif.tif
##
## Driver: GTiff
##
## Bands: 1
##
## Rows: 3525
##
## Columns: 3155
##
## Transformation:
##
##   x' = 603707.584373 + x * 28.400000 + y * 0.000000
##
##   y' = 2651670.527267 + x * 0.000000 + y * -30.800000
```

```
##
## loading: saga_temp_twi_gtif
## translation: saga_temp_twi_gtif
##
## warning: top-to-bottom and left-to-right cell sizes differ.
##  Difference: 2.400000
##
##  using cellsize: 28.400000
##
## Saving grid: saga_temp_twi_in.sgrd...
##
## SAGA Version: 6.3.0 (64 bit)
##
## library path: C:\SAGA-GIS\tools\
## library name: ta_hydrology
## library     : ta_hydrology
## tool        : SAGA Wetness Index
## identifier  : 15
## author      : J.Boehner, O.Conrad (c) 2001
## processors  : 8 [8]
##
## Loading grid: saga_temp_twi_in.sgrd...
##
## Parameters
##
## Grid system: 28.4; 3525x 3422y; 603707.584373x 2554491.127267y
## Elevation: saga_temp_twi_in
## Weights: <not set>
## Catchment Area: Catchment Area
## Catchment Slope: Catchment Slope
## Modified Catchment Area: Modified Catchment Area
## Topographic Wetness Index: Topographic Wetness Index
## Suction: 10.000000
## Type of Area: square root of catchment area
## Type of Slope: catchment slope
## Minimum Slope: 0.000000
## Offset Slope: 0.100000
## Slope Weighting: 1.000000
##
## Create index: saga_temp_twi_in
## catchment area and slope...
## pass 1 (4303541 > 0)
## pass 2 (2265316 > 0)
```

```
## pass 125 (7 > 0)
## pass 126 (0 > 0)
## post-processing...
## topographic wetness index...
## Saving grid: C:\......................\file2f14cb4434e...
## Saving grid: C:\......................\file2f141ba742ef...
## Saving grid: C:\......................\file2f146ba44733...
## Saving grid: saga_temp_twi_out.sgrd...
```

The attributes of the computed TWI raster layer can be displayed as follows.

```
>twi    #Display attributes of TWI layer
## class      : RasterLayer
## dimensions : 3422, 3525, 12062550  (nrow, ncol, ncell)
## resolution : 28.4, 28.4  (x, y)
## extent     : 603693.4, 703803.4, 2554477, 2651662  (xmin, xmax, ymin, ymax)
## crs          : +proj=utm +zone=43 +datum=WGS84 +units=m +no_defs +ellps=WGS84 +towgs84=0,0,0
## source     : memory
## names      : topoWetnessIndex
## values     : 2.622674, 10.32735  (min, max)
```

The plot of the computed Topographic Wetness Index (TWI) layer can be created by using following R-code.

```
>ggR(twi, geom_raster =TRUE) +#Plot with ggplot2
      ggtitle('Topographic Wetness Index')+#Title
      labs(x='Longitude (m)', y='Latitude (m)')+#axis labels
      theme(plot.title =element_text(hjust =0.5, #Align title in centre
      size =24), #size of title
      axis.title =element_text(size =20), #size of axis labels
      legend.key.size =unit(1, "cm"), #size of legend
      legend.title =element_text(size =20), #size of legend title
      legend.text =element_text(size =15)) +#size of legend text
      scale_fill_gradientn(colours =c('lightblue', 'darkblue'),
      name ="TWI")
```

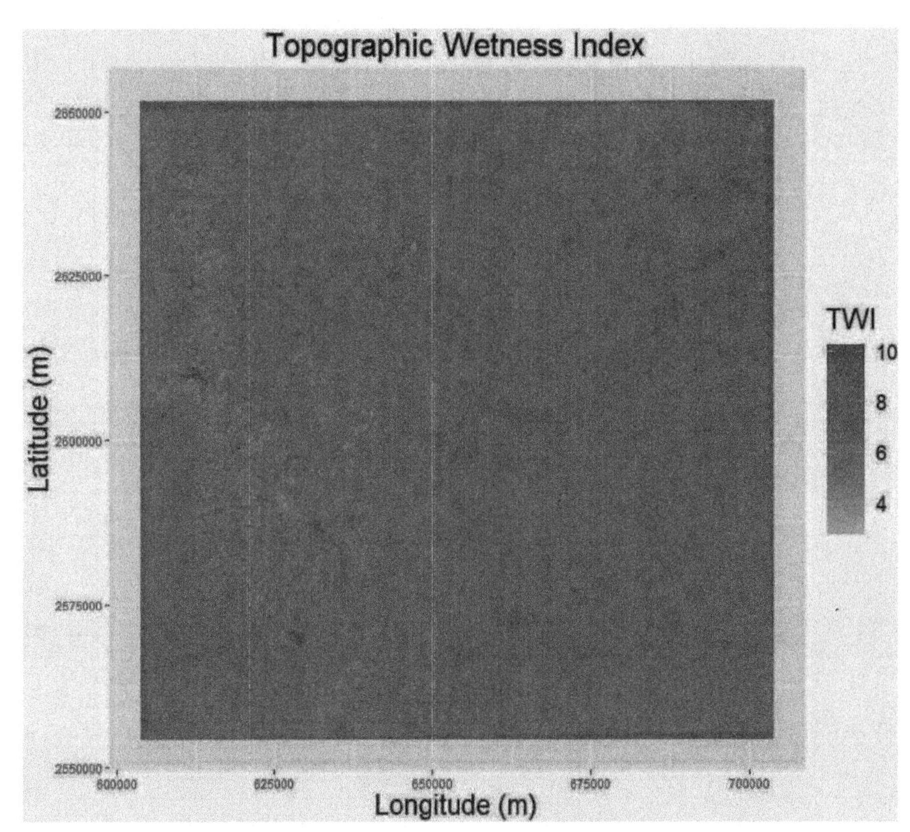

In this way, primary and secondary terrain attributes can be computed from DEM raster data in R software.

Chapter 18

Thematic Mapping

Maps make it easier to interpret and communicate the results of any GIS analysis. R has a large number of packages for making thematic maps. Static mapping is straightforward with `plot()` methods provided by core spatial packages `sf` and `raster`. The use of `plot()` function have been discussed in the previous chapters. Another package for mapping is`sp`with`spplot()` having enough options for producing good quality thematic maps.It is also possible to create advanced maps using `ggplot2`. This chapter discusses, separately, preparing thematic maps from vectors and rasters with the above mentioned packages. A dedicated map-making package `tmap`has been given special focus in this chapter.

18.1. Use of spplot in thematic mapping

18.1.1. Mapping vectors with spplot()

```
#Set the working directory
>setwd('E:/............/Chapter18/data')
>getwd() #Get the working directory
>library(rgdal)     #for reading vector layer
>library(sp)
```

Read the vector file using `readOGR`function. We will be using the shapefile named 'village.shp' in this chapter. The vector is a polygon feature of village boundaries with two attributes namely, 'total population' and 'total male population' in the villages.

```
>village=readOGR('village.shp') #get the village layer
## OGR data source with driver: ESRI Shapefile
## Source: "E:\.......\Chapter18\data\village.shp", layer:
"village"
## with 89 features
## It has 2 fields
```

```
>names(village)       #get the column names in the vector layer
## [1] "TP"   "TMP"
```

To plot the vector, use the `spplot()` function. The default spplot tries to map everything it can find in the attribute table. Sometimes, this does not work depending on the data types in the attribute table. In order to select specific values to map we can provide the spplot function with the name (or names) of the attribute variable we want to plot. In the code below we are trying to plot the 'total population' of the villages.

```
>spplot(village,'TP' )     #plot the village map with total population
```

This produces the map in default color of the `spplot()`. We will change the colour patterns now. We will be using the library RColorBrewer and select the desired color scheme and number of colour palettes based on classes identified in the legend.

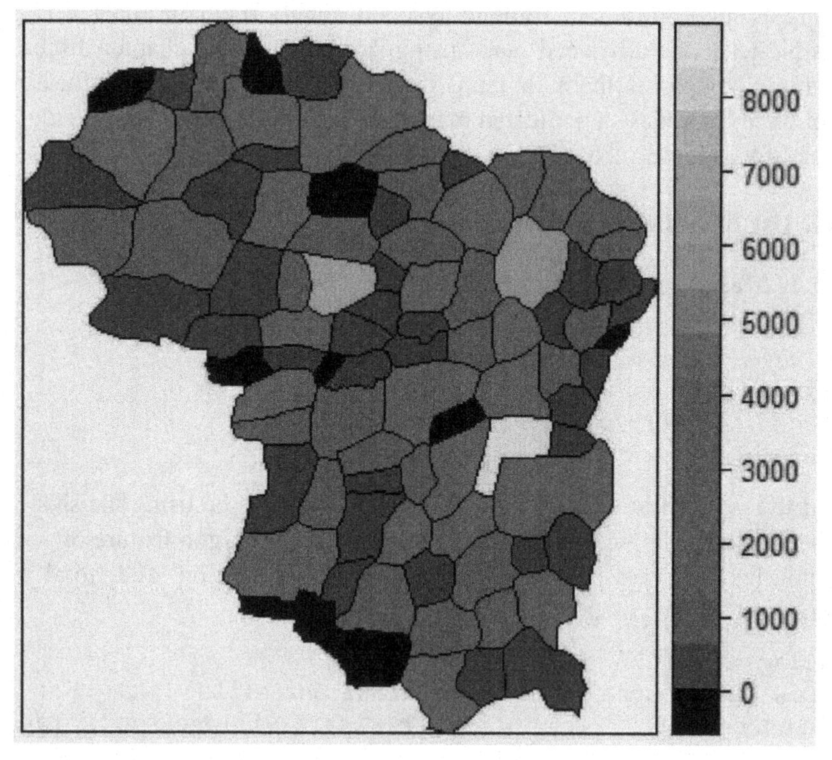


```
>library(RColorBrewer)
>display.brewer.all(type="seq")
```

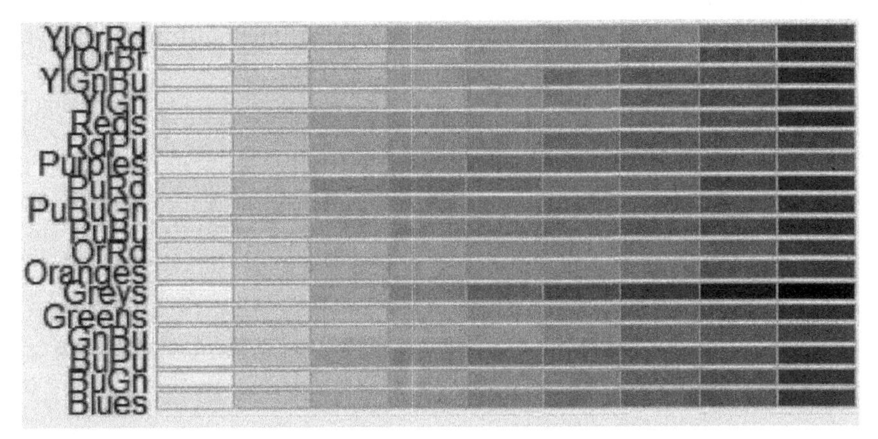

Select the number of classes and the color scheme and assign the colour to the spplot along with the number of cuts. Since we have selected 6 classes, the number of cuts will be 5 .

```
>color<-brewer.pal(6, "GnBu") # we select 6 colors from the palette
>class(color)
## [1] "character"
>spplot(village,"TP", col.regions = color, cuts =5)
```

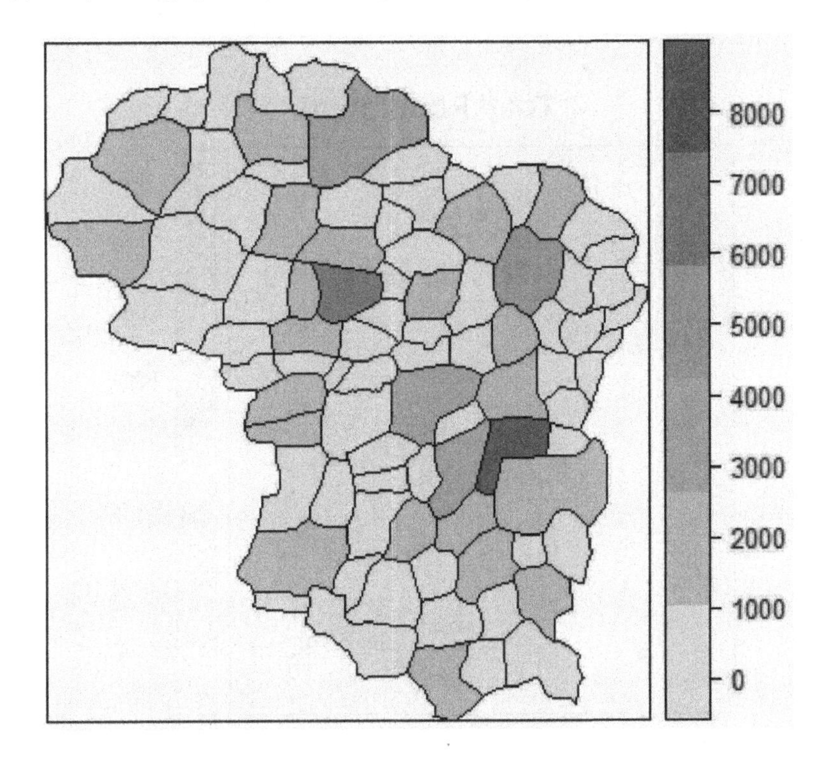

If we want to change the breaks by quantile, use classIntervals from the classInt library, something like:

```
>library(classInt)
>breaks_qt<-classIntervals(village$TP, n =6, style = "quantile")
```

We can omit the style = quantile, as it is default. See the structure of the breaks created.

```
>str(breaks_qt)
## List of 2
##  $ var : int [1:89] 3257 257 554 908 1263 508 1747 1908 723 286 ...
##  $ brks: num [1:7] 0 73.7 398 653 929 ...
##  - attr(*, "style")= chr "quantile"
##  - attr(*, "nobs")= int 81
## - attr(*, "call")= language classIntervals(var = village$TP, n = 6)
##  - attr(*, "intervalClosure")= chr "left"
##  - attr(*, "class")= chr "classIntervals"
```

Get the break values

```
>breaks_qt$brks
## [1]    0.00000   73.66667  398.00000  653.00000  929.00000 1512.00000
## [7] 8400.00000
```

Add the new break values to the plot. Also, add the title using argument, main.

```
>spplot(village,"TP", col.regions=color, at = breaks_qt$brks,
main ="Total Population")
```

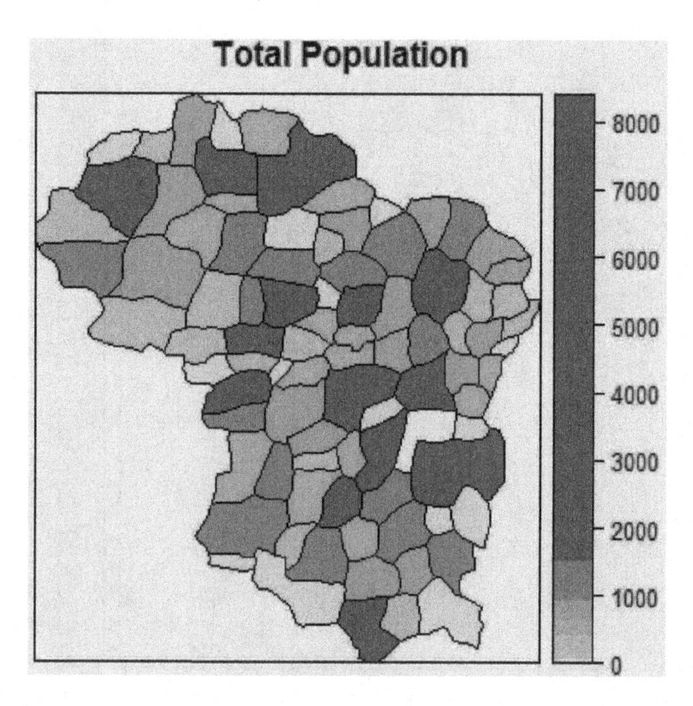

Sometimes, there may be a few blank polygons. Here are the steps to correct the breaks

```
# add a very small value to the top breakpoint, and subtract
from the bottom for symmetry
>br <-breaks_qt$brks
>offs <-0.0000001
>br[1] <-br[1] -offs
>br[length(br)] <-br[length(br)] +offs

# plot
>spplot(village, "TP", col.regions=color, at = br,  main =
"Total Population")
```

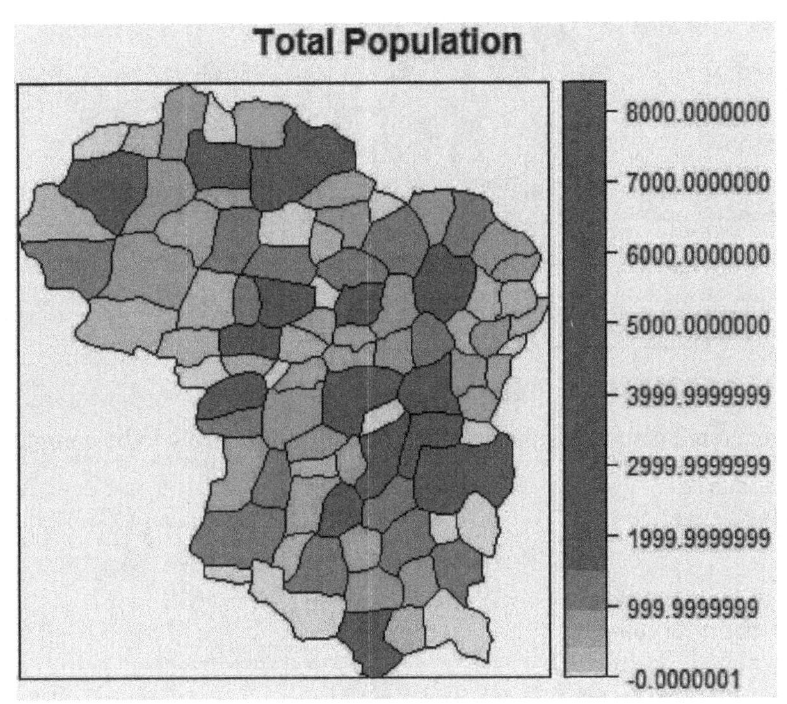

Now, to change the legend form graduated colour to continuous class intervals use the following codes.

```
>village$TP_bracket <-cut(village$TP, br)
>head(village$TP_bracket)
## [1] (1.51e+03,8.4e+03] (73.7,398]         (398,653]
## [4] (653,929]          (929,1.51e+03]     (398,653]
## 6 Levels: (-1e-07,73.7] (73.7,398] (398,653] (653,929] ...
(1.51e+03,8.4e+03]
>class(village$TP_bracket)
## [1] "factor"
```

Use this new break to the spplot. Also add the grids showing the latitude and longitude around the plot.

```
>spplot(village, "TP_bracket", col.regions=color, main ="Total Population",
scales=list(draw =TRUE))
```

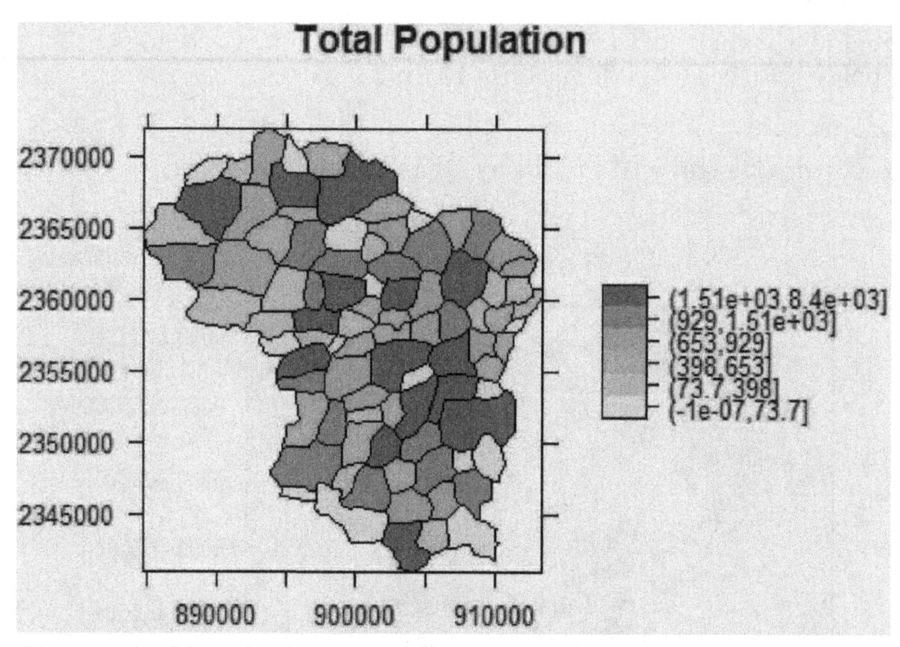

The generated map needs the scale bar and the north arrow to be completed.

```
##add scale
>scale =list("SpatialPolygonsRescale", layout.scale.bar(),
offset =c(905000, 2370000), scale =5000, fill=c("transparent", "black"))
>text1 =list("sp.text", c(905000, 2371000), "0")
>text2 =list("sp.text", c(911000, 2371000), "5 km")
##for North arrow
arrow =list("SpatialPolygonsRescale", layout.north.arrow(),
offset =c(907000, 2367000), scale =2500)
###Plotting
>spplot(village, "TP_bracket", col.regions=color, main ="Total Population",
scales=list(draw =TRUE),
sp.layout=list(scale,text1,text2,arrow))
```

The offset has two components, x and y, which shows the location of the scale, the text, and the north arrow. These are identified by the latitude and longitude from the previous map.

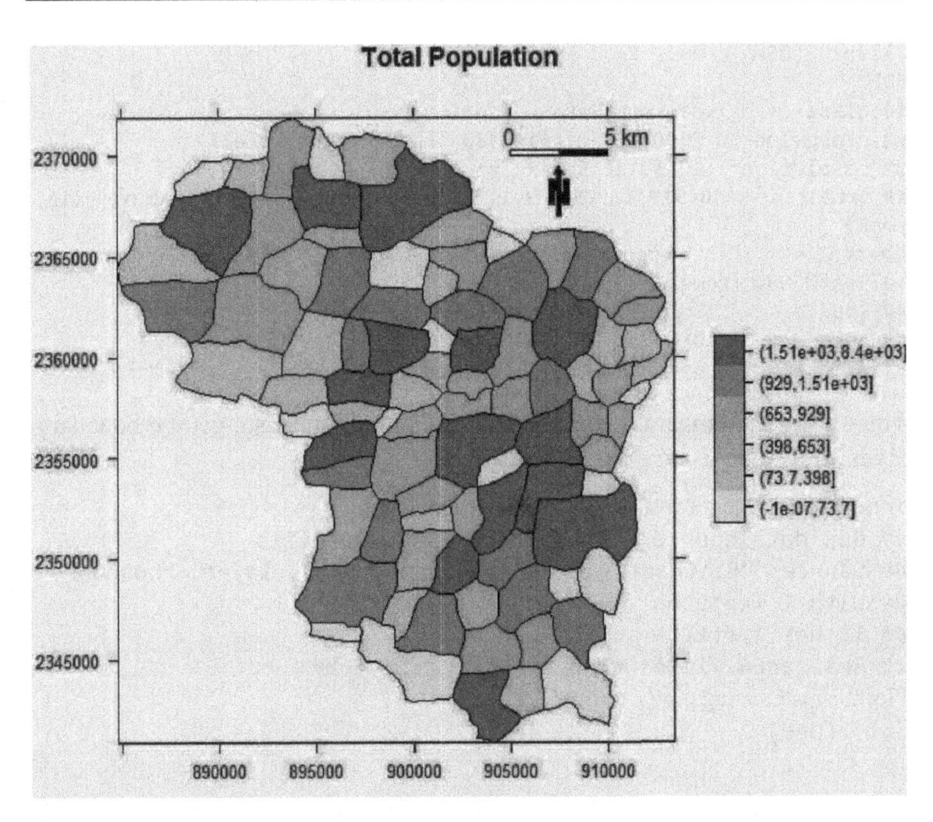

18.1.2. Mapping Rasters with spplot()

Following libraries are required for mapping rasters with spplot(). Get them installed and call. Set the working directory and import the raster named "DEM.tif".

```
>library(raster) #Raster data
## Loading required package: sp
>library(RStoolbox) #Plotting raster data
>library(RColorBrewer) #Colour pallette
>library(rgdal) #Vector data
>library(sp)
>setwd('E:/Chapter18/data')
>getwd() #Get the working directory
## [1] "E:/............/Chapter18/data"
DEM =raster('DEM.tif')
```

To avoid the null values outside the raster use the following code to define the null values

```
>NAvalue(DEM) =-9999
>DEM
## class      : RasterLayer
## dimensions : 2590, 2372, 6143480  (nrow, ncol, ncell)
## resolution : 12.5, 12.5  (x, y)
## extent     : 884519.1, 914169.1, 2340236, 2372611  (xmin, xmax, ymin, ymax)
## crs        : +proj=utm +zone=43 +datum=WGS84 +units=m +no_defs +ellps=WGS84 +towgs84=0,0,0
## source     : E:/Chapter18/data/DEM.tif
## names      : DEM
## values     : 236, 485  (min, max)
```

This shows the range of elevation from 236 to 485m. Also, get the boundary layer of the area using **readOGR**

```
>boundary =readOGR('boundary.shp')
## OGR data source with driver: ESRI Shapefile
## Source: "E:\Chapter18\data\boundary.shp", layer: "boundary"
## with 1 features
## It has 1 fields
## Integer64 fields read as strings:  T_MP
```

Plot the raster using plot()

```
>plot(DEM)
```

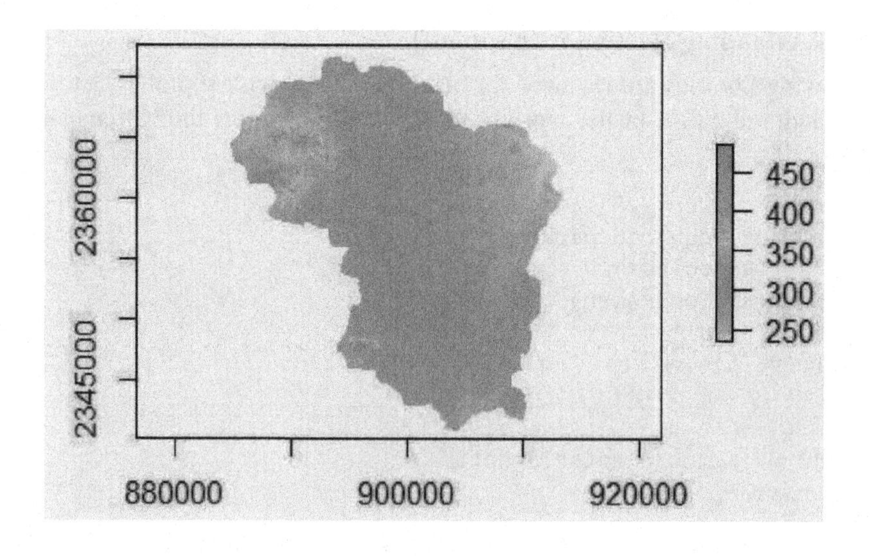

Plot the raster using spplot().

```
>spplot(DEM)
```

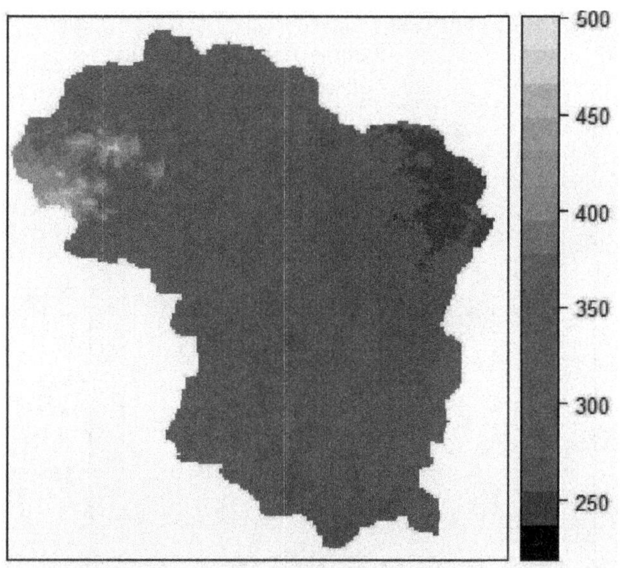

It will plot the raster with default colour pattern. We have five different colour patterns in base R namely, `rainbow,terrain.colors, heat.colors, topo.colors, and cm.colors`. We will see them one by one.

```
#select different color pattern
>spplot(DEM, col.regions =rainbow(99))
```

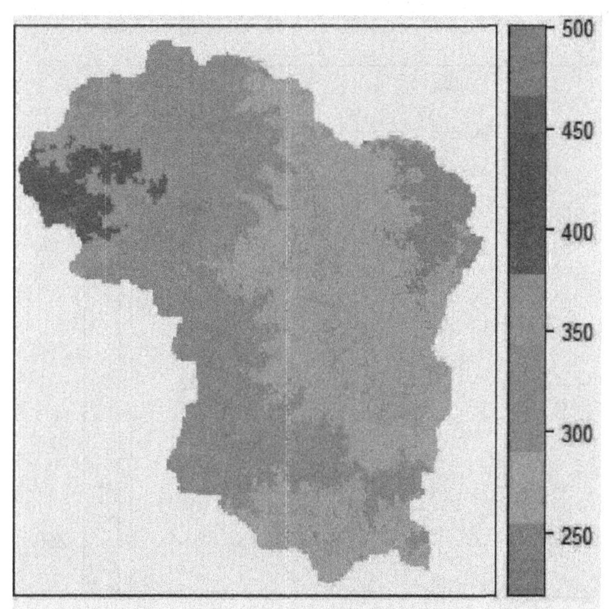

The number in the paranthesis after the color scheme shows the desired number of colours.

```
#we have five color patterns in base R
>spplot(DEM, col.regions =terrain.colors(20))
```

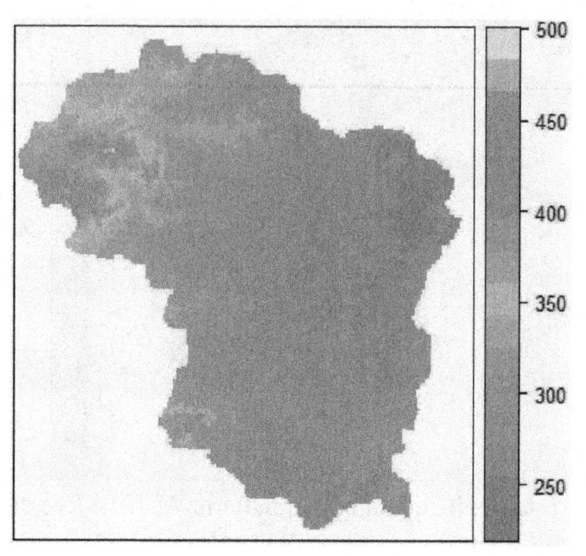

Other arguments in the colour schemes are for transparency and for order of colours.

```
#add transparency to the color
>spplot(DEM, col.regions =terrain.colors(20, alpha = .5))
```

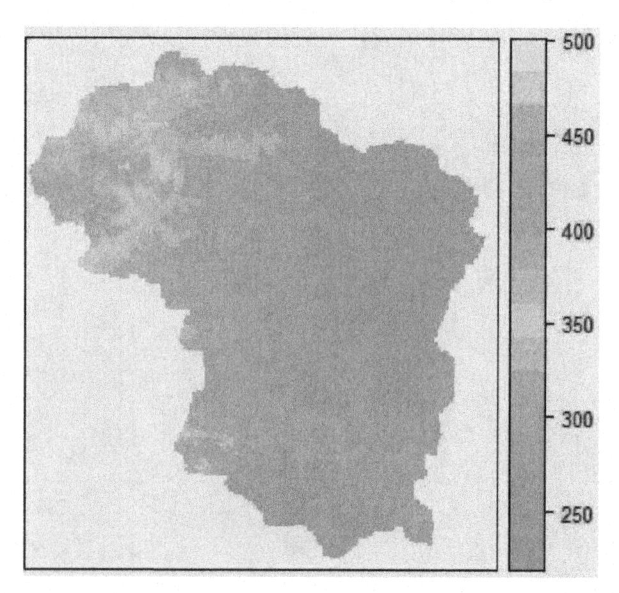

```
#reverse the color pattern
>spplot(DEM, col.regions =terrain.colors(20, rev=TRUE))
```

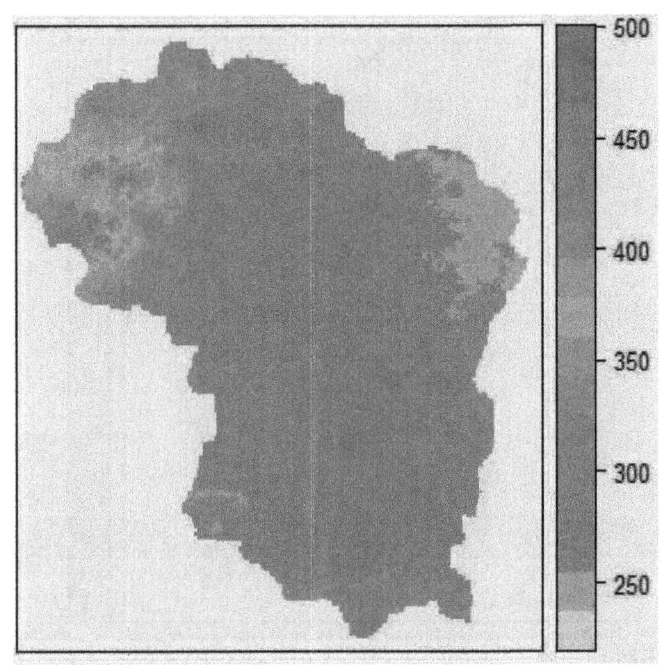


```
#other color patterns
>spplot(DEM, col.regions =heat.colors(30))
```

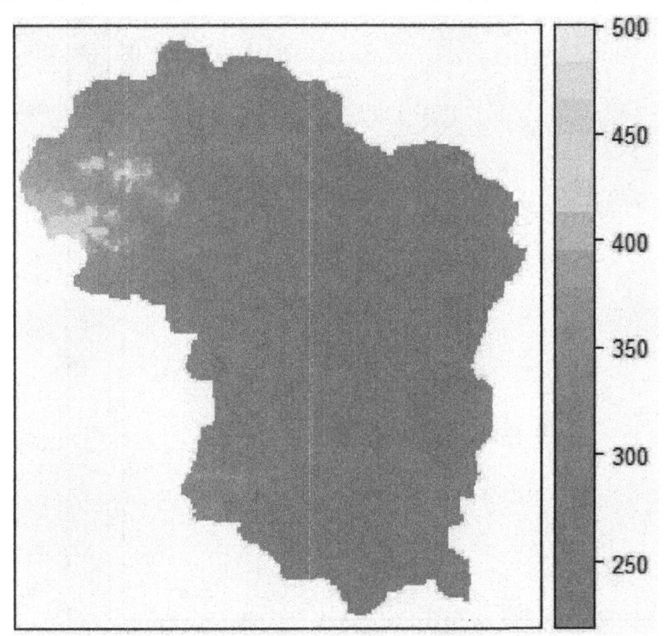

```
>spplot(DEM, col.regions =topo.colors(20))
```

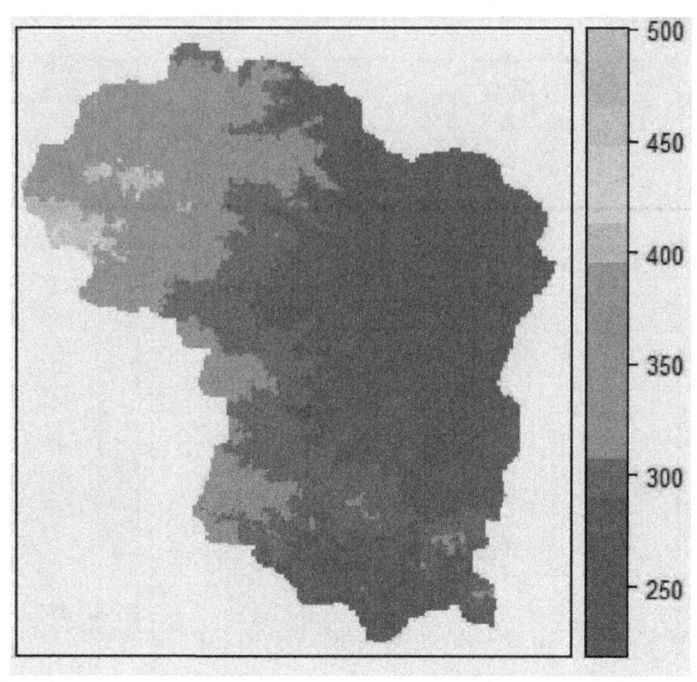


```
>spplot(DEM, col.regions =cm.colors(20))
```

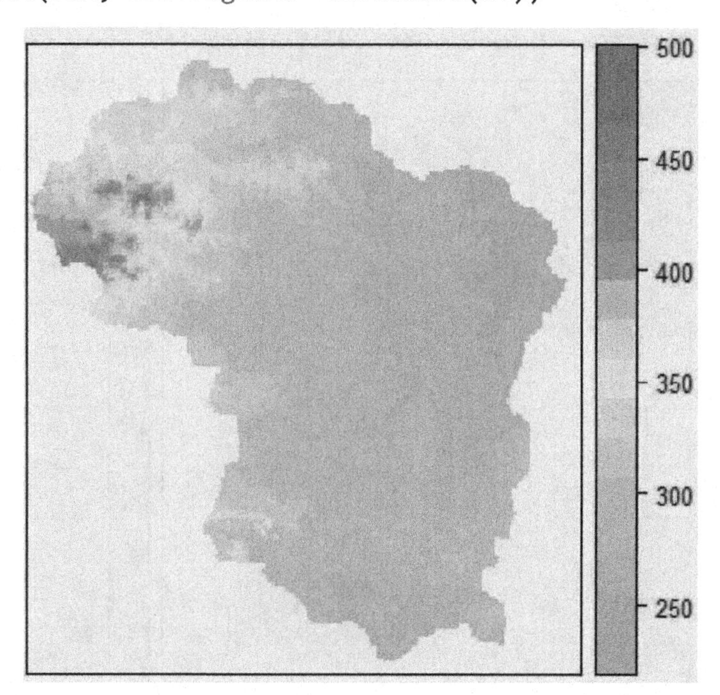

Now let's draw the scale bar and the north arrow to complete the map. First drawing the grids is needed.

```
#draw the grids
>spplot(DEM, col.regions =terrain.colors(30), scales=list(draw =TRUE))
```

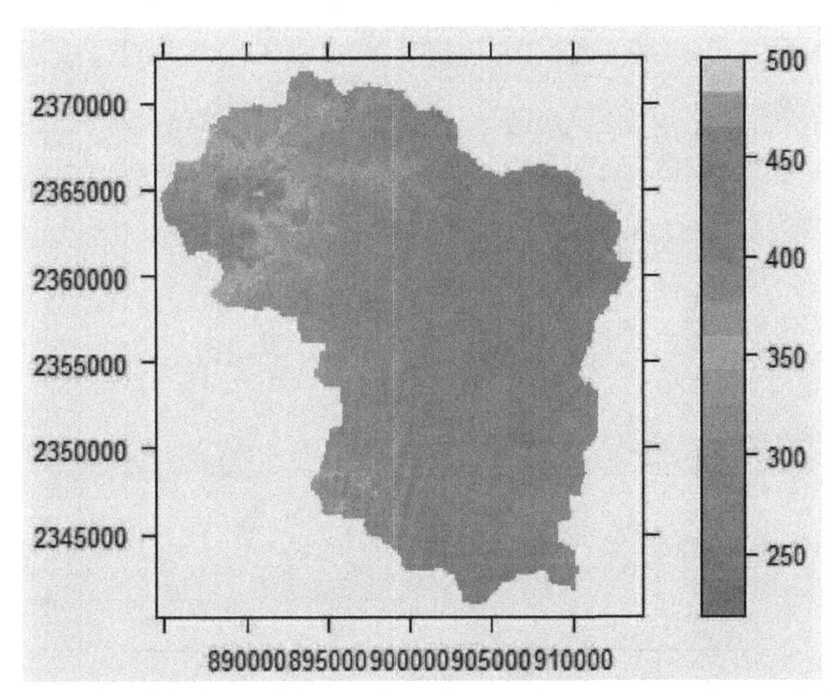

To draw the scale bar and the north arrow, use the same procedure as in the former section.

```
##add scale and north arrow
>scale =list("SpatialPolygonsRescale", layout.scale.bar(),
offset =c(905000, 2370000), scale =5000,
fill=c("transparent","black"))
>text1 =list("sp.text", c(905000, 2371000), "0")
>text2 =list("sp.text", c(910000, 2371000), "5 km")
##for North arrow
>arrow =list("SpatialPolygonsRescale", layout.north.arrow(),
offset =c(907000, 2367000), scale =2500)
###Plot with the scalebar and north arrow
>spplot(DEM, col.regions =terrain.colors(30), scales=list(draw
=TRUE),
main="Digital Elevation Model",
sp.layout=list(scale,text1,text2,arrow))
```

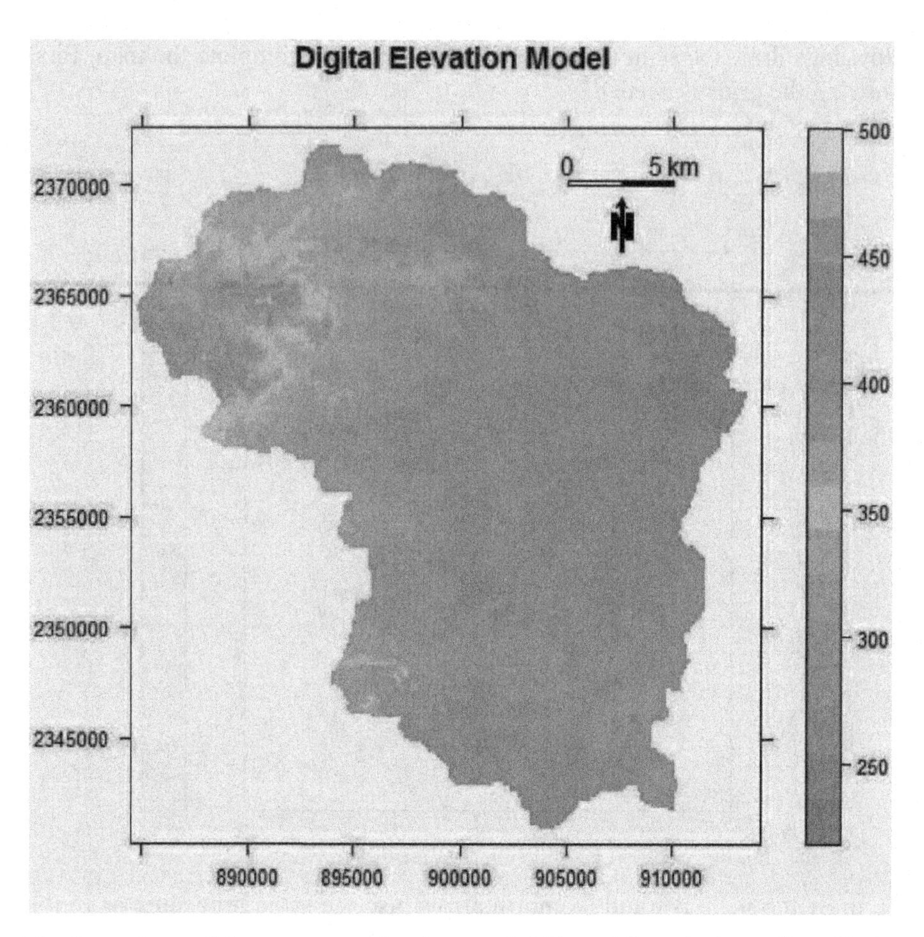

The boundary of the area may also be overlaid by adding the boundary file in the `sp.layout` function.

```
#add boundary
>spplot(DEM, col.regions =terrain.colors(30), scales=list(draw
=TRUE),
main="Digital Elevation Model",
sp.layout=list(scale,text1,text2,arrow, boundary))
```

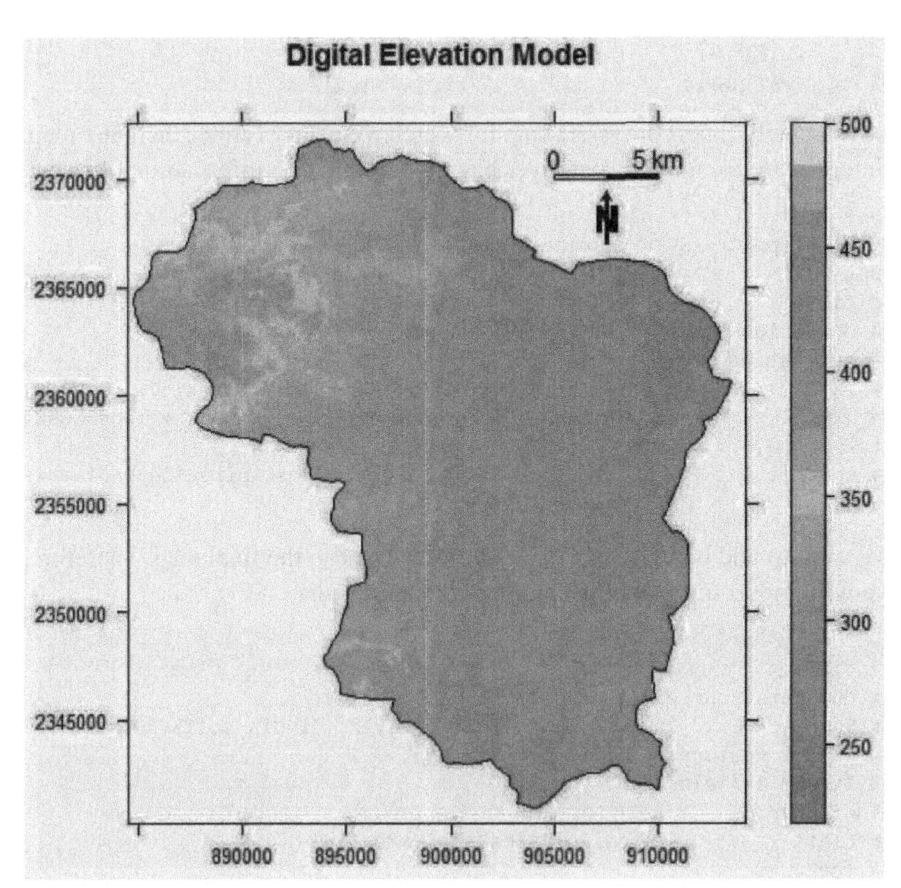

18.2. Use of ggplot2 in thematic mapping

18.2.1. Raster mapping with ggplot2

For creating thematic maps with `ggplot2` package, first of all, lets import all the related packages which are required along with `ggplot2` package to create attractive maps from the raster layer.`RStoolbox` package is required to plot raster data along with package `ggplot2`. Package `ggplot2` is required to add title, subtitle, axis, legends, etc. Package`raster` is required for importing raster data to be plotted. Package `ggsn` is required to add north arrow and scalebar in the map. Package `RColorBrewer` is required to add colour palette. Package `rgdal` is required to import shapefile.

```
#Import required packages
>library(raster)    #For importing raster data
>library(RStoolbox) #Plotting raster data with 'ggplot2'
>library(ggplot2)   #For customizing maps
>library(RColorBrewer) #For adding colour palette
```

```
>library(ggsn)          #For adding north arrow and scale bar
>library(rgdal)         #For importing shapefile
```

Now we will import the raster data for which we want to create thematic map. Here, we are using the raster layer having precipitation information of India.

```
>ppt =raster('precipitation_India.tif')      #Import raster layer
>NAvalue(ppt) =-9999  #Specify the no data value in raster
>ppt   #Display attributes of imported raster layer
## class       : RasterLayer
## dimensions  : 606, 584, 353904  (nrow, ncol, ncell)
## resolution  : 0.05, 0.05  (x, y)
## extent      : 68.2, 97.4, 6.749999, 37.05  (xmin, xmax, ymin, ymax)
## crs           : +proj=longlat +datum=WGS84 +no_defs +ellps=WGS84
+towgs84=0,0,0
## source      : E:/................................/Chapter18/data/precipitation_India.tif
## names       : precipitation_India
```

We want to add boundary of India on raster layer in the final map. Therefore, we will import the shapefile containing map of India.

```
#Import shapefile of India
>india_map =readOGR('E:\\..........\\India','india-soi154207')
## OGR data source with driver: ESRI Shapefile
## Source: "E:\..............\India", layer: "india-soi154207"
## with 1 features
## It has 1 fields
>india_map     #Display attributes of imported shapefile
## class        : SpatialPolygonsDataFrame
## features     : 1
## extent       : 68.18625, 97.41529, 6.755953, 37.07827  (xmin, xmax,
ymin, ymax)
## crs          : +proj=longlat +datum=WGS84 +no_defs +ellps=WGS84
+towgs84=0,0,0
## variables    : 1
## names        :                              Source
## value        : Survey of India State Map, Datameet
```

For adding north symbols in the map, we want to view all the north symbols available with `ggsn` package by using R-code.

```
>northSymbols() #Display north symbols available in 'ggsn' package
```

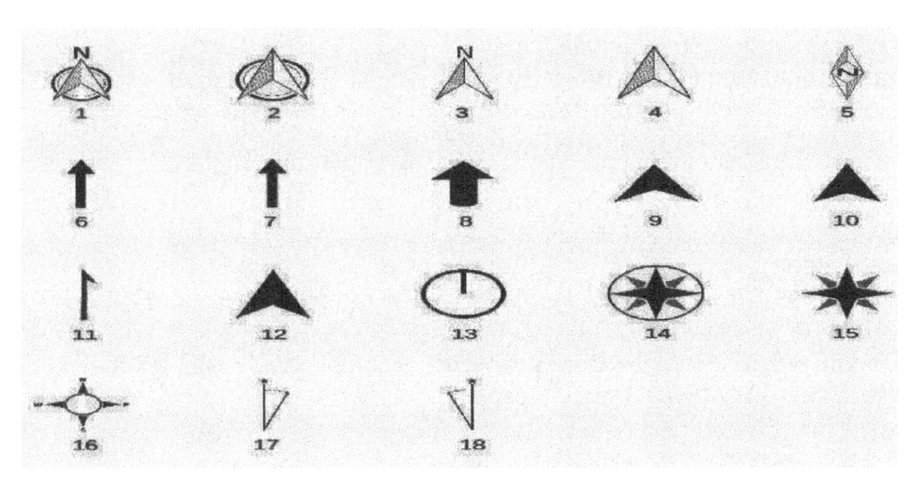

So, we can select the desired north arrow design from their respective code numbers.

We can create map from the imported raster layer with continuous stretch using following R-code. The R-code with their respective purpose is given below.

```
#Mapping with continuous stretch
>ggR(ppt, geom_raster =TRUE) +#Plot raster with 'ggplot2' package
      ggtitle('Precipitation')+#Title
      labs(x='Longitude', y='Latitude')+#Axis labels
      theme(plot.title =element_text(hjust =0.5, #Align title
in centre
      size =24), #Size of title
      axis.title =element_text(size =20), #Size of axis labels
      legend.key.size =unit(1, "cm"), #Size of legend scale
      legend.title =element_text(size =20), #Size of legend
title
      legend.text =element_text(size =15)) +#Size of legend
text
      scale_fill_gradientn(colours =brewer.pal(6,
'Blues'),#Colours
      name ="Precipitation (mm/day)",#Legend title

#Breaks in the scale bar
breaks =c(0, 300, 600, 900, 1200, 1500),
      na.value =NA)+      #No colour for missing values

#Add north arrow
      ggsn::north(location ="topright", #Location of north arrow
symbol =1, #Code number of North arrow symbol
      x.min =67, x.max =98, #Coordinates of map
      y.min =6, y.max =38) +
```

```
#Add scalebar in the map
ggsn::scalebar(location ="bottomright", #Location of scalebar
transform =TRUE,#True for coordinates in decimal degree
      dist_unit ="km", #Units of scalebar
model ="WGS84", #Ellipsoid model

#Distance to represent with each segment of scalebar
dist =500,
x.min =65, x.max =100, #Extent coordinates of map grid
y.min =5, y.max =40,
height =0.02, #Scale bar's height
st.size =3)+#Scale bar's size

#Plot shapefile of India on raster
geom_polygon(data =india_map, aes(x = long, y = lat,
      group = group),
      colour ='black', #Colour of the shapefile border
fill =NA)#No filling inside shapefile
```

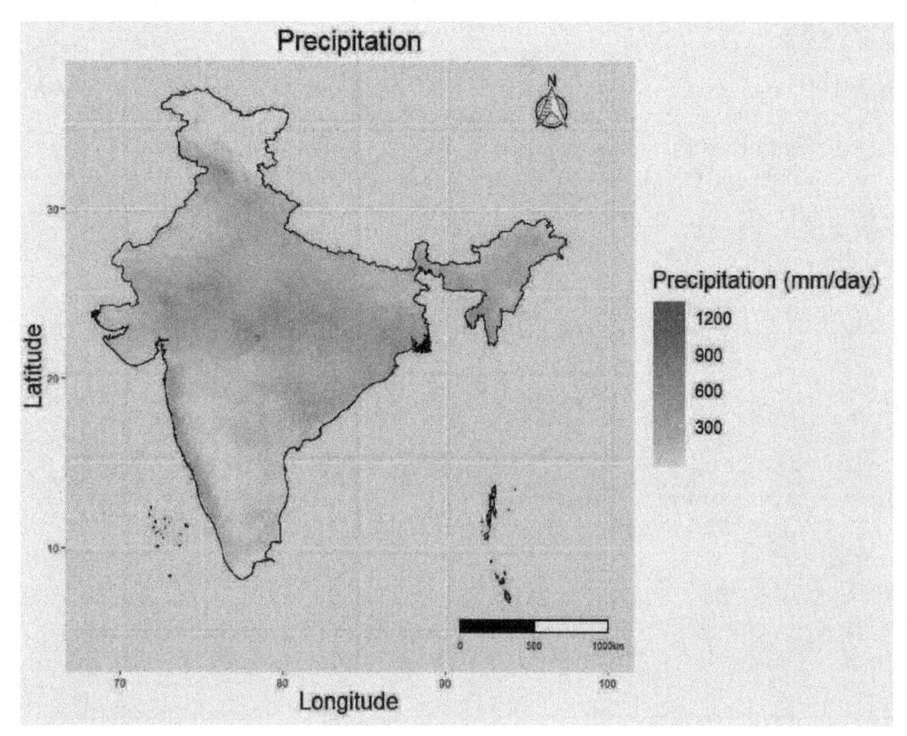

Thematic mapping with classified image

For thematic mapping with classified image, the raster data can be classified in the categories using `reclassify` function as follow.

```
>ppt_round =round(ppt, digits =0) #Round-off precipitation values
#Prepare matrix for reclassification of precipitation values
>ppt_matrix =matrix(c(0,1,0,1,300,1,300,600,2,600,900,3,
900,1200,4,1200,1500,5), ncol =3 , byrow =TRUE)
>ppt_matrix#Display matrix for reclassification
##      [,1] [,2] [,3]
## [1,] 0    1    0
## [2,]    1  300    1
## [3,]  300  600    2
## [4,]  600  900    3
## [5,]  900 1200    4
## [6,] 1200 1500    5
# Reclassification of precipitation raster
>ppt_class =reclassify(ppt_round, ppt_matrix)
>ppt_class#Display attributes of classified output layer
## class      : RasterLayer
## dimensions : 606, 584, 353904  (nrow, ncol, ncell)
## resolution : 0.05, 0.05  (x, y)
## extent     : 68.2, 97.4, 6.749999, 37.05  (xmin, xmax, ymin, ymax)
## crs        : +proj=longlat +datum=WGS84 +no_defs +ellps=WGS84
+towgs84=0,0,0
## source     : memory
## names      : layer
## values     : 0, 5  (min, max)
```

After reclassification of the precipitation values into categories, the thematic map can be prepared using following R-code. The purpose of each code is also given below.

```
#Mapping with classified labels
>ggR(ppt_class, geom_raster =TRUE) +#Plot raster with 'ggplot2'
package
      ggtitle('Precipitation')+#Title
      labs(x='Longitude', y='Latitude')+#Axis labels
      theme(plot.title =element_text(hjust =0.5, #Align title
in centre
      size =24), #Size of title
      axis.title =element_text(size =20), #Size of axis labels
      legend.key.size =unit(1, "cm"), #Size of legend scale
      legend.title =element_text(size =20), #Size of legend
title
      legend.text =element_text(size =15)) +#Size of legend
text
      #Add customized colours to the raster categories
scale_fill_gradientn(colours =c('red', 'orange', 'yellow',
'lightblue','blue', 'darkblue'),
```

```
name =”Intensity of precipitation”,#Legend title
labels =c(‘No precipitation’, ‘Very low’, “Low”,
“Medium”, “High”, “Very High”),#Legends
na.value =NA)+#No colour for missing values

#Add north arrow
      ggsn::north(location =”topright”, #Location of north arrow
symbol =1, #Code number of North arrow symbol
      x.min =67, x.max =98, #Coordinates of map
      y.min =6, y.max =38) +

#Add scale bar in the map
ggsn::scalebar(location =”bottomright”, #Location of scale bar
transform =TRUE,#True for coordinates in decimal degree
dist_unit =”km”, #Units of scale bar
model =”WGS84", #Ellipsoid model

#Distance to represent with each segment of scale bar
dist =500,
x.min =65, x.max =100, #Extent coordinates of map grid
y.min =5, y.max =40,
height =0.02, #Scale bar’s height
st.size =3)+#Scale bar’s size

#Plot shapefile of India on raster
      geom_polygon(data =india_map, aes(x = long, y = lat,
group = group),
      colour =’black’, #Colour of the shapefile border
fill =NA)    #No filling inside shapefile
```

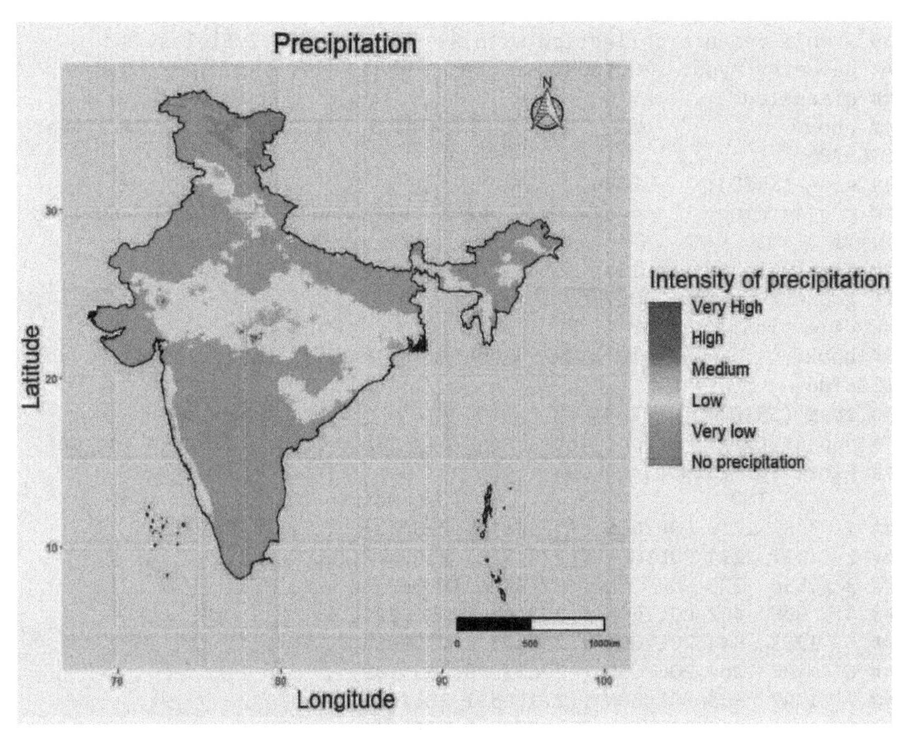

18.2.2. Vector mapping with ggplot2

The maps from shapefile representing attribute data can be created using `ggplot2` package. For creating thematic map from vector data, first of all, import the required packages which are required along with `ggplot2` package using following R-codes. Package `sf` is required for reading shapefile. Package `RColorBrewer` is required for assigning pre-defined colour palettes. Package `ggsn` is required for adding north arrow and scalebar to the plot.

```
#Import required packages
>library(sf) #To read shapefile
>library(ggplot2)  #To create map
>library(RColorBrewer) #Colour palette
>library(ggsn)      #For adding north arrow and scalebar
```

Now we will import the shapefile of a village having population data as an attribute using following R-code.

```
#Import shapefile
>vill =st_read('village/village.shp')
## Reading layer 'village' from data source
'E:\.......\Chapter18\data\village\village.shp' using driver
'ESRI Shapefile'
```

```
## Simple feature collection with 89 features and 2 fields
## geometry type:  POLYGON
## dimension:      XY
## bbox:           xmin: 262402.9 ymin: 2337437 xmax: 290834.7 ymax:
2368706
## epsg (SRID):    32644
## proj4string:    +proj=utm +zone=44 +datum=WGS84 +units=m +no_defs
>vill          #Display metadata of imported shapefile
## Simple feature collection with 89 features and 2 fields
## geometry type:  POLYGON
## dimension:      XY
## bbox:           xmin: 262402.9 ymin: 2337437 xmax: 290834.7 ymax:
2368706
## epsg (SRID):    32644
## proj4string:    +proj=utm +zone=44 +datum=WGS84 +units=m +no_defs
## First 10 features:
##      TP  TMP                       geometry
## 1  3257 1710 POLYGON ((275081.4 2364424,...
## 2   257  134 POLYGON ((270173.7 2365404,...
## 3   554  279 POLYGON ((271502 2360052, 2...
## 4   908  467 POLYGON ((272479.9 2366324,...
## 5  1263  644 POLYGON ((275163.2 2359912,...
## 6   508  264 POLYGON ((276933.5 2365588,...
## 7  1747  903 POLYGON ((271508.1 2364701,...
## 8  1908  987 POLYGON ((265802.9 2362584,...
## 9   723  380 POLYGON ((268287.2 2361866,...
## 10  286  150 POLYGON ((267204.9 2361413,...
```

From the metadata of shapefile, we can see that it has two associated attributes *i.e.* 'TP' and 'TMP'. We will create a thematic map of total population data using the sets of following R-code. The respective purpose of each R-code is also given below.

```
#Create Map of population data
>ggplot(data =vill) +#Plot shapefile
       geom_sf(aes(fill = TP)) +#Fill regions with value of 'TP'
       labs(title ="Total Population",                #Title
       x ="Longitude", y ="Latitude") +               #Axis labels
       theme(plot.title =element_text(hjust =0.5, #Align title in centre
              size =24), #Size of title
       axis.title =element_text(size =20), #Size of axis labels
       legend.key.size =unit(1, "cm"),                #Size of legend
       legend.title =element_text(size =20),          #Size of legend
title
       legend.text =element_text(size =15)) +         #Size  of  legend
text
       scale_fill_gradientn(colours =brewer.pal(6, "YlOrBr"), #Colour
       name ="Population") +#Legend Title
```

```
#Add North Arrow
        ggsn::north(vill, location ="topright", #Location of North arrow
        symbol =1) +#Code number of North arrow symbol

#Add Scalebar
ggsn::scalebar(vill, location ="bottomleft", #Location of scale bar
transform =FALSE, #False for coordinates not in degree
dist_unit ="m", #Units of scale bar
#Distance to represent with each segment of scale bar
dist =5000,
st.size =3) #Scale bar's size
```

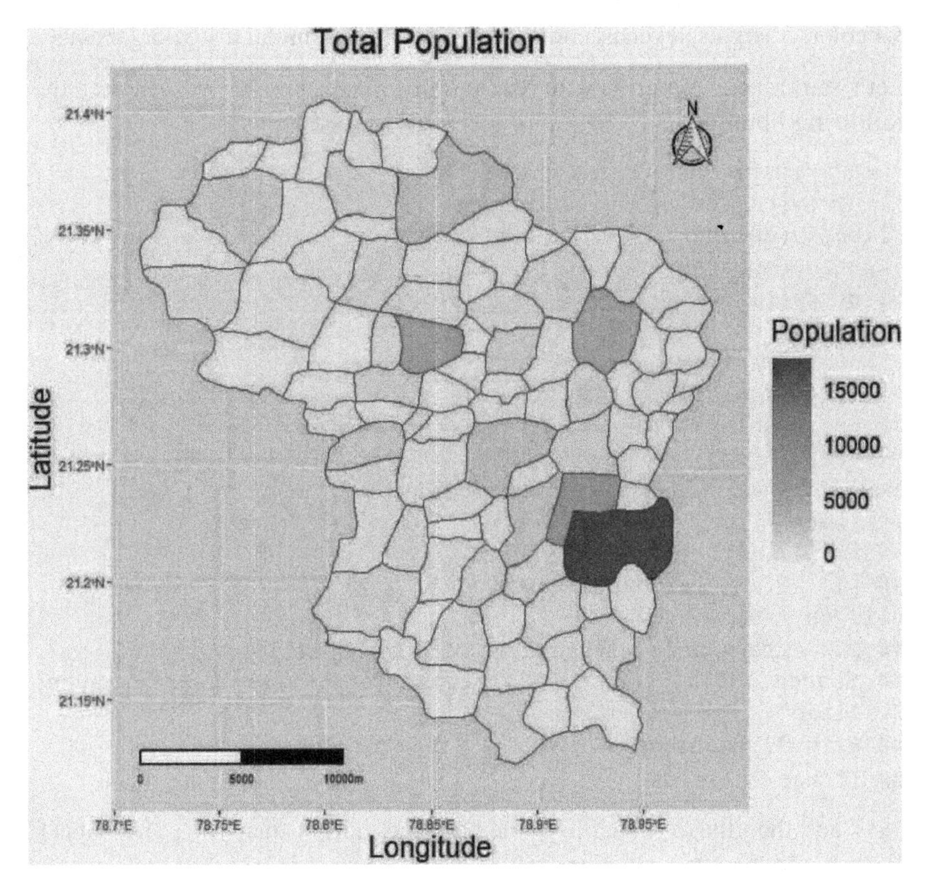

18.3. Use of tmap in thematic mapping

18.3.1. Mapping vectors with tmap

tmap is a dedicated package for thematic mapping which works grammatically, as the ggplot2. tm_shape() is the base function for spatial data, and tm_…() for layers, similar to ggplot2 geom_…(…), where you can add geometry, mapping and scaling. For example,

- tm_symbols()
- tm_raster()
- tm_polygons()
- tm_squares()
- tm_bubbles()
- tm_fill()

tm_facets() for small multiples, similar to ggplot2 facet_wrap() and tm_layout() for colors, margins, legends, compass, grid lines, etc. similar to ggplot2 theme().

Let's start map preparation with reading the village.shp file. We will need the following libraries:

```
>library(sf)
>library(raster)
>library(dplyr)
>library(spData)
>library(tmap)    # for static and interactive maps
>library(leaflet) # for interactive maps
>library(mapview) # for interactive maps
>library(ggplot2) # tidyverse data visualization package
>library(shiny)   # for web applications
>library(rgdal) #Vector data
```

Now set the working directory and read the shapefile.

```
>setwd('E:/............../Chapter18/data')
>getwd() #Get the working directory
## [1] "E:/........../Chapter18/data"
village =readOGR('village.shp')
## OGR data source with driver: ESRI Shapefile
## Source: "E:\..........\Chapter18\data\village.shp", layer: "village"
## with 89 features
## It has 2 fields
```

Now add the village shapefile to the base function of tmap, i.e., tm_shape(). We can add the other functions by '+' symbol as:

```
# Add fill layer to village shape
>tm_shape(village) +
tm_fill()
```

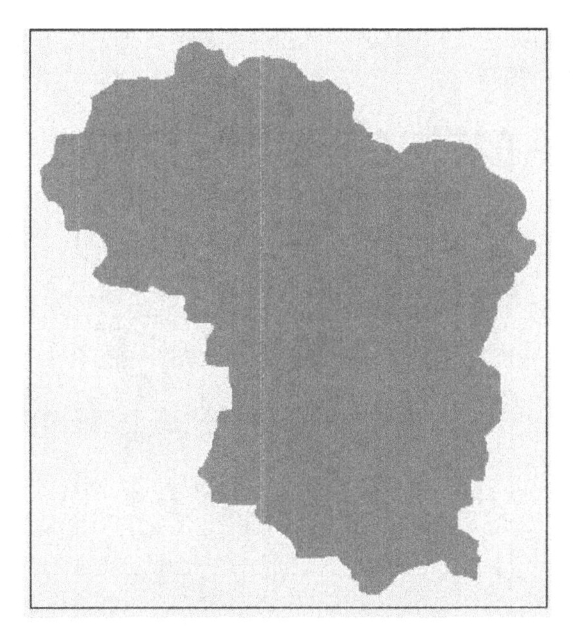

We can add the borders of the villages by tm_border() function. Further, we can add both the fill and the border to the map. tm_dots()can be used to create points in the centre of the polygons.

```
# Add border layer to village shape
>tm_shape(village) +
tm_borders()
```

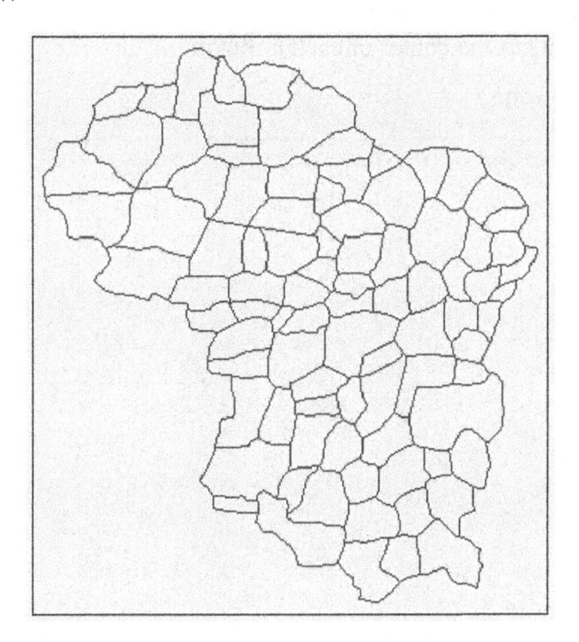

```
# Add fill and border layers to village shape
>tm_shape(village) +
tm_fill() +
tm_borders()
```

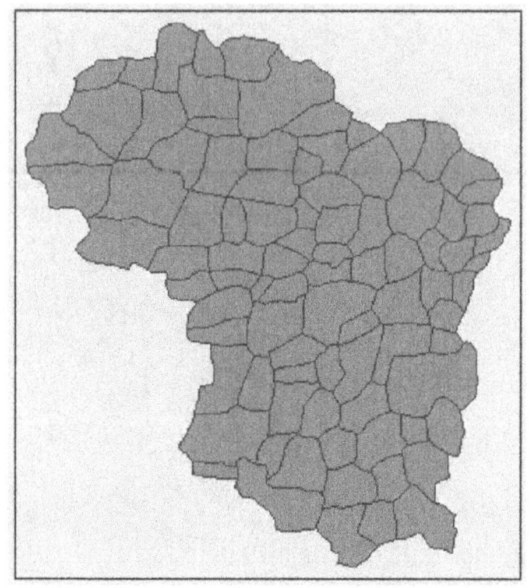

These three steps may be done by qtm() function in one go

```
>qtm(village)
```

Or we can use the tm_polygons() function with tm_shape()

```
>tm_shape(village) +tm_polygons()
```

Images are similar to the previous one. Further, a function tm_dots(), developes a map with points in the center of each polygon

```
>tm_shape(village) +
tm_dots() +
tm_borders()
```

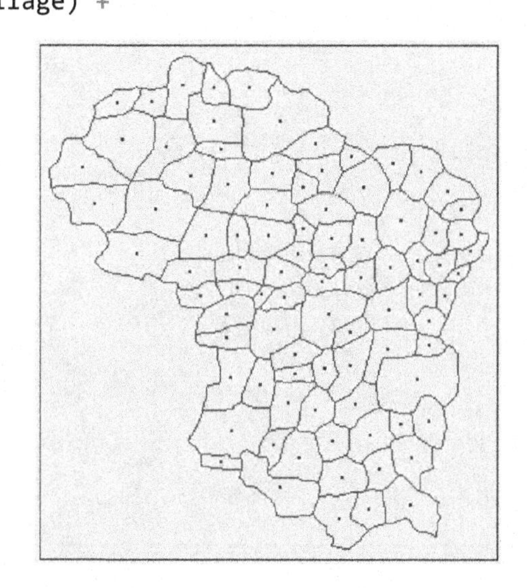

tmap stores objects representing maps. This can be verified by saving the map by the name of 'villagemap'

```
#tmap store objects representing maps
>villagemap=tm_shape(village) +tm_polygons()
#get the class of the tmap object, villagemap created above
>class(villagemap)
## [1] "tmap"
```

The following codes define the line width, line type and color. We will save these maps as objects to show the use of tmap_arrange() function for arranging the maps. First, we will create the maps and save them as objects and then arrange them in one frame.

```
>map1 =tm_shape(village) +tm_fill(col =”red”) #col defines the
fill color
>map2 =tm_shape(village) +tm_fill(col =”red”, alpha =0.3) #alpha
defines the transpirancy
>map3 =tm_shape(village) +tm_borders(col =”blue”)
>map4 =tm_shape(village) +tm_borders(lwd =3)  #lwd defines the
line width
>map5 =tm_shape(village) +tm_borders(lty =2)  #lty defines the
line type. here 2 difines dashed lines
```

The line types can either be specified as an integer (0=blank, 1=solid (default), 2=dashed, 3=dotted, 4=dotdash, 5=longdash, 6=twodash) or as one of the character strings "blank", "solid", "dashed", "dotted", "dotdash", "longdash", or "twodash", where "blank" uses 'invisible lines' (i.e., does not draw them).

```
>map6 =tm_shape(village) +tm_fill(col =”red”, alpha =0.3) +
tm_borders(col =”blue”, lwd =2, lty =3)
>tmap_arrange(map1, map2, map3, map4, map5, map6)
```

We can assign the fill color according to any attribute associated with the polygon. We will now give the color according to total population. It will be provided by the argument *col=* in the tm_fill() function

```
>tm_shape(village) +tm_fill(col ="TP")
```

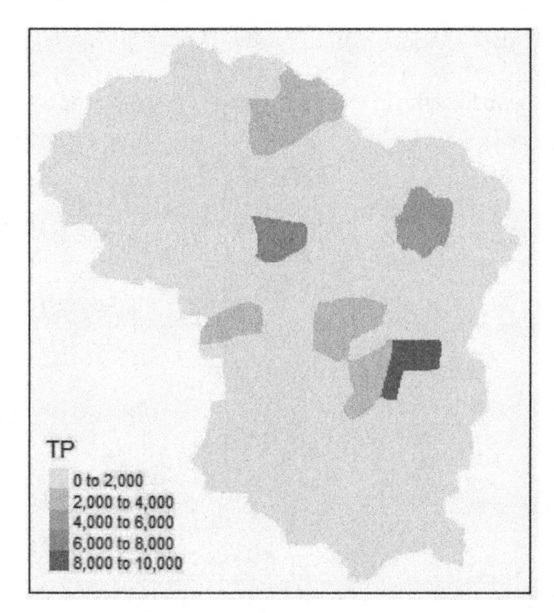

We can change the label of the legend by argument *title*= in the tm_fill(0 function. Now we will give the label of the legend as "Total Population"

```
>tm_shape(village) +
tm_fill(col ="TP", title ="Total Population") +tm_borders()
```

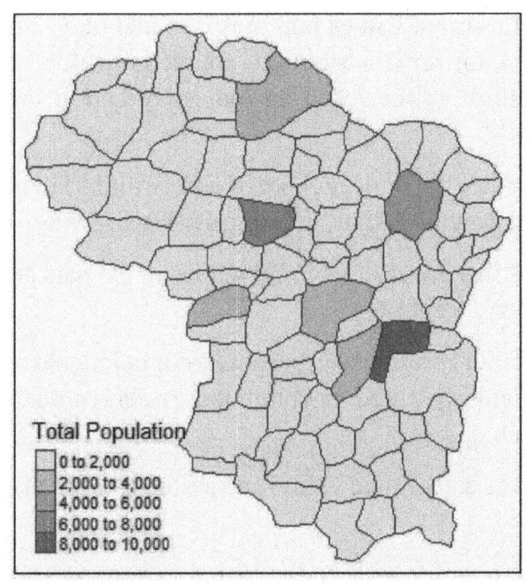

Now we can change the breaks and define colors for each category. We can define default breaks by providing number of classes by argument *n*= in the tm_fill() function.

```
>tm_shape(village) +
tm_fill(col ="TP", title ="Total Population", n=10) +tm_borders()
```

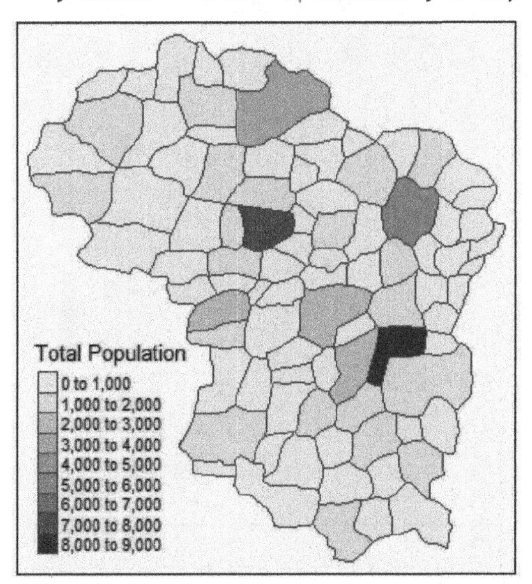

Other ways for assigning default breaks is to use the argument 'style'. Six types of styles are available.

style = pretty, the default setting, rounds breaks into whole numbers where possible and spaces them evenly;

style = equal divides input values into bins of equal range and is appropriate for variables with a uniform distribution (not recommended for variables with a skewed distribution as the resulting map may end-up having little color diversity);

style = quantile ensures the same number of observations fall into each category (with the potential downside that bin ranges can vary widely);

style = jenks identifies groups of similar values in the data and maximizes the differences between categories;

style = cont (and order) present a large number of colors over continuous color fields and are particularly suited for continuous rasters (order can help visualize skewed distributions);

style = cat represents categorical values and assures that each category receives a unique color.

We will assign the quantile style to the map for example.

```
>tm_shape(village) +
tm_fill(col ="TP", title ="Total Population", style ='quantile')
+tm_borders()
```

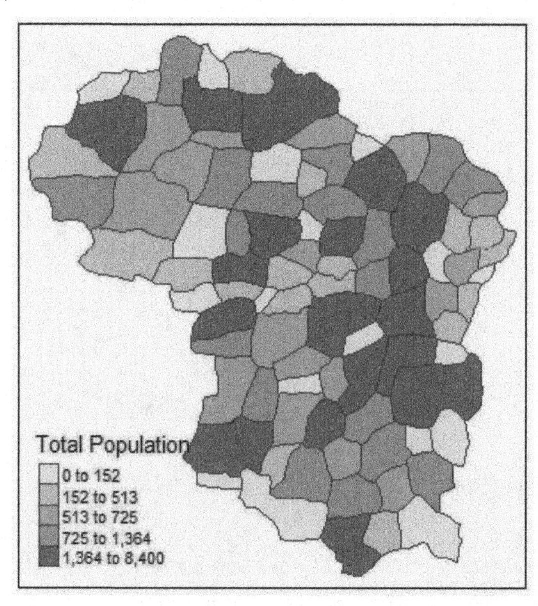

For assigning user defined class breaks, first break object is needed to be made. If you have 6 classes, give 7 break values. The break created is assigned to the map by argument *breaks=* in the function tm_fill().

```
#define your own breaks
>breaks =c(0, 100, 300, 500, 1000, 5000, 18000)
>tm_shape(village) +
tm_fill(col ="TP", title ="Total Population", breaks = breaks)
+tm_borders()
```

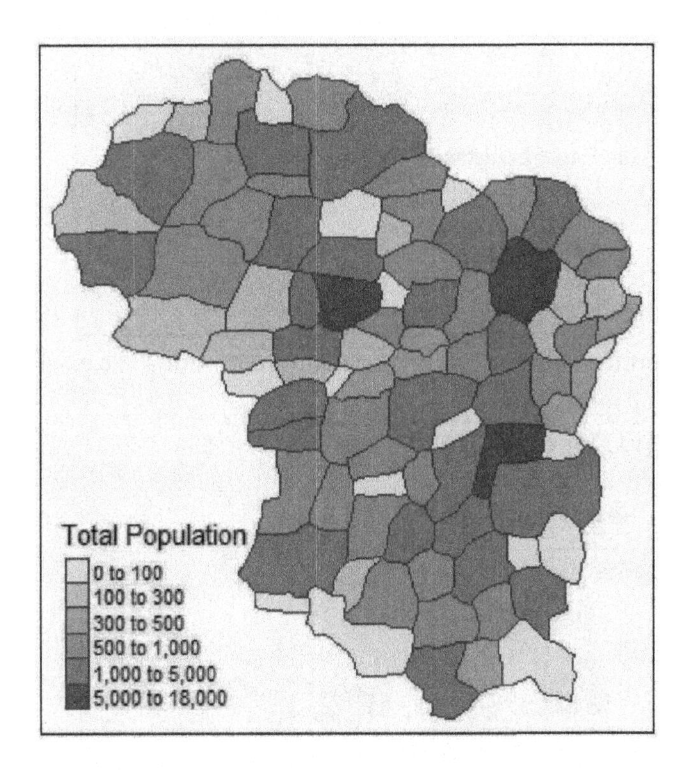

Now the user defined color is to be assigned. We can use the library, "RColorBrewer". Here in example, we select "BuGn" palette from the ColorBrewer.

```
>library("RColorBrewer")
#we select the "BuGn" palette for example
>breaks =c(0, 100, 300, 500, 1000, 5000, 18000)
>tm_shape(village) +
tm_fill(col ="TP", title ="Total Population", breaks = breaks,
palette ="BuGn") +tm_borders()
```

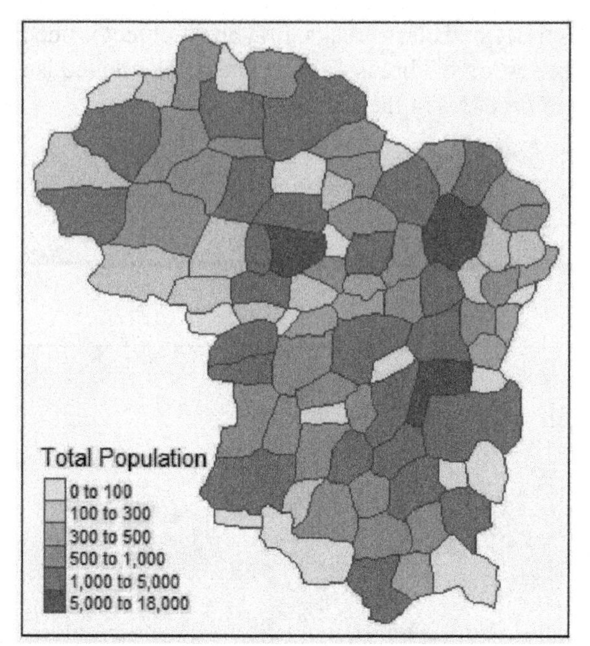

We can invert the color by prefixing palette name with"-" . e.g., -BuGn

```
>breaks =c(0, 100, 300, 500, 1000, 5000, 18000)
>tm_shape(village) +
tm_fill(col ="TP", title ="Total Population", breaks = breaks,
palette ="-BuGn") +tm_borders()
```

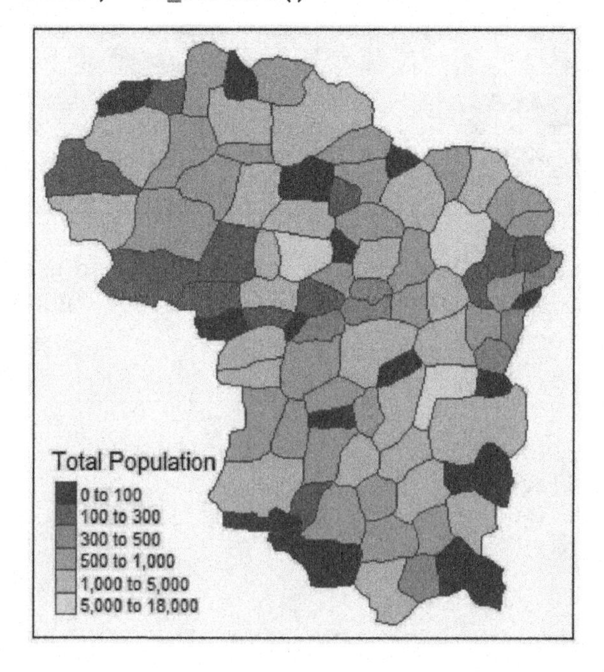

We can add another shape file to the map by using '+'. Here we will be adding the boundary file to the map and will keep the line type of the polygon as dotted.

```
#read the boundary file
>boundary=readOGR('boundary.shp')
## OGR data source with driver: ESRI Shapefile
## Source: "H:\Chapter18\data\boundary.shp", layer: "boundary"
## with 1 features
## It has 1 fields
## Integer64 fields read as strings:  T_MP
>tm_shape(village) +
tm_fill(col ="TP", title ="Total Population", breaks =c(0, 100,
300, 500, 1000, 5000, 18000),
palette ="BuGn") +tm_borders(lty =3)+
tm_shape(boundary)+
tm_borders()
```

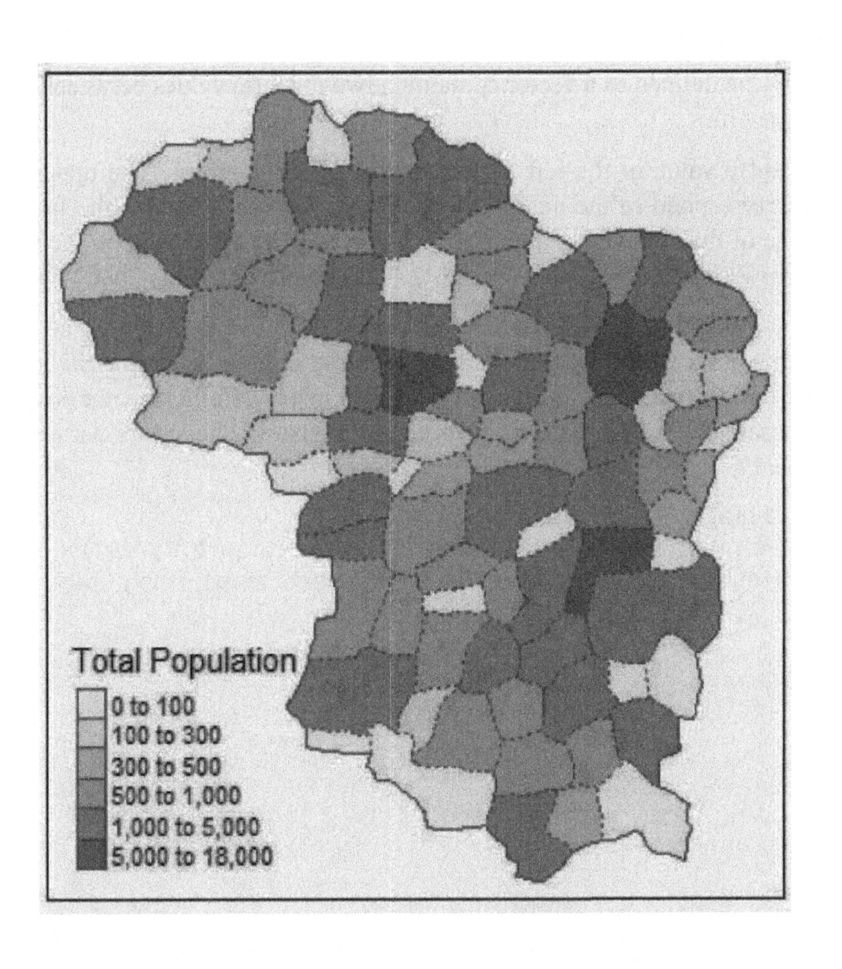

Now we will define the layout i.e., the scale bar, north arrow, and legend position etc.

For that we will be saving the map built till now as an object called 'map_village'

```
>map_village=tm_shape(village) +
tm_fill(col =”TP”, title =”Total Population”, breaks =c(0, 100,
300, 500, 1000, 5000, 18000),
palette =”BuGn”) +tm_borders(lty =3)+
tm_shape(boundary)+
tm_borders()
```

in tmap package, the north arrows and scale bars have their own functions: tm_compass() and tm_scale_bar(). The type of north arrow available are "arrow", "4star", "8star", "radar", and "rose". The position of the compass is defined as a vector of two values, specifying the x and y coordinates. Either this vector contains "left", "LEFT", "center", "right", or "RIGHT" for the first value and "top", "TOP", "center", "bottom", or "BOTTOM" for the second value,

or it may be defined as a vector containing two numeric values between 0 and 1 that specifies

the x and y value of the left bottom corner of the compass. The uppercase values correspond to the position without margins (so tighter to the frame). The size of the compass is a number. The default values depend on the type: for "arrow" it is 2, for "4star" and "8star" it is 4, and for "radar" and "rose" it is 6.

The scale bar's breaks are defined by a vector of numbers defining the initial (0), the middle (e.g., 2.5), and the last (e.g., 5) numbers defining km. the position of the scale bar is defined similar to the north arrow. The size is defined by argument *text.size=*.

```
>map_village+
tm_compass(type =”8star”,  size =2, position=c(.8, .87)) +
tm_scale_bar(breaks =c(0, 2.5, 5), text.size = .5, position
=c(.75, .8))
```

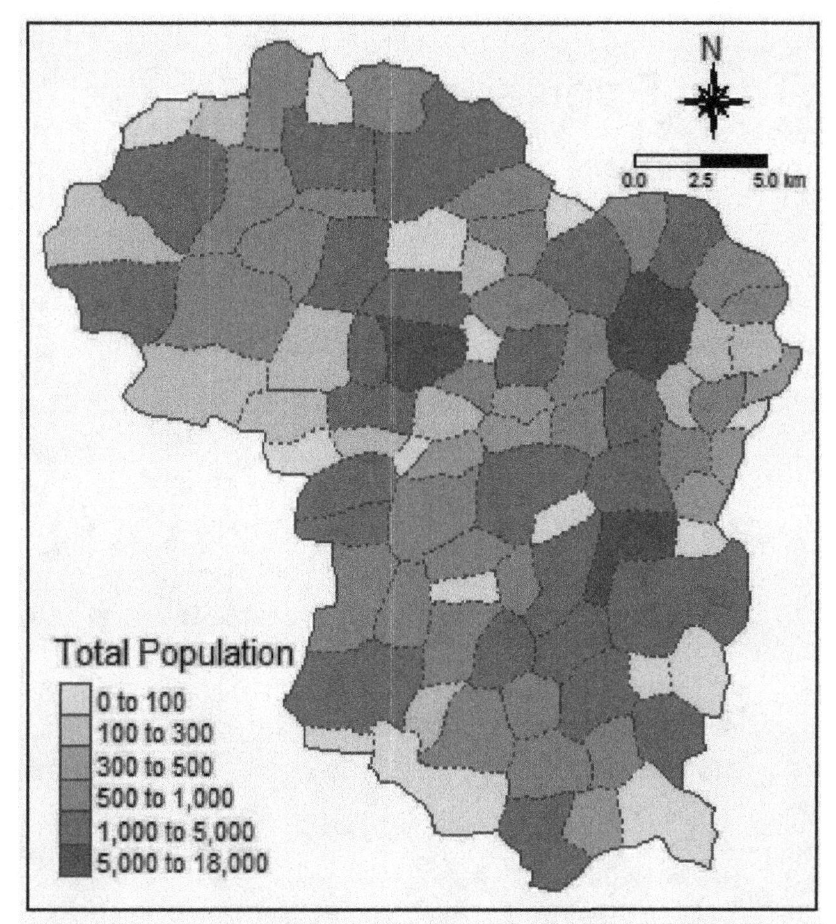

To control title, margins, aspect ratio, colors, frame, legend, and many other specifications of the map layout tm_layout() function is used. The *inner.margin* argument is useful to accommodate the title, legend, and other map elements. It is the relative margins inside the frame defined by a vector of four values specifying the bottom, left, top, and right margin. Values are between 0 and 1. By default, it is 0 for each side if master shape is a raster, otherwise 0.02. The *bg.color* argument gives the background color of the map.

```
>map_village+
tm_layout(title ="Total Population", bg.color ="grey",
inner.margins =c(.04, .03, .1, .01))
```

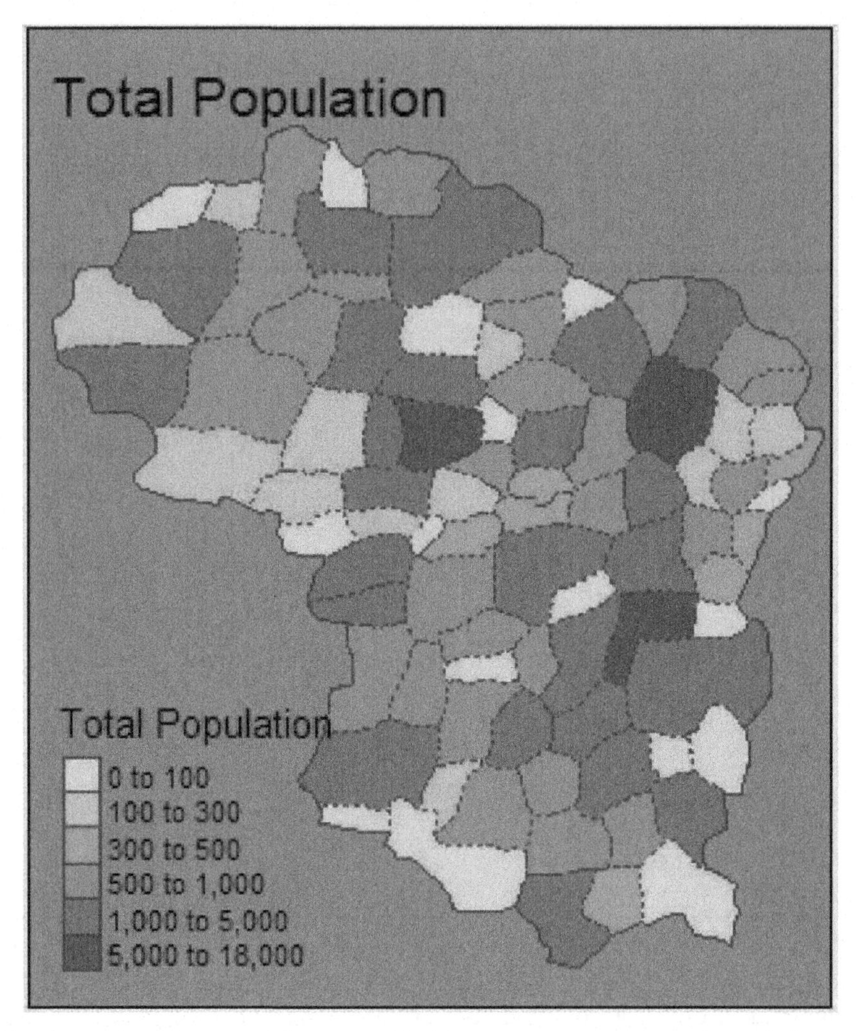

The legend position can be controlled by many arguments. *legend.show*, a logical argument will show legend if TRUE. Similarly, the argument *legend.outside* plots legend outside the frame if TRUE. Let's try to remove the legend.

```
map_village+
tm_layout(title ="Total Population", bg.color ="grey",
inner.margins =c(.04, .03, .1, .01), legend.show =FALSE)
```

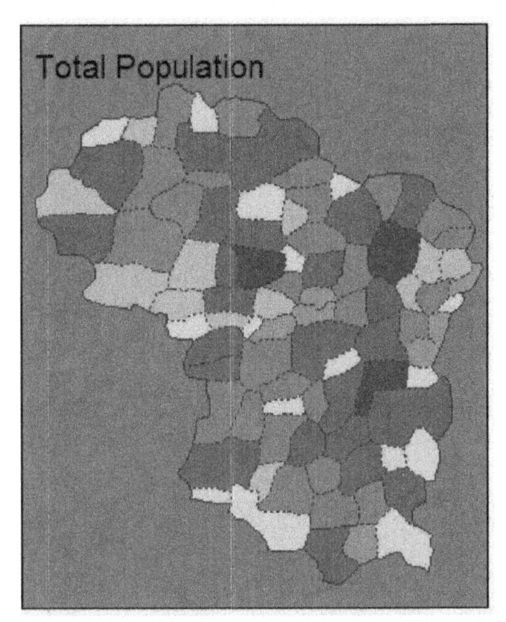

To put the legend in the left outside the frame, try this code:

```
>map_village+
tm_layout(title ="Total Population", bg.color ="grey",
inner.margins =c(.04, .03, .1, .01), legend.outside =TRUE,
legend.outside.position ="left")
```

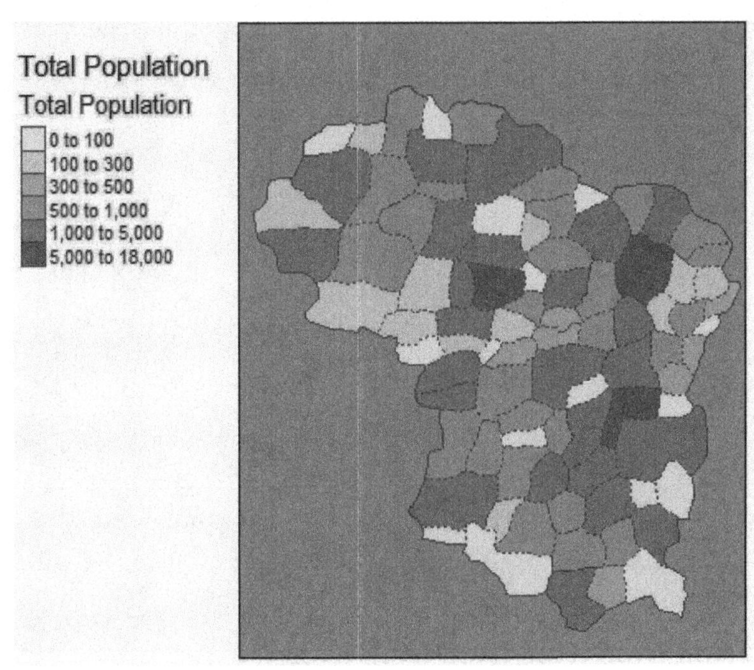

The final code for mapping a vector is

```
>tm_shape(village) +
tm_fill(col ="TP", title ="Total Population", breaks =c(0, 100,
300, 500, 1000, 5000, 18000),
palette ="BuGn") +tm_borders(lty =3)+
tm_shape(boundary)+
tm_borders(lwd =2)+
tm_compass(type ="8star",  size =2, position=c(.8, .87)) +
tm_scale_bar(breaks =c(0, 2.5, 5), text.size = .5, position
=c(.75, .8))+
tm_layout(title ="Total Population", bg.color ="grey",
inner.margins =c(.04, .1, .15, .01)) #this inner.margin
accommodates the north arrow, scale, title, and legend well.
```

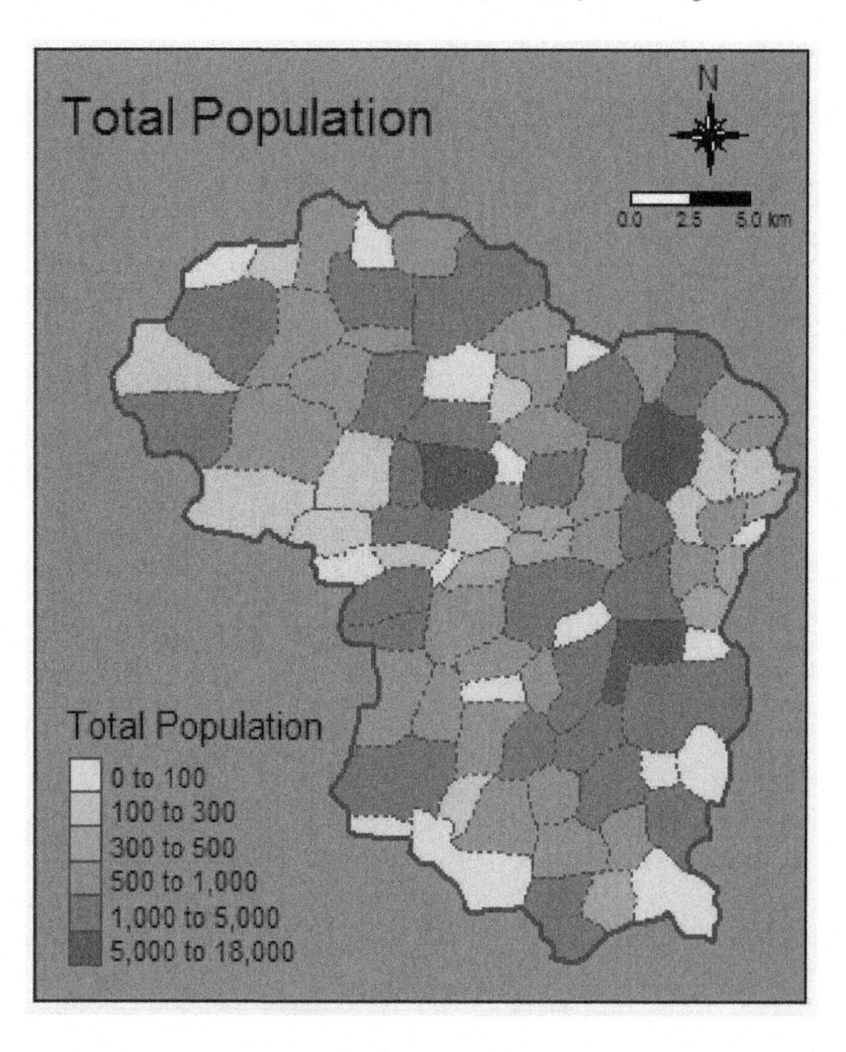

18.3.2. Mapping Rasters with tmap

Mapping rasters with tmap is similar to mapping the vector. Mapping raster in tmap requires the raster to be read in by tm_shape(). tm_raster() is used for defining color, style of color, number of classes, title of legend etc. First read the raster file:

```
>DEM =raster('DEM.tif')
>NAvalue(DEM) =-9999
>DEM
## class      : RasterLayer
## dimensions : 2590, 2372, 6143480  (nrow, ncol, ncell)
## resolution : 12.5, 12.5  (x, y)
## extent    : 884519.1, 914169.1, 2340236, 2372611  (xmin, xmax, ymin, ymax)
## crs        : +proj=utm +zone=43 +datum=WGS84 +units=m +no_defs +ellps=WGS84 +towgs84=0,0,0
## source     : E:/........../Chapter18/data/DEM.tif
## names      : DEM
## values     : 236, 485  (min, max)
names(DEM)              #get the names of values in the raster
## [1] "DEM"
```

Now add the DEM to the tm_shape() function, and define the color pallete, style, and legend label in the tm_raster() function as was done in tm_fill() function in case of vector. Additionally, define the legend position by tm_legend() function and *legend.position=* argument. It is a vector of two characters as shown in the code below, i.e., ("left", "bottom"). It may be ("right", "bottom") or ("left", "top"), etc. This function may also be used in case of vector mapping.

```
>tm_shape(DEM) +
tm_raster(col ="DEM",
palette ="-Spectral", style ="quantile", n=15,
title="Elevation (m)") +
tm_legend(legend.position =c("left", "bottom"))
```

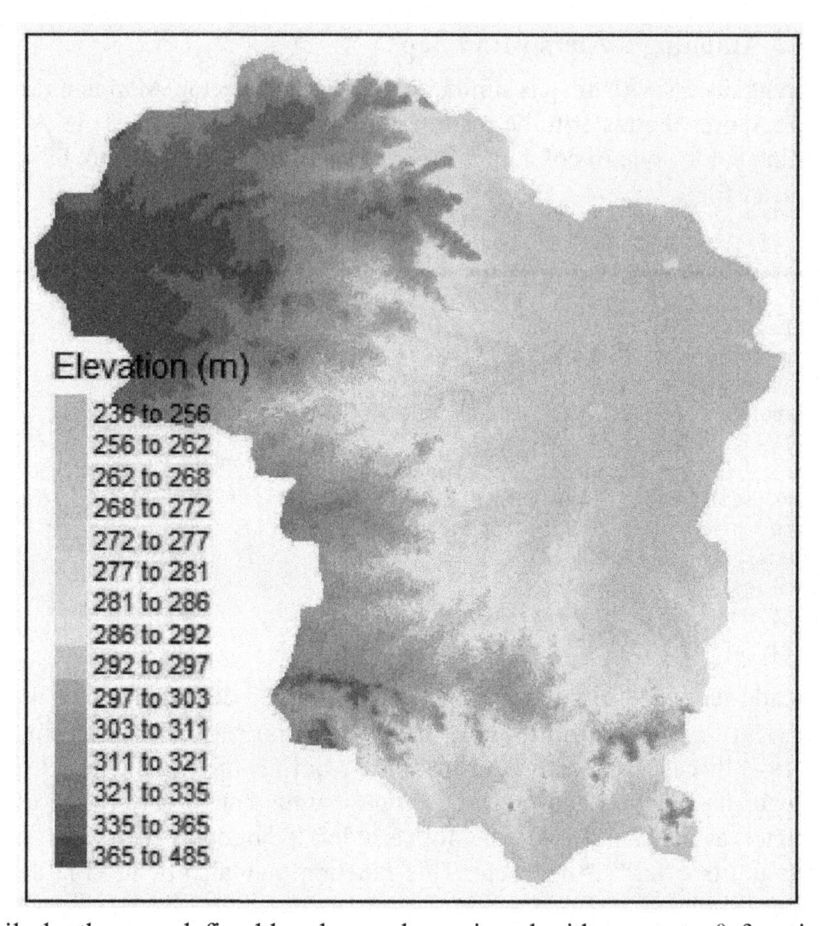

Similarly, the user defined breaks can be assigned with tm_raster() function, as in case of tm_fill().

```
#define desired breaks
>tm_shape(DEM) +
tm_raster(col =”DEM”,
palette =”-Spectral”, style =”fixed”, breaks=c(200,250, 300,
350, 400, 450, 500),
title=”Elevation (m)”) +
tm_legend(legend.position =c(“left”, “bottom”))
```

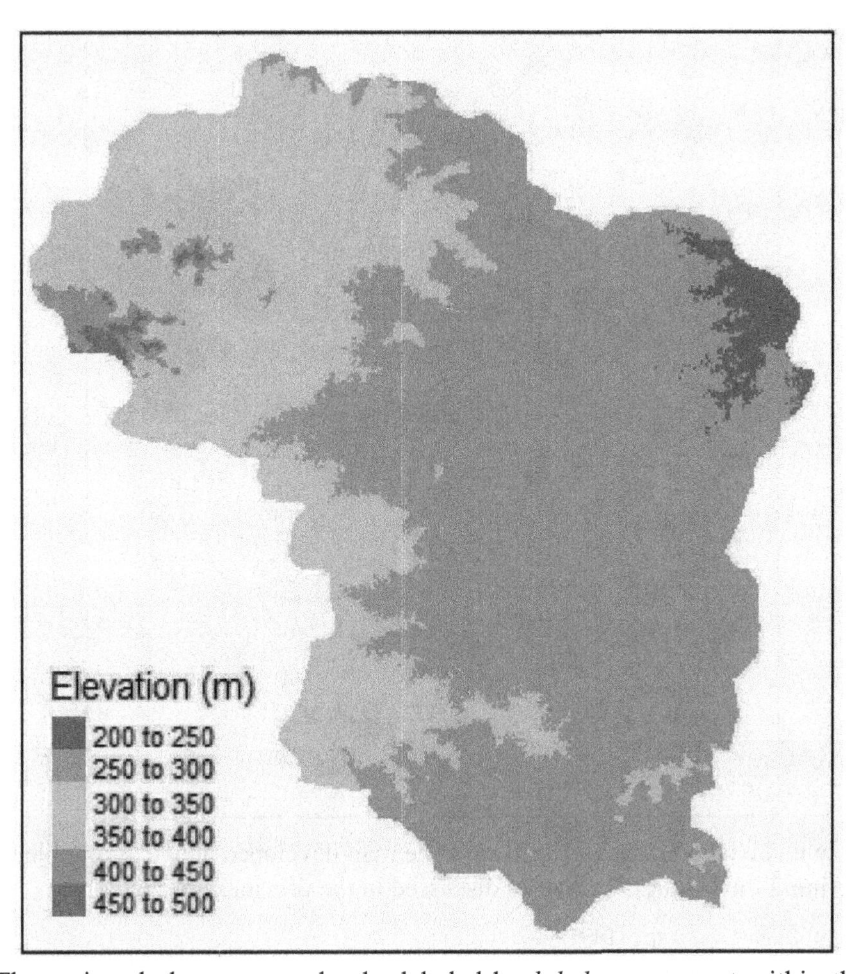

The assigned classes may also be labeled by *labels*= argument within the tm_raster() function.

```
# give class name to the defined classes
>tm_shape(DEM) +
tm_raster(col =”DEM”,
palette =”-Spectral”, style =”fixed”, breaks=c(200,250, 300,
350, 400, 450, 500),
labels=c(“Very low”, “Low”, “Moderate”, “Moderately high”, “High”,
“Very high”) ,
title=”Elevation Classes”) +
tm_legend(legend.position =c(“left”, “bottom”))
```

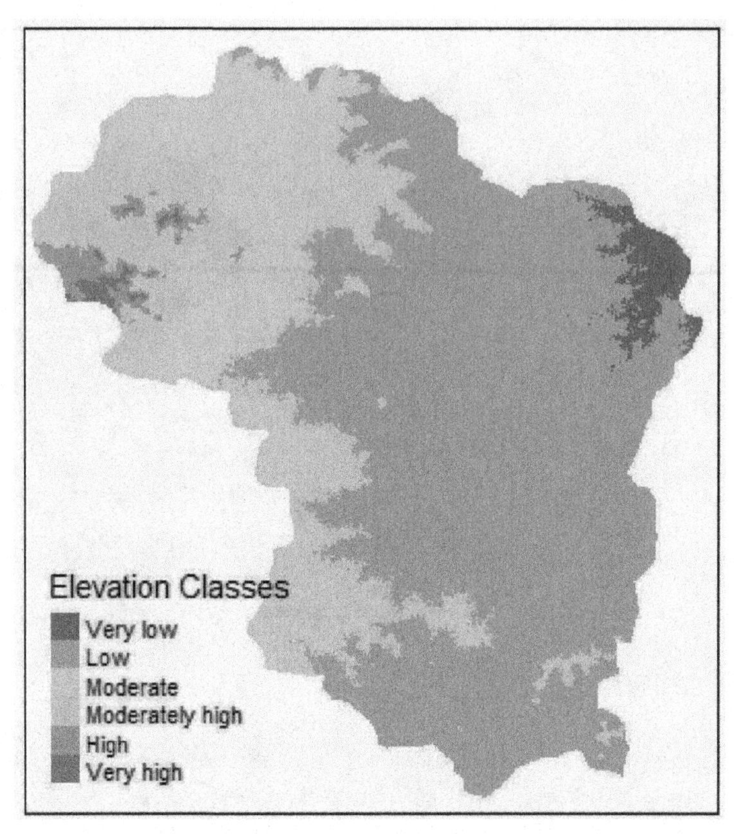

Now make layout. First we will save the map developed above as an object. Defining the layout is similar as discussed in the previous section.

```
>map_DEM=tm_shape(DEM) +
tm_raster(col =”DEM”,
palette =”-Spectral”, style =”fixed”, breaks=c(200,250, 300, 350, 400, 450, 500),
labels=c(“Very low”, “Low”, “Moderate”, “Moderately high”, “High”, “Very high”) ,
title=”Elevation Classes”) +
tm_legend(legend.position =c(“left”, “bottom”))

>map_DEM+
tm_shape(boundary)+
tm_borders(lwd =2)+
tm_compass(type =”8star”,  size =2, position=c(.8, .87)) +
tm_scale_bar(breaks =c(0, 2.5, 5), text.size = .5, position =c(.75, .8))+
tm_layout(title =”Digital Elevation Model”, bg.color =”lightblue”, inner.margins =c(.04, .1, .15, .01))
```

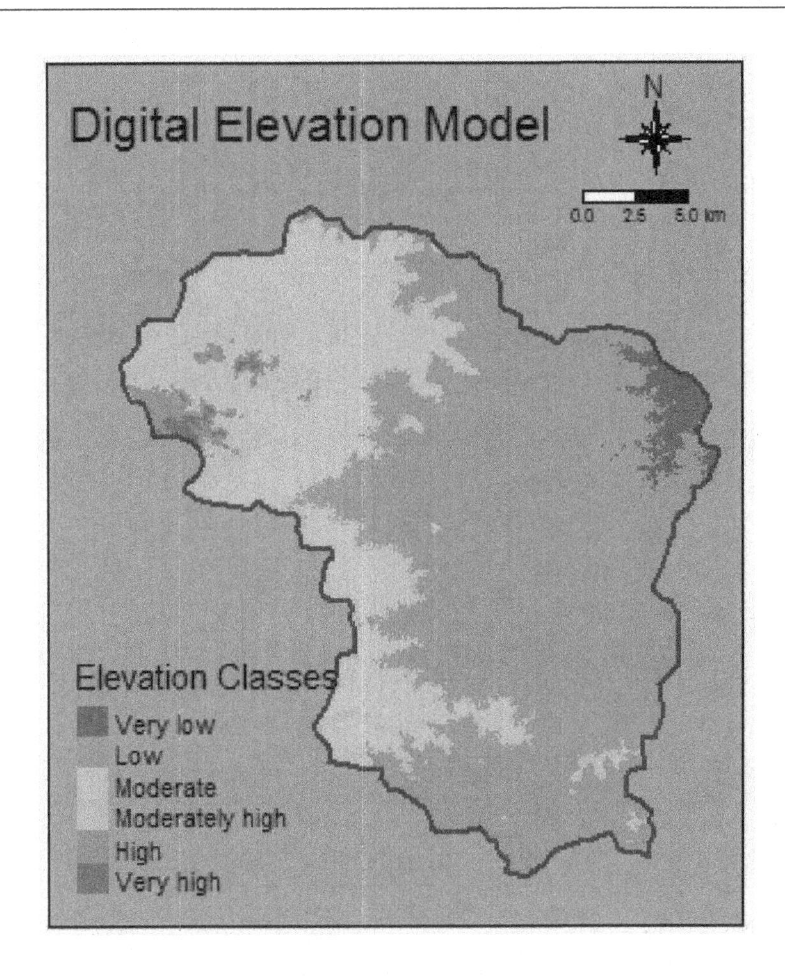
Digital Elevation Model
N
0.0 2.5 5.0 km
Elevation Classes
Very low
Low
Moderate
Moderately high
High
Very high

Bibliography

If you need to cite R, there is a very useful function called citation().

```
> citation()
To cite R in publications use:
  R Core Team (2019). R: A language and environment for statistical
  computing. R Foundation for Statistical Computing, Vienna,
  Austria. URL https://www.R-project.org/.
A BibTeX entry for LaTeX users is
  @Manual{,
    title = {R: A Language and Environment for Statistical Computing},
    author = {{R Core Team}},
    organization = {R Foundation for Statistical Computing},
    address = {Vienna, Austria},
    year = {2019},
    url = {https://www.R-project.org/},
  }
We have invested a lot of time and effort in creating R, please cite it
when using it for data analysis. See also ‘citation(“pkgname”)’ for
citing R packages.
```

If you want to cite just a package, just pass the package name as a parameter, e.g.:

```
> citation(package=”shiny”)

To cite package ‘shiny’ in publications use:

  Winston Chang, Joe Cheng, JJ Allaire, Yihui Xie and Jonathan
  McPherson (2019). shiny: Web Application Framework for R. R
  package version 1.3.2. https://CRAN.R-project.org/package=shiny

A BibTeX entry for LaTeX users is

  @Manual{,
    title = {shiny: Web Application Framework for R},
    author = {Winston Chang and Joe Cheng and JJ Allaire and Yihui Xie
      and Jonathan McPherson},
    year = {2019},
    note = {R package version 1.3.2},
    url = {https://CRAN.R-project.org/package=shiny},
  }
```

In this book, we have used several packages of R. These are listed below in the form of R code to know their citations. The citations of these packages are listed after the package list.

List of packages

```
citation(package = “raster”)
citation(package=”RStoolbox”)
citation(package=”sp”)
citation(package=”rgdal”)
citation(packagc=”ggplot2")
citation(package=”rgeos”)
citation(package=”colorspace”)
citation(package=”rasterVis”)
citation(package=”cluster”)
citation(package=”randomForest”)
citation(package=”doParallel”)
citation(package=”nnet”)
citation(package=”caret”)
citation(package=”kernlab”)
citation(package=”rpart”)
citation(package=”rpart.plot”)
citation(package=”RSAGA”)
citation(package=”spatialEco”)
citation(package=”envirem”)
citation(package=”RColorBrewer”)
citation(package=”classInt”)
citation(package=”ggsn”)
citation(package=”sf”)
citation(package=”dplyr”)
citation(package=”spData”)
citation(package=”tmap”)
citation(package=”leaflet”)
citation(package=”mapview”)
citation(package=”shiny”)
```

Citations of packages

A. Liaw and M. Wiener (2002). Classification and Regression by randomForest. R News 2(3), 18—22.

Alexander Brenning, Donovan Bangs and Marc Becker (2018). RSAGA: SAGA Geoprocessing and Terrain Analysis. R package version 1.3.0. https://CRAN.R-project.org/package=RSAGA

Alexandros Karatzoglou, Alex Smola, Kurt Hornik, Achim Zeileis (2004). kernlab - An S4 Package for Kernel Methods in R. Journal of Statistical Software 11(9), 1-20. URL http://www.jstatsoft.org/v11/i09/

Benjamin Leutner, Ned Horning and Jakob Schwalb-Willmann (2019).RStoolbox: Tools for Remote Sensing Data Analysis. R package version 0.2.6. https://CRAN.R-project.org/package=RStoolbox

Erich Neuwirth (2014). RColorBrewer: ColorBrewer Palettes. R package version 1.1-2.https://CRAN.R-project.org/package=RColorBrewer

Evans JS (2019). _spatialEco_. R package version 1.2-0, <URL: https://github.com/jeffreyevans/spatialEco>.

H. Wickham. ggplot2: Elegant Graphics for Data Analysis. Springer-Verlag New York, 2016.

Hadley Wickham, Romain François, Lionel Henry and Kirill Müller (2019). dplyr: A Grammar of Data Manipulation. R package version 0.8.3. https://CRAN.R-project.org/package=dplyr

Joe Cheng, Bhaskar Karambelkar and Yihui Xie (2018). leaflet: Create Interactive Web Maps with the JavaScript 'Leaflet' Library.R package version 2.0.2. https://CRAN.R-project.org/package=leaflet

Maechler, M., Rousseeuw, P., Struyf, A., Hubert, M., Hornik, K.(2019). cluster: Cluster Analysis Basics and Extensions. R package version 2.1.0.

Max Kuhn. Contributions from Jed Wing, Steve Weston, Andre Williams, Chris Keefer, Allan Engelhardt, Tony Cooper, Zachary Mayer, Brenton Kenkel, the R Core Team, Michael Benesty, Reynald Lescarbeau, Andrew Ziem, Luca Scrucca, Yuan Tang, Can Candan and Tyler Hunt. (2019). caret: Classification and Regression Training. R package version 6.0-84. https://CRAN.R-project.org/package=caret

Microsoft Corporation and Steve Weston (2019). doParallel: Foreach Parallel Adaptor for the 'parallel' Package. R package version 1.0.15. https://CRAN.R-project.org/package=doParallel

Oscar Perpinan Lamigueiro and Robert Hijmans (2019), rasterVis. R package version 0.46.

Oswaldo Santos Baquero (2019). ggsn: North Symbols and Scale Bars for Maps Created with 'ggplot2' or 'ggmap'. R package version 0.5.0. https://CRAN.R-project.org/package=ggsn

Pebesma, E., 2018. Simple Features for R: Standardized Support for Spatial Vector Data. The R Journal 10 (1), 439-446, https://doi.org/10.32614/RJ-2018-009

Pebesma, E.J., R.S. Bivand, 2005. Classes and methods for spatial data in R. R News 5 (2), https://cran.r-project.org/doc/Rnews/.

R Core Team (2019). R: A language and environment for statistical computing. R Foundation for Statistical Computing, Vienna,Austria. URL https://www.R-project.org/.

Robert J. Hijmans (2019). raster: Geographic Data Analysis and Modeling. R package version 3.0-2.https://CRAN.R-project.org/package=raster

Roger Bivand (2019). classInt: Choose Univariate Class Intervals. R package version 0.4-1. https://CRAN.R-project.org/package=classInt

Roger Bivand and Colin Rundel (2019). rgeos: Interface to Geometry Engine - Open Source ('GEOS'). R package version 0.5-1. https://CRAN.R-project.org/package=rgeos

Roger Bivand, Jakub Nowosad and Robin Lovelace (2019). spData: Datasets for Spatial Analysis. R package version 0.3.2. https://CRAN.R-project.org/package=spData

Roger Bivand, Tim Keitt and Barry Rowlingson (2019). rgdal: Bindings for the 'Geospatial' Data Abstraction Library. R package version 1.4-4. https://CRAN.R-project.org/package=rgdal

Roger S. Bivand, Edzer Pebesma, Virgilio Gomez-Rubio, 2013. Applied spatial data analysis with R, Second edition. Springer, NY. http://www.asdar-book.org/

Stauffer R, Mayr GJ, Dabernig M, Zeileis A (2009). "Somewhere over the Rainbow: How to Make Effective Use of Colors in Meteorological Visualizations." _Bulletin of the American Meteorological Society_, 96 (2), 203-216. doi: 10.1175/BAMS-D-13-00155.1 (URL: https://doi.org/10.1175/BAMS-D-13-00155.1).

Stephen Milborrow (2019). rpart.plot: Plot 'rpart' Models: An Enhanced Version of 'plot.rpart'. R package version 3.0.8. https://CRAN.R-project.org/package=rpart.plot

Tennekes M (2018). "tmap: Thematic Maps in R." _Journal of Statistical Software_, 84 (6), 1-39. doi: 10.18637/jss.v084.i06 (URL: https://doi.org/10.18637/jss.v084.i06).

Terry Therneau and Beth Atkinson (2019). rpart: Recursive Partitioning and Regression Trees. R package version 4.1-15. https://CRAN.R-project.org/package=rpart

Tim Appelhans, Florian Detsch, Christoph Reudenbach and Stefan Woellauer (2019). mapview: Interactive Viewing of Spatial Data in R. R package version 2.7.0. https://CRAN.R-project.org/package=mapview

Title P.O., Bemmels J.B. 2018. ENVIREM: an expanded set of bioclimatic and topographic variables increases flexibility and improves performance of ecological niche modeling. Ecography. 41:291–307.

Venables, W. N. & Ripley, B. D. (2002) Modern Applied Statistics with S. Fourth Edition. Springer, New York. ISBN 0-387-95457-0

Winston Chang, Joe Cheng, JJ Allaire, Yihui Xie and Jonathan McPherson (2019). shiny: Web Application Framework for R. R package version 1.3.2. https://CRAN.R-project.org/package=shiny.

Zeileis A, Fisher JC, Hornik K, Ihaka R, McWhite CD, Murrell P, Stauffer R, Wilke CO (2019). "colorspace: A Toolbox for Manipulating and Assessing Colors and Palettes." arXiv 1903.06490, arXiv.org E-Print Archive. <URL: http://arxiv.org/abs/1903.06490>.

Zeileis A, Hornik K, Murrell P (2009). "Escaping RGBland: Selecting Colors for Statistical Graphics." _Computational Statistics \& Data Analysis_, 53 (9), 3259-3270. doi: 10.1016/j.csda.2008.11.033 (URL: https://doi.org/10.1016/j.csda.2008.11.033).

Online resources

Apart from some text books and research articles, several online resources ahve been consulted for preparation of this book. These are listed below:

http://amsantac.co/blog/en/2015/11/28/classification-r.html

http://desktop.arcgis.com/en/arcmap/10.3/manage-data/raster-and-images/tasseled-cap-transformation.htm

http://desktop.arcgis.com/en/arcmap/10.3/manage-data/raster-and-images/curvature-function.htm

http://desktop.arcgis.com/en/arcmap/10.3/tools/spatial-analyst-toolbox/how-principal-components-works.htm

http://desktop.arcgis.com/en/arcmap/10.3/tools/spatial-analyst-toolbox/flow-direction.htm

http://dst-iget.in/assets/pdf/tutorial/IGET_RS_010_Terrain_Analysis.pdf

http://gadm.org/data.html.

http://gsp.humboldt.edu/OLM/Courses/GSP_216_Online/lesson4-2/radiometric.html

http://gsp.humboldt.edu/olm_2015/Courses/GSP_216_Online/lesson4-1/radiometric.html

http://knightlab.org/rscc/legacy/RSCC_Contrast_Enhancement.pdf

http://northstar-www.dartmouth.edu/doc/idl/html_6.2/Filtering_an_Imagehvr.html

http://pages.stat.wisc.edu/~loh/treeprogs/guide/wires11.pdf

http://paper.ijcsns.org/07_book/201002/20100222.pdf

http://remote-sensing.eu/unsupervised-classification-with-r/

http://spatial-analyst.net/ILWIS/htm/ilwisapp/principal_component_analysis_functionality.htm

http://statisticstimes.com/demographics/population-of-indian-states.php

http://webhelp.esri.com/arcgisdesktop/9.3/index.cfm?TopicName=Flow%20Direction

http://www.dl-c.com/Temp/downloads/Whitepapers/Resampling.pdf

http://www.geol-amu.org/notes/rs12-4-1.htm

http://www.guru-gis.net/tag/raster/

http://www.jennessent.com/downloads/tpi-poster-tnc_18x22.pdf

http://www.naturalearthdata.com/downloads/.

https://bleutner.github.io/RStoolbox/rstbx-docu/coregisterImages.html

https://bleutner.github.io/RStoolbox/rstbx-docu/rasterPCA.html

https://cran.r-project.org/mirrors.html

https://datacarpentry.org/r-raster-vector-geospatial/06-vector-open-shapefile-in-r/
https://datascienceplus.com/random-forests-in-r/
https://docs.qgis.org/2.8/en/docs/user_manual/processing_algs/gdalogr/gdal_analysis/triterrainruggednessindex.html
https://earthexplorer.usgs.gov/
https://ecosystems.psu.edu/research/labs/walter-lab/manual/chapter-6-three-dimensional-analysis/6.2-three-dimensional-exploration-of-digital-elevation-models-dems
https://eden.dei.uc.pt/~ruipedro/publications/Tutorials/slidesML.pdf
https://en.wikibooks.org/wiki/Data_Mining_Algorithms_In_R/Clustering/CLARA
https://geoawesomeness.com/list-of-top-10-sources-of-free-remote-sensing-data/
https://geocompr.robinlovelace.net/adv-map.html
https://gisgeography.com/flow-direction/
https://homepages.inf.ed.ac.uk/rbf/HIPR2/filtops.htm
https://johnwilson.usc.edu/wp-content/uploads/2016/09/geoinfoscience_n209.pdf
https://mgimond.github.io/Spatial/coordinate-systems-in-r.html
https://nature.berkeley.edu/~penggong/textbook/chapter6/html/home65.htm
https://nptel.ac.in/courses/105108077/module5/lecture21.pdf
https://paginas.fe.up.pt/~ec/files_1112/week_06_Clustering_part_II.pdf
https://pro.arcgis.com/en/pro-app/help/data/imagery/curvature-function.htm
https://proj.org/usage/projections.html
https://rdrr.io/cran/RStoolbox/man/spectralIndices.html
https://rpubs.com/huanfaChen/ggplotShapefile
https://rstudio-pubs-static.s3.amazonaws.com/72662_b874adb74477422ca4464e5e542a41fc.html
https://segoptim.bitbucket.io/docs/index.html
https://subscription.packtpub.com/book/big_data_and_business_intelligence/9781783982103/5/ch05lvl1sec53/the-clara-algorithm
https://topepo.github.io/caret/train-models-by-tag.html#Support_Vector_Machines
https://uc-r.github.io/random_forests
https://webapps.itc.utwente.nl/librarywww/papers_2003/misca/hengl_digital.pdf
https://www.datanovia.com/en/lessons/clara-in-r-clustering-large-applications/
https://www.diva-gis.org/gdata, http://download.geofabrik.de/asia.html
https://www.eanswers.net/wiki/en/Positive_and_negative_predictive_values/
https://www.eanswers.net/wiki/en/Topographic_wetness_index/?ext=t&cid=5036
https://www.esri.com/about/newsroom/arcuser/create-amazing-hillshade-effects-quickly-and-easily-in-arcgis-pro/
https://www.hydrol-earth-syst-sci.net/10/101/2006/hess-10-101-2006.pdf
https://www.latlong.net/category/states-102-14.html
https://www.leg.state.mn.us/docs/2016/mandated/160722.pdf
https://www.listendata.com/2014/11/random-forest-with-r.html
https://www.machinelearningplus.com/machine-learning/evaluation-metrics-classification-models-r/
https://www.nceas.ucsb.edu/~frazier/RSpatialGuides/OverviewCoordinateReferenceSystems.pdf
https://www.neonscience.org/dc-open-shapefiles-r
https://www.neonscience.org/image-raster-data-r
https://www.neonscience.org/raster-data-r
https://www.neonscience.org/vector-data-series#scrollNav-6
https://www.neonscience.org/vector-data-series#scrollNav-7
https://www.r-bloggers.com/how-to-implement-random-forests-in-r/

https://www.rdocumentation.org/packages/caret/versions/6.0-84/topics/negPredValue
https://www.rdocumentation.org/packages/randomForest/versions/4.6-14/topics/partialPlot
https://www.rdocumentation.org/packages/raster/versions/2.9-23/topics/terrain
https://www.rdocumentation.org/packages/raster/versions/3.0-2/topics/terrain
https://www.rdocumentation.org/packages/RStoolbox/versions/0.2.4/topics/panSharpen
https://www.rdocumentation.org/packages/spatialEco/versions/1.2-0/topics/curvature
https://www.r-exercises.com/2018/02/28/advanced-techniques-with-raster-data-part-1-unsupervised-classification/
https://www.r-project.org/
https://www.rstudio.com/products/rstudio/download/
https://www.slideshare.net/BinteAdaam/resampling-gis
https://www.srbc.net/pennsylvania-lidar-working-group/docs/twi-srbc.pdf
https://www.tutorialspoint.com/dip
https://www.usna.edu/Users/oceano/pguth/md_help/html/geomorph_curvature.htm
https://www.usna.edu/Users/oceano/pguth/md_help/html/topo_rugged_index.htm
https://www.wrc.umn.edu/sites/wrc.umn.edu/files/terrain_analysis_slides.pdf
https://zia207.github.io/geospatial-data-science.github.io/support-vector-machine.html

Books consulted

Kaufman, Leonard, and Peter Rousseeuw. 1990. *Finding Groups in Data: An Introduction to Cluster Analysis*.

Lillesand, T. M., Kiefer, R. W., & Chipman, J. W. (2008). *Remote sensing and image interpretation*. Hoboken, NJ: John Wiley & Sons.

Panda, B.C. (2005). *Remote Sensing: Principles and Applications*. Viva Books Private Limited.

Printed in the United States
by Baker & Taylor Publisher Services